Praise for James Donovan's

SHOOT FOR THE MOON

"Exceptionally researched, this exciting, sometimes harrowing book highlights the work not only of the pioneering astronauts but also of thousands of technicians and engineers. This is a perfect volume to commemorate the fiftieth anniversary of the first lunar landing and all that led up to it." —*Publishers Weekly* (starred review)

"Donovan's account of Apollo 11 is a breath-stopping page-turner."
—Michael Barnes, *Austin American-Statesman*

"Donovan slips behind the shiny curtain of astronaut perfection to portray the men as they really were. He saves the best for last: a meticulous, almost minute-by-minute re-creation of Apollo 11's round trip to the moon, especially the landing and the two and a half hours that Armstrong and Aldrin spent on the moon."
—Edward Kosner, *Wall Street Journal*

"This is the best book on Apollo that I have read. Extensively researched and meticulously accurate, it successfully traces not only the technical highlights of the program but also the contributions of the extraordinary people who made it possible."
—Mike Collins, command module pilot, Apollo 11

"Donovan's narrative is a well-crafted one. It is one of the rarest of Apollo books that manages to weave together the political, the technical, and the heroic... One of the best in print."

—Asif Siddiqi, *Science*

"It was one of humankind's greatest achievements, and here, perhaps for the first time, is the whole story, fastidiously reported and elegantly told. With *Shoot for the Moon,* James Donovan captures it all—the science, the engineering, the clashing egos, the Cold War politics. But what's even more impressive, he does it without depriving us of the essential magic that was Apollo, this Promethean program that dared to aim as high and as far as man could go."

—Hampton Sides, author of *Ghost Soldiers, In the Kingdom of Ice,* and *On Desperate Ground*

"The Apollo 11 astronauts come to life in Donovan's vividly readable pages, and each is a vigorously drawn personality."

—Steve Donoghue, *Christian Science Monitor*

"Donovan combines his masterful research skills and narrative gifts in recounting the full story of the most famous Apollo trip... A powerfully written and irresistible celebration of the Apollo missions."

—*Booklist* (starred review)

"A gripping yet wonderfully detailed account of one of humanity's greatest achievements. *Shoot for the Moon* gives a fascinating insight into the golden age of space exploration."

—Tim Peake, head of astronaut operations, European Space Agency

"*Shoot for the Moon*'s breadth and detail will give you a new appreciation for just how complex and dangerous this mission was. You'll come away marveling that, against all odds, we put people on the motherfuckin' moon!" —Robert Faires, *Austin Chronicle*

"Showcasing a brilliant eye for detail and an elegant sense of historical narrative, Donovan's *Shoot for the Moon* is sure to be a space-race classic." —Annie Jacobsen, author of *Operation Paperclip* and the Pulitzer Prize finalist *The Pentagon's Brain*

ALSO BY JAMES DONOVAN

The Blood of Heroes:
The 13-Day Struggle for the Alamo—and the Sacrifice That Forged a Nation

A Terrible Glory:
Custer and the Little Bighorn—the Last Great Battle of the American West

Custer and the Little Bighorn:
The Man, the Mystery, the Myth

SHOOT FOR THE MOON

THE SPACE RACE AND THE EXTRAORDINARY VOYAGE OF APOLLO 11

JAMES DONOVAN

BACK BAY BOOKS
Little, Brown and Company
New York Boston London

For my sister and brothers,
in memory of a Brooklyn room with four beds:
"Whenever the moon and stars are set..."

Back Bay Books / Little, Brown and Company
Hachette Book Group
1290 Avenue of the Americas, New York, NY 10104
littlebrown.com

Originally published in hardcover by Little, Brown and Company, March 2019
First Back Bay trade paperback edition, March 2020

Back Bay Books is an imprint of Little, Brown and Company, a division of Hachette Book Group, Inc. The Back Bay Books name and logo are trademarks of Hachette Book Group, Inc.

All photographs courtesy NASA unless otherwise stated.

ISBN 978-0-316-34178-3 (hardcover) / 978-0-316-45443-8 (Canadian paperback) / 978-0-316-34181-3 (U.S. trade paperback)
LCCN 2018947611

10 9 8 7 6 5 4 3 2 1

LSC-C

Printed in the United States of America

CONTENTS

Neither the sun nor death can be looked at steadily.

François de La Rochefoucauld

SHOOT FOR THE MOON

PROLOGUE

CAPE KENNEDY, JULY 16, 1969, 4:15 A.M.

The beast—what some early missile men called a rocket—stood on the launchpad, dozens of spotlights bathing its milky-white skin, hissing, groaning, gurgling, thick umbilical hoses pumping fuel into it, sheets of ice sliding down its sides from the super-chilled liquid oxygen inside, some of it boiling off in thick white clouds of breath, and looking as if it might shake off the arms of its gantry, rip itself from its moorings, and stalk off down the Florida coast.

Eight miles away, on the third floor of Kennedy Space Center's Building 24, Deke Slayton walked down the hall of the crew quarters and rapped on three doors. With each knock, he said cheerily, "It's a beautiful day," and he meant it.

Raised on a farm, the unpretentious Slayton was fiercely protective of his charges—part den mother, part dictator—while at the same time envious of every one of them. A former World War II bomber pilot and test pilot and an original Mercury Seven astronaut, he'd been made chief of NASA's astronaut office when a minor heart problem grounded him before he had a chance to fly a mission into space. Slayton understood better than anyone that to a certain extent, the men's fates were in his hands, since he selected the crew for each mission and could make or break their careers. He was scrupulously fair in his choices—for the most part.

The men behind the three doors were astronauts Neil Armstrong, Edwin "Buzz" Aldrin, and Michael Collins. They constituted the crew of Apollo 11 scheduled for launch that morning. In three hours, they would climb into a small chamber atop the 363-foot, three-stage Saturn V, the most powerful machine ever built, and blast off into space. A few days later, two of them would attempt to do something that had never been done before: pilot a small, fragile craft down to another world, 239,000 miles from Earth, and walk on its surface.

These three men and others, most of them culled from the ranks of the nation's top test and fighter pilots, had committed their lives to this goal and worked tirelessly toward this moment.

When the American space program began, in 1958, no human had journeyed into the hostile environment of space—an airless, low-gravity vacuum with temperatures of extreme cold and intense heat that no living being could withstand. Without an artificial life-support system, a man would die instantly. Even with one, he might die; the effects of weightlessness, radiation, meteors, and the enormous forces accompanying launch and reentry were largely unknown. Each astronaut had trained for years to overcome these dangers and others.

Along the way they also spent thousands of hours learning how to use the machines that would carry them into space, machines far more sophisticated and complex than any previously invented, machines designed by a cadre of visionary scientists and engineers united by a dream of space travel, an insatiable curiosity, and the determination to make that dream come true. They all worked insanely long hours, often at the expense of their personal lives and relationships. Along with the four hundred thousand other men and women who actually built the machines, the astronauts devoted themselves to helping their country triumph against the Communist threat. At stake was not just supremacy in space but quite possibly America's survival as a democracy.

They had not reached this point without major setbacks and great tragedies. Rockets exploded. Systems malfunctioned. Men died. The murder of a visionary president whose bold challenge had fired the program only reaffirmed their dedication to finishing the job.

But in October 1957, still flying high just a dozen years after their victory in World War II, Americans had no idea how a small metallic ball with a radio transmitter would change the world.

I.

UP

CHAPTER ONE

COSSACKS IN SPACE

Our aim from the beginning was to reach infinite space.

MAJOR-GENERAL WALTER DORNBERGER,
COORDINATOR OF GERMANY'S V-2 PROGRAM

ONE SATURDAY MORNING in October 1957, a fourteen-year-old boy in the small farming town of Fremont, Iowa, woke up to find the world a different place. The Soviet Union had launched a beach-ball-size silver sphere into orbit around the Earth. They called it Sputnik—literally, "fellow traveler." The Russians, those steppe-riding, vodka-swilling Cossacks who were widely seen as a second-rate technological power, had beaten the United States into space.

The boy's name was Steve Bales, and he was of average height with thick brown hair and glasses. His mother worked in a beauty parlor, and his father, who at the age of thirty-nine had been drafted into the U.S. Army and served with the 102nd Infantry Division in World War II, owned a hardware store. Steve told his parents, his three younger brothers, and anyone else who'd listen how angry he was that America hadn't launched a satellite first. He'd been interested in space ever since he was ten, when he and his father and brothers spent many a summer night sleeping outside on well-worn gray blankets in the field behind their house on the edge of town. As darkness fell, their dad would point out the Big Dipper, Cassiopeia, and other constellations, and nothing seemed as wonderful as the universe and its mysteries. That excitement spiked when the boy watched a 1955 Walt Disney TV special that featured an intense rocket scientist with a slight German accent describing how one day man would reach the moon.

Now there was an artificial satellite, and it belonged to the Soviets—the enemy in this Cold War. But it was only a matter of time, the boy knew, before the United States would launch its own, and there would be more space exploration. And he wanted to be a part of it.

Lyndon Johnson, Senate majority leader, was relaxing with some friends at his family ranch in the Texas Hill Country that Saturday, October 5, when he heard about Sputnik. After dinner, they took a walk down a dark road, and everyone looked up at the sky. "In some new way," Johnson remembered, "the sky seemed almost alien." He spent most of that evening calling aides and colleagues, mobilizing them to begin an inquiry into the nation's satellite and missile programs. Johnson knew more about this new frontier than any other elected official in Washington—he had been spearheading congressional hearings and inquiries into America's space programs since the late 1940s—and he didn't like the feeling of being second to America's greatest enemy. He wanted to respond immediately to the Soviet challenge, and he began plans to chair a Senate Preparedness Subcommittee. It was clear to him that a comprehensive space program was necessary. That the Eisenhower administration's ineptitude in space provided an opportunity for political gain—so much the better.

In the dozen years since the end of World War II, America's onetime ally had become its greatest enemy. Some 418,000 Americans had lost their lives in the war, but that number paled in comparison to the twenty-seven million Russians who had died. While the United States emerged from the struggle as the world's most powerful nation, a wary USSR viewed America's constant meddling in the affairs of other countries—some overt, but much of it covert—as imperialism and believed the Western powers might finish off what the Nazis had started: their country's conquest. (After all, in 1941, Senator Harry S. Truman had publicly equated Nazi Germany with Stalinist Russia.) Increasing the tension between the two countries was the fact that the long-term goal of the totalitarian form of Communism, as frequently enunciated, was the elimination of the Western capitalist system. The Soviets' espoused desire for world domination (albeit preferably by a series of national revolutions, "small wars of liberation"), their nuclear and ICBM capability, and their frequent saber-rattling (for instance, in 1956 they threatened England, France, and Israel with hydrogen bombs if those countries did not end their war against Egypt, a Soviet ally) created a constant state of paranoia in the United States. That mentality was reflected and refracted in a burgeoning wave of science fiction and fact concerning rocketry and space travel in movies, TV shows, books, and magazines. All this was magnified by the realization that nuclear bombs were completely capable of annihilating mankind.

And there was most certainly a war going on, regardless of temperature, and the stakes involved were serious. Most Americans expected smaller countries around the world to gradually succumb, one by one, to the creeping menace of Communism. In November 1956, at a party at the Polish embassy in Moscow, Nikita Khrushchev, the tough, blustery Soviet premier, told Western ambassadors, "We will bury you," and while the phrase as translated might not have represented his precise meaning, it was accurate as to his intent. That declaration jibed with his publicly stated prediction that President Dwight Eisenhower's grandchildren would live under socialism, since capitalism was in its death throes. And the Soviets weren't the only ones rattling their nukes. In 1953, Eisenhower had threatened to use a hydrogen bomb against China, and

U.S. senators had often called publicly for an atomic bomb to be dropped on Russia.

Soon after the Soviet Union had detonated its own nuclear device, the two superpowers had begun to coexist under an unwritten but clearly understood doctrine known as "mutually assured destruction," meaning that everyone was aware that the full-scale use of nuclear weapons would cause the almost complete annihilation of both the attacker and the defender. This knowledge—and each country's fear of a massive, preemptive nuclear strike from the other—was all that kept the Cold War from becoming hot. Both developed massive forces of nuclear weapons and long-range bombers and missiles, more than a thousand on each side, though the Russian Tu-4 bomber, a direct copy of the American B-29, was inferior to most of the U.S. fleet.

Americans had always taken comfort in the fact that their country was superior to Russia in every way, including in the realm of science and technology. It was Americans, after all, who had split the atom and created the nuclear monster that had ended the war—even if the Russians had developed their own A-bomb in 1949 and then a much more powerful hydrogen bomb in 1953, a year after the United States.

So when Americans awoke on October 5, 1957, to the news that a shiny, 184-pound steel ball with four trailing antennas and two radio transmitters was *beep-beep-beep*ing its way around the world and that this Russian moon was traveling right over the United States, hundreds of miles above their heads, seven times a day, most people in the country were aghast.

But leaders of the U.S. satellite program were not surprised by the announcement. In 1955 both Russia and the United States had announced their intention to launch a satellite during the International Geophysical Year, a multinational agreement to share scientific information that would run from July 1, 1957, to

December 31, 1958, and involve sixty-seven nations in various earth-science projects. And just weeks before the launch, the Soviets had gone so far as to provide the frequencies at which Sputnik's telemetry—its electronic data transmission—could be monitored.

Not all of America's military leaders were intimidated by the satellite; one U.S. admiral called Sputnik a "hunk of iron almost anyone could launch." A few others, including President Eisenhower, tried to play down its significance, a position somewhat weakened two days later when the United States offered the United Nations "a plan to control outer-space arms." Ike had no intention of entering into an expensive space race, especially when it appeared that the Soviets were so far in front. And Sputnik, it was noted, was not large enough to carry a nuclear weapon.

But the press, Johnson, and Johnson's Democratic colleagues, sensing an opportunity to do damage to the opposition, disagreed. "The Soviet has taken a giant leap into space," proclaimed the *New York Times*, and Senator Henry M. Jackson, a Democrat from Washington, called the launching "a devastating blow to the United States' scientific, industrial, and technical prestige in the world." The industry monthly *Missiles and Rockets* made it even more Manichaean: "The nation that controls space will control the world. The choice is democracy or slavery." The magazine also revealed Soviet plans to land a small tank on the moon that would constantly move about, film the Earth, and relay the images back to Russia. Over six weeks, Johnson's Senate Preparedness Subcommittee hearings paraded an impressive array of aviation, military, and rocketry experts before the public. Each one emphasized the dangers of ignoring the Sputnik achievement and what it portended. When the committee finished the hearings on January 8, Johnson's no-holds-barred summary statement threw down the gauntlet. "Control of space means control of the world," declared Johnson, echoing the *Missiles and Rockets* editorial. He linked the fate of the free world to it and virtually dared anyone to argue cost—"bookkeeping concerns of fiscal officers," he made clear, were "irrelevant."

Prestige—a code word for political strength and one that would pop up frequently in speeches and reports—was something statesmen worried about. In the global tug-of-war between the democratic free world, led by the United States, and the Communist countries, headed by the USSR, dozens of less advanced, recently decolonized nations were still undecided as to which side of the ideological rope they would grab onto; in a sense, they were waiting to see which team had the advantage. Neither superpower knew exactly what would persuade them, though, clearly, superiority in science—particularly its military applications—would play a large part. So advances in these crucial areas were trumpeted loudly, with Western Europe a rapt audience. If America's North Atlantic Treaty Organization (NATO) allies could be shaken loose from that coalition, the Soviets felt, Sputnik and its descendants would be worth every ruble.

Until Sputnik launched—an event that followed the Russian announcement six weeks earlier that a Soviet intercontinental ballistic missile had been fired a great distance—the United States had appeared comfortably in the lead. "The Soviet Union is thought to be making a conscious effort to persuade people, especially in Asia and Africa, that Moscow has taken over world leadership in science," warned the *New York Times* on its front page in October 1957, and another op-ed in the paper was even bleaker: "The neutral nations may come to believe the wave of the future is Russian; even our friends and allies may slough away." The struggle for the undecideds' hearts and minds—and pocketbooks, since another worry was the international market for American goods and tools—was a very real part of the Cold War. A top secret government report just a week after Sputnik launched seemed to agree; it concluded that American prestige had suffered a severe blow and cited several examples of how the satellite had enhanced the Soviet Union's prestige and damaged America's. Another reported that "within weeks there was a perceptible decline in enthusiasm among the public in West Germany, France and Italy for 'siding with the west' and the North Atlantic Treaty Organization."

But if *prestige* was a politician's word, it was also a politician's worry. The average American was more concerned about survival than prestige, and Johnson's statements hit a nerve. Clearly the United States was no longer safe from nuclear attacks. What was next? Soviet military bases on the moon or orbiting the Earth? A 1956 article in *Collier's* magazine had vividly suggested just that, with color illustrations of New York City being firebombed from above. Yes, that might be next, and it seemed even more likely after the Soviets launched Sputnik 2 just thirty-two days after its predecessor. That satellite carried a small mongrel dog named Laika into space, though she died from overheating after only a few hours in orbit. (At the time, the Soviets claimed that she survived for a week; the truth would not be revealed for decades.) Sputnik 2's 1,121-pound payload—heavier than a Soviet nuclear warhead—made it abundantly clear that the Soviets could now reach America with an A-bomb. And there was only one reason to send a dog into orbit around the Earth. It was just a matter of time before there were Russians in space—and that meant right above America.

After a few more weeks of teeth-gnashing by Democrats in DC and nonstop fear-stoking by the press—"Why is the U.S. still lagging in a race that may decide whether freedom has any future?" asked *Time* magazine—even those Americans initially unimpressed by Sputnik were whipped into near hysteria. The United States trailed the leader in a race that could end in the destruction of the American way of life…or destruction, period. Paranoia increased; not surprisingly, UFO sightings quadrupled immediately after Sputnik. Martians? Russians? Both were postulated.

How could these primitive Cossacks have accomplished such a thing? Much ink was spilled and time spent lamenting how soft America had become in the little more than a decade since the end of World War II. While op-ed writers bemoaned an "education gap," and a government report showed that Soviet children took far more science and math in high school than the fun-loving, sock-hopping American kids, who were getting dumber every day, it was clear that Soviet triumphs would continue. All this self-recrimination would result in the National Defense Education Act, signed into law by Eisenhower in September 1958, which was designed to kick-start the U.S. educational system with grants, low-interest loans, and the like. Eisenhower, in a press conference a few days after Sputnik 1 launched, insisted that it wasn't the Russians who had built the satellite—it was "all of the German scientists" they had captured at the end of the war.

This could not have been further from the truth. Almost all the top German rocket scientists and engineers had been captured and co-opted not by the Soviets but by the United States under the auspices of a secret project, and they had been living in America since 1945. Their leader, Wernher von Braun, was a handsome, charming ex-SS officer who had been the chief architect of an ambitious rocket program that had killed thousands during the war—and who now spread the gospel of space exploration to Americans in Walt Disney TV specials.

This group of so-called Nazi scientists—a misnomer, since less than half of them had been members of the Nazi Party and all but a few were engineers and technicians, not scientists—were capable enough; their forty-six-foot-long V-2 rocket, a tremendous feat of science and engineering, had been the first man-made object to enter space. The trouble was, its creators had been virtually handcuffed by their American superiors for almost a decade, despite the best efforts of their persuasive director, and relegated to low-priority upper-atmosphere experiments, since the military saw long-distance bombers, not rockets, as superior weapons. Von Braun had striven for permission to construct a satellite and launch it before the Soviets did their own, and he had grandiose plans for an orbiting space platform that would work as either a refueling stop halfway to the moon or a battle station from which to rain down nuclear missiles. The satellite project was denied due to budget restrictions, and the space-platform idea wasn't even considered. In the late 1940s, anybody who talked of going anywhere in space was considered a crackpot. Von Braun was determined to change that thinking, but it wouldn't happen overnight.

Wernher was born into German nobility (which accounted for the *von* in his name) in 1912; his father served as minister of agriculture during the Weimar Republic of the 1920s, and his mother could trace her ancestry to medieval kings in four countries. Von Braun began dreaming of traveling to other heavenly bodies during his pampered childhood. His mother, an amateur astronomer, gave him a telescope for his confirmation instead of his first pair of long pants, which was the standard gift for Lutheran boys, and he devoured the space-travel novels of H. G. Wells and Jules Verne. In his teens he became familiar with the writings of the great early theoreticians of rocketry—an obscure Russian schoolteacher, Konstantin Tsiolkovsky; the German mathematician Hermann Oberth; and the American physicist and inventor Robert Goddard—and graduated from fireworks to his own inventions and plans, not only for rockets, but for spaceships and their pilots. At sixteen, he organized an astronomy club at school and managed to acquire funds to buy a good-size refractor telescope. A German spaceflight movement was growing in the 1920s, and von Braun became a part of it.

He was soon working with Oberth and other leading figures in German rocketry. By the time he entered university, the tall, blue-eyed, dark-haired young man was a science and math prodigy committed to a career in rockets and determined to pursue his dream of journeying to the moon and the planets.

He graduated at twenty from the Berlin Institute of Technology with a degree in mechanical engineering and began postgraduate work in physics at the University of Berlin, but his studies were curtailed by larger forces. In the summer of 1932, some months before the Nazi Party came to power, the German army was casting about for some sort of long-range weaponry not banned by the 1919 Treaty of Versailles, which had severely restricted the country's war-making abilities. They were not unaware of the destructive potential of a large, long-range rocket and asked von Braun to participate in secret research for military applications. He was initially impressed with the Nazis' leader, Adolf Hitler, because of his nationalist ideas and his "astounding intellectual capabilities." Above all, von Braun was an opportunist, and the idea of war seemed highly unlikely at the time. The army did not share his dream of spaceflight, but if they could help him build rockets, he would build rockets for them—a tempting prospect, considering the cost of the rockets he envisioned, and somewhat inevitable, since civilian rocket activity was soon banned. He accepted the offer and became the army's top civilian specialist at their new (and only) rocket station, an artillery range hidden in a pine forest near Kummersdorf, sixty miles south of Berlin. His laboratory was half a concrete pit, his staff a single mechanic. But he continued his studies, and by the end of 1934, he was awarded a doctorate in physics in aerospace engineering, though his full dissertation was kept classified by the army. By that time, his group, which had grown to include several men who would work closely with him over the next decade and beyond, had successfully launched two small liquid-fueled rockets. The director of the program was only twenty-two.

Over the next several years, his greatest talent would emerge—the managing of massive engineering projects. Few visionaries could mesh their dreams with the practical realities of administration on a grand scale; von Braun had a genius for it. His blend of charisma, enthusiasm, and knowledge inspired loyalty and hard work, and he became a superb leader of a large group of scientists, engineers, and technicians. In 1937 he was made technical director of the German army's long-range ballistic missile program after it moved to Peenemünde, a facility on an island off the coast of northern Germany. Three years later, with World War II in full swing, he led a staff of more than a thousand. In the spring of 1940, he was pressured by Reichsführer Heinrich Himmler to join the SS, the elite paramilitary organization in charge of enforcing Nazi racial policies and, by extension, policing and intelligence. After some deliberation, von Braun joined, since not doing so might have damaged his career and his rocket work; he was given the rank of lieutenant. He participated in very few SS activities besides the monthly meetings, and even those he attended only half the time.

By 1942, the Peenemünde team had developed the world's first long-range guided ballistic missile, the forty-six-foot A-4 (soon renamed the V-2, the V for *vengeance*). It was a monster of a rocket, unlike anything ever seen before, and a huge leap forward in technology, from its powerful engine to its advanced guidance system. The V-2 could carry a 2,200-pound payload of explosives two hundred miles and do it fairly accurately, or at least it could toward the end of the war, after its guidance system was improved. The missile reached a speed of thirty-five hundred miles an hour, making it virtually impossible to shoot down—when it didn't explode during liftoff.

But all of that was still in the future. In October 1942, von

Braun and his colleagues were under extreme pressure. The first two launch attempts had failed, and if the third wasn't successful, the program would likely be shut down and its members shipped to combat zones—probably the Russian front, a virtual death sentence. On October 3, a black-and-white A-4 sat on a simple five-foot-high frame, frost from its super-chilled liquid-oxygen propellant coating the hull. The firing command was given, a switch was thrown, and as hundreds of tense spectators watched, the rocket lifted off the stand with a deafening roar, slowly at first and then gaining speed, finally reaching fifty miles beyond the stratosphere into the mesosphere. Von Braun and his team danced and wept for joy. It had taken almost a decade of hard work, but they had achieved their lofty goal. Even the army general overseeing the rocket development, Walter Dornberger—a short, bald engineer and former artillery officer who had fought in World War I and who, after working twelve years with von Braun, had become sympathetic to his dreams of interplanetary travel—knew what it meant to them. "This afternoon the *spaceship* has been born!" he said.

For the German army, desperate for help in a losing war, the rocket's only raison d'être was the destruction of the country's enemies, and work continued in that direction. Hitler had been lukewarm about the technology at first, and funding for manpower and materials had been tight. By the summer of 1943, however, with the war going badly for Germany, the führer decided to give the program the highest national priority, and its budget was increased dramatically. In September 1944, hundreds of V-2 rockets, each capable of carrying a one-ton warhead, began raining down on targets in England, France, and Belgium. The V-2's destructive effect could be significant; one November 1944 hit on London killed 160 people and injured 108. By the war's end, V-2s would kill 2,754 British civilians and injure 6,523—though it would fail to become the supreme instrument of terror, the *Wunderwaffe*, or miracle weapon, that Hitler had hoped for.

In August 1943, the Royal Air Force bombed Peenemünde and killed about five hundred civilian workers, most of them Polish and Russian prisoners of war involved in fabrication work. Only two of von Braun's top scientists died, but the damage to the facility was extensive, and Hitler ordered production of the rocket moved to the Mittelwerk, an underground site carved out of an abandoned gypsum-mine shaft located two hundred and fifty miles south, in the Harz Mountains. Ten thousand slave laborers—again, most of them Polish and Russian prisoners, men from the nearby Dora-Mittelbau concentration camp—built the plant's extensive series of large tunnels a mile into the hillside. Those workers and fifty thousand more endured hellish conditions and inhumane treatment to meet the factory's target of nine hundred V-2s a month, an ambitious goal that was never quite achieved. By the time production was halted, in March 1945, twenty thousand of those workers had died as a result of starvation, disease, beatings, or execution.

Despite his long working hours, von Braun, now tall and broad-shouldered with thick brown hair, found time for pleasure. He was given a civilian-adapted Messerschmitt for his own personal use, and on weekends he would fly it to Berlin, a hundred and sixty miles away, to visit a girlfriend—when he wasn't romancing one of the Peenemünde secretaries. Sometimes his younger brother, Magnus, whom von Braun had hired as his assistant to keep him out of combat, would accompany him. There was the occasional bicycle ride with von Braun's own attractive young secretary, and he was rarely alone on his sailboat in the Baltic Sea. And every so often, he and his team would let off steam at a weekend gathering.

At one such party in March 1944, an inebriated von Braun and two colleagues discussed the fact that the war was going badly for Germany as well as their disappointment that they weren't working on a spaceship. Von Braun's near-constant comments to almost anyone within earshot about going to the moon finally caught up with him. A local female physician who was a spy for the Gestapo heard their remarks and reported them to the SS. Ten days later, von Braun and four others were arrested by the Gestapo on suspicion of treason and sabotage, charges that could have resulted in imprisonment or execution. The rocketeers were detained for a week, until General Dornberger, with the help of the minister of war armaments and production Albert Speer, a relatively sane Nazi despite being one of the führer's favorites, engineered their release and the dropping of all charges; they argued that von Braun, especially, was too valuable to the war effort. Even Hitler admitted to Speer that von Braun was "indispensable." But von Braun had gotten on the bad side of Heinrich Himmler and would never again be trusted by the SS Reichsführer or his minions. It was about then that von Braun began planning what he would do when the war was over.

By the end of 1944, it was clear that the V-2 would not significantly alter the outcome of the war—and that outcome, von Braun knew, would not be a German victory. It was also clear that his plans for a transatlantic armed rocket able to reach New York, Washington, DC, and other cities on the U.S. mainland, part of an operation known as Projektil Amerika, would never be realized. With the Allied troops just days away from toppling the Nazi regime, von Braun had only one question: Which conquering nation was more likely to help him further his goal of space exploration—England, France, the United States, or the Soviet Union? The answer did not require long agonizing. He dismissed the first two for lacking the necessary funding, so the choice was the USSR or the United States; the U.S., with its relatively benign democracy, its booming economy, and its reputation for considerate treatment of prisoners of war, won hands down. The New World, with its siren song of freedom and a better life, had maintained a romantic hold on Germans for a century and a half—six million of them had immigrated there between 1820 and World War I—and the just-ended conflict had not quashed those dreams. Russia inspired no such imaginings. Few of the rocketeers preferred the Russians, who were rapidly advancing westward; stories of their murder, rape, and pillage in revenge for Germany's invasion of their country were rampant. Besides, von Braun's country had lost two wars during his lifetime. "The next time, I wanted to be on the winning side," he said later. He and the members of his inner circle discussed the options, and then they took a vote. All but one were in favor of the Americans. When he broached the subject with the rest of his team, most agreed, though some decided to cast their lots with the Russians.

Early in 1945, von Braun and his staff were relocated to central Germany for their safety—or, some of them suspected, for a mass execution by the SS that would deny their valuable knowledge to the fatherland's enemies. Von Braun oversaw the move south, but first he fabricated documents—orders to himself that he wrote on official SS stationery—and employed no small amount of subterfuge, including creating an acronym, V2BV, that he said stood for a top secret agency answerable only to Himmler himself; he had V2BV stenciled on boxes, crates, and vehicles. Thousands of personnel and tons of equipment, V-2 parts, and documents traveled by mule, horse, train, and truck convoy. Weeks later, with the technicians settled into empty factories and buildings in towns near the Mittelwerk, von Braun and a couple dozen of his closest colleagues ensconced themselves at a Bavarian Alps ski resort. After hiding the most valuable materials and papers in an abandoned gypsum mine, he began engineering the surrender of five hundred of his staff to the U.S. Army. When news of Hitler's death on April 30 reached the area the next day, it became easier for von Braun to carry out his plans, as the SS troops guarding the missile team gradually disappeared.

On the morning of May 2, 1945, Magnus von Braun—chosen because he spoke passable English, a language he'd learned in childhood from his British nanny—biked down a mountain road with a white handkerchief tied to the bicycle's handlebars. When he ran into an American antitank platoon, he stopped his bike and walked it up to a private named Fred Schneikert. He told the soldier that the inventors of the V-2, including his brother Wernher von Braun, were up the mountain in a hotel, and they wanted to be taken to Ike—General Dwight Eisenhower, supreme commander of the Allied forces in Europe. "You are a nut," the private said, but after some confusion, Magnus was sent back up the mountain to fetch the V-2 scientists. Later that afternoon, Wernher and six others drove in three cars and were escorted to U.S. Army intelligence. They were treated well and given a meal of scrambled eggs, bread and butter, and real coffee, a rarity in Germany at the time. Von Braun posed for photos with GIs and acted like a visiting dignitary. The Americanization of Wernher von Braun had begun.

It took a few weeks before von Braun and his team came to an agreement with their American captors, but the two sides finally did. The Germans were at least twenty years ahead of the United States in the field of rocketry, and the rocketeers understood that their knowledge and their V-2 hardware and their priceless papers were valuable bargaining chips. By the terms of the Yalta Conference agreement, that part of Germany would be incorporated into the Soviet occupation zone. The Russians hadn't arrived yet, but they would any day. Racing against the clock to beat them, the Americans absconded with about a hundred unfinished V-2s and enough parts to fill hundreds of railroad cars. They also located the mine shaft where von Braun had hidden the program's most important plans and blueprints—fourteen tons of crates—and at the last minute they found another cache of valuable V-2 documents that Dornberger had hidden himself. The final treasures were shipped out just two days before the June 1 handover of the area to the Soviets.

A few weeks later, the transfer of von Braun and a hundred and twenty-six of his top rocketeers to the United States was officially approved as part of Operation Paperclip, a quickly planned and executed evacuation of thousands of German scientists, engineers, and technicians. They were soon granted security clearances and began arriving in September as "wards of the Army," requiring no entry permits—but first they were provided with false employment histories, and their Nazi Party affiliations were expunged from their records. The U.S. military claimed they had not imported any "ardent Nazis," but there seemed to be plenty of ways to avoid being classified as "ardent."

The original idea was to bring the Germans over to America for six months to help defeat Japan—a costly invasion of the Japanese home islands had seemed unavoidable, since the Japanese had ignored Allied demands for unconditional surrender. The V-2, some thought, might be of use in the Pacific. But a U.S. Army Air Forces B-29 dropped a nuclear bomb on Hiroshima on August 6 and another on Nagasaki on August 9, resulting in at least a hundred and thirty thousand deaths and the complete obliteration of each city; Japan raised the white flag on August 15. Meanwhile, the Soviet Union was shaping up to be a formidable postwar enemy, and the Germans began signing five-year contracts with the U.S. Army.

In September of 1945, von Braun and company arrived in Fort Bliss, just north of El Paso, Texas, and were subjected to a loose form of house arrest in surplus barracks—"prisoners of peace," as they referred to themselves, half jokingly. They were not allowed to leave the base without a military escort. With their families set to arrive a year later, the scientists had time to explore the exotic landscapes around them. The Texas desert was unappealing to most, though von Braun found it beautiful in an Old West kind of way—he and many Germans had an obsession with cowboys and Indians, chiefly due to Karl May, a late-nineteenth-century German novelist who had written dozens of hugely popular Westerns without ever visiting America.

Over the next five years, von Braun and his colleagues spent most of their time assembling V-2s, launching them at the White Sands Proving Ground in New Mexico, seventy-eight miles north, and training personnel in the use of rockets and guided missiles, all the while avoiding the subject of Nazi war crimes and adjusting to their new lives. In 1947, the thirty-four-year-old von Braun was granted permission to return to Germany to marry his beautiful blond first cousin Maria von Quistorp, who was eighteen; first-cousin marriages were not uncommon among the old European aristocracy. (He was accompanied by American agents to prevent the Russians from abducting him, and U.S. Army MPs went along on the honeymoon.) A few weeks later he returned to America with his bride and his parents, the Baron and Baroness von Braun, who had lost almost everything in the war, including their ancestral estate, which had been confiscated by the Russians.

During the next few years, with rare exceptions, any new rocketry ideas of von Braun's were dismissed, and his dreams of space exploration remained just that. He and his team could take only meager satisfaction in the fact that their sixty-odd V-2 launches (most of which were successful) were furthering high-altitude research, as their increasingly accurate rockets penetrated the upper atmosphere carrying various scientific experiments. But in the spring of 1950, the U.S. Army—alarmed at the deteriorating world situation, particularly the escalating tensions between South Korea and the Communist countries of China and North Korea—moved them to the superior facilities of two adjacent, shuttered arsenals at Huntsville, in northern Alabama. The former chemical weapons factory and depot were combined to make the Redstone Arsenal.

There, with larger budgets, they would develop a rocket and missile center. The Tennessee River ran along the arsenal's southern edge, and the foothills of the Appalachians lay to the west. The Germans appreciated the lush green hills of Huntsville more than they had arid West Texas, and they settled into a project worthy of their talents: developing the Redstone, a large missile with a two-hundred-mile range—a kind of "super V-2." When Chinese intervention in North Korea that winter spiked fears of an imminent Soviet invasion of Western Europe, American rocket development became a high priority.

For his part, von Braun set out to educate the American public—and the world—about space exploration and legitimize a subject widely perceived as Buck Rogers silliness. In 1951, a paper he authored on organizing a manned mission to Mars was read at an astronautics symposium. The next year, he wrote eight articles for *Collier's* magazine discussing manned rockets, space stations, space shuttles, and moon expeditions—"Man Will Conquer Space Soon!" blared the series' title. Vivid, detailed color illustrations by respected artists such as Chesley Bonestell accompanied the stories. Despite the U.S. government's lack of interest in von Braun's true passion, he continued to write and speak about his extravagant ideas. He also began working with Walt Disney on three TV specials. The first, which aired on March 9, 1955, was seen by forty-two million viewers, an impressive 25 percent of the U.S. population. His zeal was infectious. With von Braun as their professor, Americans were acing Space Exploration 101—and they understood, or thought they understood, the implications of those Sputniks in 1957.

One American not so excited about space exploration was President Eisenhower, who refused to encourage the army's increasingly ambitious plans for military rocket applications. As a result, the program was modest and underfunded. But von Braun was permitted to develop a multistage version of the Redstone, an advance that allowed rocket stages to be jettisoned after use, resulting in a lighter craft that accelerated more easily. Dubbed the Jupiter-C, it could deliver warheads to distant targets. By September 1954, von Braun's Huntsville team had expanded to a thousand employees. The navy also began developing its own rocket, the science-oriented Vanguard, and the air force produced a much larger military booster, the Atlas.

In July 1955, the Pentagon chose the Vanguard, a superior rocket but one that at the time existed only on paper, to be first in space. Von Braun was incensed, declaring that the Vanguard—which some suggested should be renamed Rearguard for its slipping schedule—would fail. The Redstone had been deemed operational, and his Jupiter-C, which was capable of boosting a satellite into orbit, was close to completion. Von Braun had been pushing to launch a satellite for years, but he was allowed to prepare the Jupiter-C as a backup to the Vanguard. On September 20, 1956, von Braun and his team fired a successful Jupiter-C that reached a world-record altitude of 682 miles and a speed of 12,800 miles per hour. When he asked for permission to use it to put a satellite into orbit, he was refused. Neither Eisenhower nor the Pentagon believed that it was important to beat the Soviets into space. The backup Jupiter-Cs went into storage. Sputnik's success infuriated and depressed him.

On December 6, 1957, at Cape Canaveral, the navy prepared to send a small four-pound satellite into orbit using their Vanguard rocket, an event that would be broadcast live. The Cape, as it was known—officially, the Florida Missile Test Range, on Patrick Air Force Base—was the military's new missile-launch facility, built several years earlier on an abandoned naval station over a desolate stretch of gator- and mosquito-infested sand and palmetto scrub arcing out from the eastern coast of Florida. In truth, the launch was only a test, the missile's first, and the navy was not happy with the pressure of having it broadcast live on national TV. But millions watched, eager to see America's answer to Sputnik, and they were appalled when the seventy-foot rocket lifted about four feet off the ground and then exploded in a huge orange-and-yellow fireball. The grapefruit-size satellite fell from the nose section and rolled away into some bushes, where it began transmitting signals. "Kaputnik" and "Flopnik" were just two of the derisive names the press came up with for the disaster. American humiliation was complete when Russia's UN delegate suggested that the United States take advantage of a Soviet program that offered technical assistance to underdeveloped nations.

Another Vanguard misfire—this one in secret—would occur two months later. But von Braun had already been given the green light. He and about forty of his German colleagues and their families had officially been sworn in as American citizens in 1955, and they had long since become civil service employees. They had spent years refining their V-2 into the much superior Redstone, a short-range missile originally built to carry an atomic warhead and then converted into a three-stage rocket capable of escaping Earth's gravity. On the night of January 31, 1958, von Braun journeyed to the Cape by train—the base was remote, and air travel there was infrequent—to witness the culmination of almost thirty years of work, but upon arrival, he was disappointed to find that he would not be allowed to watch. Instead, he was flown to Washington, DC, for what the army hoped would be a triumphant announcement. It was. The white-bodied, black-tipped Jupiter-C, with a fourth stage added and renamed Juno, lifted off and launched an eight-foot-long, thirty-one-pound tube called Explorer into the sky, though it took tracking stations an hour and a half to confirm that the satellite had been successfully boosted into orbit. A large crowd of reporters and radio and TV broadcasters showed up for a 1:30 a.m. press conference at the Great Hall of the National Academy of Sciences to hear the former Nazi who had saved America's honor speak. They didn't leave for two hours.

America was finally in space, and in the race.

After the successful launch of Explorer, von Braun was feted like a war hero. Unlike Sputnik, Explorer had accomplished something beyond the simple fact of its orbit. The satellite had carried the same radiation sensors as the Vanguard, and they had detected the belts of radiation ringing the Earth that were later named after the scientist who directed the experiment, James Van Allen. Von Braun appeared on the cover of *Time* magazine and was invited to the White House to be congratulated by President Eisenhower, who didn't much like him—von Braun had not only publicly disagreed with the president when he downplayed the importance of Sputnik but also campaigned constantly and loudly for space funding. It was vindication of the sweetest sort. Von Braun was the right man at the right time, and he knew it. His speaking fees went up dramatically, and a movie about his life was announced.

At a prestigious gathering in Chicago a few weeks later, von Braun waxed eloquent—by this time, his written English was as good as the best speechwriters'. He delineated the grave danger to "free men everywhere" from the "Red menace" and asked whether America could "meet the total competition of aggressive communism and still preserve its way of life." In a speech that sounded like a presidential State of the Union address, he spoke of the importance of education in meeting the challenge and of the many years it would take to catch up to and pull ahead of the Russians. He finished with a call to arms: "We have stepped into a new, high road from which there can be no turning back."

If the challenge had not been formally accepted before, it was now.

CHAPTER TWO

OF MONKEYS AND MEN

It doesn't really require a pilot, and besides, you'd have to sweep the monkey shit off the seat before you could sit down.

CHUCK YEAGER

UNLIKE MOST OTHER INTERNATIONAL contests, those involving space required massive expenditures of money, and a president determined to balance the budget was in no mood to grant them. From top secret intelligence based on reports from U-2 spy planes, which had been operating since 1956, Eisenhower knew that the United States was comfortably ahead of the USSR in guided missile development—the opposite of what most Americans assumed. He continued to deny or minimize the importance of the Sputniks, insisting there was no value in a space race with the Soviets. He even went so far as to make several TV appearances in an effort to convince the American people of this. He took great pains to point out the difference between satellites and rockets designed for scientific purposes and those intended for military use.

But neither the public nor the press seemed to care about the distinction, and the May 15, 1958, successful launch into orbit of Sputnik 3, a 2,926-pound research satellite with a large array of instruments, only increased the nation's anxiety. The United States and the USSR had yet to engage in full-scale combat, but each side was heavily armed, and the doctrine of mutually assured destruction was of little comfort. Further fueled by an almost constant barrage of opinion pieces and articles on the imminent dangers of Soviets in space, not to mention speeches and statements by Senator Lyndon Johnson and other Democratic congressmen eager to exploit the purported missile gap for their own political gain, Americans quickly began demanding a full-fledged space program.

The president grudgingly conceded, though he dreaded adding another bureaucracy and the expenditures it would create. To Eisenhower, fed up with the endless territorial squabbling of the army, navy, and air force over space—he would later warn Americans of the "military-industrial complex"—it had become increasingly clear that the program needed to be nonmilitary. (In January 1958, he had even proposed to Russia that the two superpowers agree that "outer space should be used for peaceful purposes." The Soviets rejected the offer.)

Despite Ike's insistence that there was no cause for alarm, the Democrat-controlled Congress chartered the National Aeronautics and Space Administration in July 1958, and on October 1, NASA became operational. The new civilian organization would incorporate the National Advisory Committee for Aeronautics (NACA), an agency devoted to aeronautical R and D, and its far-flung research and test centers (most significantly Langley Memorial Aeronautical Laboratory, on the banks of the Chesapeake Bay in Virginia, to be renamed Langley Research Center). It would also include other important facilities, such as the Jet Propulsion Laboratory in Pasadena, California. Transferred to the fledgling agency were ongoing projects from the three branches of the military, each of which had been developing its own missiles and edging closer to space; the army had its Explorer, the navy had the Vanguard, and the air force had its massive F-1 rocket engine.

The NACA had been founded by Congress during the early years of World War I when the government realized that the United

States' meager air military force was at a severe disadvantage to other nations'. Just prior to the war, the U.S. had about thirty planes; Russia, England, and Germany together had more than a thousand. By 1915, the Germans were using fighter aircraft—a plane with a machine gun timed to fire through the propeller. Two years later, the U.S. responded by establishing its first civilian aeronautical research facility at present-day Langley Field. With a mandate to study the problems of flight "with a view to their practical solution, and to direct and conduct experiments in aerodynamics," the agency had improved aviation techniques dramatically by the mid-1930s and played a huge part in developing the superior aircraft that helped win World War II.

But the NACA produced research, not products; any ideas with potential were turned over to others—sometimes the military, other times a private aircraft company—to develop and produce. The agency was run by committee and operated by consensus—and did it all slowly and carefully. Employees were encouraged to work only from eight to five; Security locked the doors at five p.m. every day. A regular lunchtime activity was a paper-airplane contest. After Sputnik, that environment—and particularly those hours—would change completely. In the years following World War II, the agency had been withering away, its budget slashed repeatedly. Now it would help to win another war. Some far-seeing NACA engineers had even begun to research the task of putting a man into space—and they were eager to take on the challenge.

But that would require powerful rockets, and NASA had no rocket program. The NACA's specialty had been applied research; it didn't build aircraft, only told military and industrial entities how to make them better and safer. NASA would need von Braun and his ballistic-missile team in Huntsville. That presented a problem; the army, still basking in the glow of launching the first American satellite, refused to part with von Braun and the other Germans and the massive booster they were developing, the Saturn. Von Braun had an eye toward using it to transport components of a military space station, one of his pet projects. A three-way tug-of-war for his services would emerge between the army, the air force, and NASA.

Meanwhile, since the successful Jupiter-C launch of Explorer, von Braun had become even more of a national hero, and he had capitalized on his fame. In its cover story about him, *Time* magazine dubbed von Braun "the Seer of Space," and he was well paid for his articles and speeches. The Disney movie about his life was produced and titled *I Aim at the Stars.* The film was neither dramatically effective nor factually accurate—the former SS major was depicted as being persecuted by the Nazis, and his Peenemünde secretary became an Allied spy, two of many perversions of the truth. The film's most lasting legacy came from comedian Mort Sahl, who suggested that it should have been subtitled *But Sometimes I Hit London.* The public's tepid response to the film, however, didn't damage von Braun's reputation a whit. In a July 1957 article he had written for *Missiles and Rockets*—he was on the magazine's advisory board—he waxed messiah-like on the promise of space travel. "Space flight," he wrote, "will free man from his remaining chains, the chains of gravity which still tie him to this planet. It will open to him the gates of heaven."

For several months after the launch of Explorer, the American and Soviet programs launched—or attempted to launch—rockets into orbit almost every week. More Sputniks went up, most of them successfully, though at least one, launched three days after Explorer, failed to reach orbit and crashed back to Earth. (Because the Soviets didn't announce planned launches, no one in the West would know about it for decades.) The navy's Vanguards continued to malfunction, and von Braun's next Jupiter-C, launched on March 5, failed as well. Two weeks later, a Vanguard finally reached orbit with a three-pound test satellite. Nine days after that, on March 26, another Jupiter-C lifted Explorer 3 into orbit. Then three consecutive Vanguards failed. Each nation also attempted to land rockets on the moon; the Soviets tried three times, the Americans four. All seven were unsuccessful.

In March 1958, six months before the official conversion of the NACA to NASA, an ad hoc committee of some thirty-five NACA engineers had started planning for what they suspected would be the inevitable: sending a man into space. The committee was named the Space Task Group. Soon after NASA's creation, the goal was made official, and the group was charged with putting a man into orbit as quickly as possible.

A man in space; to most Americans, it sounded like science fiction. Space, or outer space, was generally considered to begin at around a hundred kilometers, or sixty-two miles, above the Earth, where the planet's atmosphere ended. No human had so much as approached that rarefied air, or lack of it. The airplane, introduced in 1903, had yet to reach half that altitude. On September 7, 1956, test pilot Iven Kincheloe had flown a Bell X-2 rocket plane to a height of 23.92 miles—an impressive distance above the Earth but still a long way from space.

Chosen to helm the Space Task Group was forty-five-year-old Robert Gilruth, a large, mostly bald man with thick, dark eyebrows who possessed an enduring and infectious passion for flight. He had been working in the NACA's test-engineering area since 1937 and in aeronautics even before that. As a young boy, he'd constructed his own rubber-band-powered model airplanes, and as an aeronautical engineering student at the University of Minnesota, he had helped design the world's fastest airplane. After eight years as a "dirty-hands" engineer, he had made a name for himself in the area of high-speed flight, and he was appointed director of his own fiefdom, the Pilotless Aircraft Research Division, whose purview included guided missiles and, eventually, supersonic flight. He was soft-spoken and, like von Braun, had a knack for inspiring those around him to do their best—he hired good people and mostly left them alone. His management style owed something to the Socratic method. "He never once told me what I should do," an employee would later remember about meetings with Gilruth, but nonetheless, he said, "I never once left not knowing what I should do." He believed in a bottom-up organization with as few bureaucratic layers as possible, where the lowest technician could talk to the highest-level manager. He inspired unequivocal loyalty, and it led to impressive results.

His chief designer was thirty-seven-year-old Maxime Faget, a brilliant engineer, small in physical stature, who had been working for Gilruth since 1946 and designing rocket-powered aircraft almost as long. Another young flight-research engineer, thirty-four-year-old Christopher Columbus Kraft Jr., was slim, dark-haired, and intense; he was a local boy, born and bred in the small town of Phoebus, just seven miles from Langley. He'd tried to enlist as an eighteen-year-old early in World War II but was rejected due to his damaged right hand, which had been badly burned in a fire when he was three. After graduating from Virginia Polytechnic Institute in 1944 with a degree in aeronautical engineering, he'd spent fifteen impressive years with NACA in flight-test operations. It didn't take him long to start contributing—in the last year of the war, he'd helped solve problems in the P-47 Thunderbolt and the P-51 Mustang and was made project engineer on the P-80 Shooting Star, the first American-made jet fighter. Along the way he developed a palpable competence and confidence, an aura of leadership that inspired those he worked with.

Kraft's orders from his boss were to write the flight plan for the first program missions—launch, orbital, reentry, and recovery. "Chris, you come up with a basic mission plan," he was told. "You know, the bottom-line stuff on how we fly a man from a launchpad into space and back again. It would be good if you kept him alive." Kraft's flight-operations division would control the spacecraft and monitor every aspect of its progress—and every biomedical detail of its occupant—in real time, through telemetry and other methods. Kraft realized that this quantum leap in ground support would require more than the traditional reinforced-concrete blockhouse; the concept of the mission control center was born and would quickly become reality. Someone would have to coordinate all this and make final decisions. Kraft wanted to be that person, and so the role of flight director was created. Soon he and everyone else were working sixty-hour weeks and bringing work home. No one at NASA complained; most didn't even think of it as work. This was fine for single employees but hard on the married ones, who spent less time with their families.

Faget, Kraft, and forty-four other NACA employees—thirty-five engineers, eight female secretaries and "computers" (professional mathematicians adept at operating mechanical calculators), and one male file clerk—would form the nucleus of the NASA division that would handle manned spaceflight.

On December 17, 1958—the fifty-fifth anniversary of the Wright brothers' flight at Kitty Hawk—America's first manned space program, Project Mercury, was announced. Named for the Roman messenger of the gods, its mission was simple: to put a man into orbit and return him safely to Earth. They had the rocket. They had the flight plan. But they were missing one important part: the man. But who should they hire? What were the requirements for a job that had never before existed?

The men in the program would need to be accustomed to danger—they had to know how to cope, and cope well, in high-pressure situations. The initial idea was to find individuals with experience in high-risk or stressful jobs, people who had developed the requisite level of toughness: test pilots, deep-sea divers, mountaineers, race-car drivers, balloonists, submariners, polar explorers, parachutists. Professionals in these fields were all considered, along with acrobats and contortionists; NASA, it appeared, would soon be recruiting from the country's circuses. (Other candidates that were suggested, half seriously, were midgets, since they took up less room; women, for their greater fortitude; Eskimos, because they were smaller and accustomed to different solar cycles; and Buddhist monks, who were less "time-oriented" and could put themselves in a trancelike state.)

But before the ranks of those stressful professions could be scrutinized, and before the halls of NASA's medical-testing facilities began resembling circus auditions, President Eisenhower was advised against choosing from such a broad pool of candidates. Not only would the process threaten to become the butt of jokes, but security clearances—since there would be classified aspects of the program—would complicate hiring. Another factor played a part in his decision not to cast such a wide net. "Ike felt it would be embarrassing," remembered a psychologist involved in the selection process, "for us to be going out to select astronauts when the best we could do at the time was a grapefruit-sized satellite." So in December 1958, the president decreed that the applicant pool would be limited to the nation's test pilots. (Of course, that meant men only, since there were no female military test pilots.) This would simplify both security concerns and the process itself, because their detailed personnel records were already on file. They would also be accustomed to pressure suits, complex cockpits, and the discipline of military life, since even civilian test pilots were former military men. And maybe, just maybe, their extensive flying experience would play an important part in the success of a mission, despite the automated capsule planned.

So in January 1959, NASA published the qualifications for the job of "Research Astronaut-Candidate." The starting salary would be between $8,330 and $12,770, a quite comfortable amount for the time. The minimum requirements for the job included fifteen hundred hours of flying time, graduation from test-pilot school, excellent physical condition, being between twenty-five and forty years old and no taller than five eleven (to fit into the small one-man capsule that was on the drawing boards), and having a bachelor's degree in engineering or the equivalent. The "equivalent"—as in, a technical degree other than engineering—would make an important difference to the future of U.S. space exploration.

Some worried that this new program was just another of the many man-in-space projects that already existed or had already been canceled. The air force had the X-15, Dyna-Soar, Man in Space Soonest (the final stage of which would be a moon landing), and another lunar-landing program, the top secret Project Lunex; the army had Project Adam; and the navy had a plan for a Manned Earth Reconnaissance spacecraft. There was also the concern that being part of this program might sidetrack a military man's career. Despite these qualms, the records of 508 test pilots were pulled. By January, that number had been whittled down to 110 who met the minimum standards. They were divided into three groups; the first two groups were invited to Washington for a top secret briefing about the Mercury project. The third group was kept in reserve in case not enough pilots in the first two groups decided to apply.

Though the pilots were assured that they would have a significant role in operating the Mercury capsule, some of the men thought it sounded like a stunt, not a serious research program. Put a man in orbit? For what purpose? And if and when that was accomplished, what then? But others were intrigued by the engineering challenge. They were test pilots, men addicted to flying the highest, the fastest, the farthest—and this venture offered the distant promise of exceeding the known limits in each of these categories. It was a test pilot's dream, and as one of the men chosen would point out, "This was a chance for immortality."

Some men declined, but most in the first two groups agreed to apply, so the third group wasn't brought in. That left almost seventy applicants. That number was soon shaved down to fifty-six, and a month later, after in-depth interviews and psychological tests, thirty-two candidates were invited to undergo various examinations that would result in six winners of the spaceman sweepstakes.

The physical exams came first. For eight days at the private Lovelace Clinic in Albuquerque, New Mexico, which specialized in aerospace medicine, doctors and scientists conducted more than sixty tests on their subjects. Every organ, orifice, system, and body part was thoroughly scrutinized. Many of the exams were based on physicals administered to potential U-2 spy pilots, but since the medical corps had little idea of what would happen to a human in the weightless vacuum of space, with its extremes of heat and cold, some of the tests were speculative. Most had their origins in air force space-medicine research collected over the previous decade. (Doctors were particularly concerned by the high incidence of "anal problems"—mostly hemorrhoids—in the fliers, so after a cleansing enema, each candidate was subjected to an especially thorough proctosigmoidoscopy with a tool the subjects dubbed the "steel eel.") The candidates were tested from early morning through late evening and spent hours on stationary bikes, on treadmills, on tilt tables, and in centrifuges. (Though none of them approached the astounding record of eighty-three and a half g's that an air force officer named Eli Beeding would reach a few years later, resulting in spinal bruising and temporary paralysis and blindness. As a comparison, an automobile crashing into a brick wall at forty-five miles an hour reaches only sixty g's.) They also delivered innumerable stool, urine, and semen samples. At the end of the eight days, only one of the thirty-two was deemed unhealthy enough to be eliminated, and that was solely for a high bilirubin level in his blood (a usually temporary finding).

Psychological and stress evaluations were next, so the men were immediately flown to the Wright Aeromedical Laboratory in Dayton, Ohio. More than two dozen exams were administered. Some were classics, such as the Rorschach ink-blot test and the 566-question Minnesota Multiphasic Personality Inventory, which assessed psychopathology. Others resembled parlor games, such as Draw-a-Person and Who Am I?, which required the candidate to write twenty answers to that question. (The first several answers—"I am a father," "I am a naval officer," "I am an American"—were easy, but after a while the subject was forced to a deeper level of self-examination and, hopefully, self-revelation.) Another favorite test was showing a blank card to a subject and asking what he saw in it. These and other tests evaluated personality, intellect, aptitude, peer sociability, and motivation, among many other traits. Psychiatrists delved into each man's adolescence to root out any feelings of sexual inadequacy that might have fueled his interest in high-performance aircraft. Then came half a dozen psychological-stress tests designed to evaluate each man's capacity to tolerate conditions expected in space. Some involved the subject's reactions to isolation, low pressure, severe vibrations, loud noises, extreme heat and cold, or blinking strobe lights; others had no discernible goal. Indeed, since no one knew precisely what the men would face during a mission, many of the tests had been invented for these evaluations and seemed bizarre—for instance, blindfolding the subject, sticking a hose in an ear, and pumping cold water into his ear canal until he was dizzy, or submerging a man's feet in a bucket of ice-filled water until they went numb or he couldn't take it anymore. Some tests might have been added just to measure the participants' determination. "They were free to be deliberately brutal," said one pilot, calling the doctors "sadists to a man." One doctor admitted as much: "We did our best to drive them crazy," he said. Throughout, the candidates were constantly photographed for medical records, sometimes when they were naked, often in embarrassing or undignified poses, the cameras pointing into virtually every orifice.

They were all active-duty pilots in excellent health, so none of the thirty-one washed out. But the barrage of tests and interviews helped rank which were more psychologically and physically suited for the job. When the dust—and heart rates and blood pressure—had settled, the doctors recommended eighteen men "without medical reservations." After a three-man committee reevaluated their interviews and reviewed their engineering acumen, seven men were rated ever so slightly higher than the rest, though the final results were kept strictly confidential. Bob Gilruth, director of the Space Task Group, decided to accept all seven. (Apparently none of the seven had been rejected for either of two little-known causes found in "Medical Standards for Selection of Crews": "extreme ugliness" or "any deformity which is markedly repulsive.") An especially important consideration was each man's ability to relate and interact effectively with others, because success in the program would depend on the men working with thousands of people toward an ultimate goal years down the road.

On the first two days of April, the chosen seven were called and invited to join the program. For some of them, the decision wasn't easy. They all had promising careers and likely promotions in their futures—and, possibly, fulfillment of dreams they'd spent their entire adult lives chasing. But this new aspiration had taken root, and every man accepted.

At two in the afternoon on April 9, 1959, at a press conference in a small ballroom at NASA's temporary headquarters in the Dolley Madison House in downtown Washington, DC, the "astronaut volunteers" were introduced to the world: Scott Carpenter, Gordon Cooper, John Glenn, Virgil "Gus" Grissom, Walter "Wally" Schirra, Alan Shepard, and Donald "Deke" Slayton. They were all married with children, each with an apparently perfect family, though a few of the marriages were far from it, a fact that would not be revealed until years later.

The Mercury Seven represented the air force and the navy equally—three men from each—with a Marine (Glenn) thrown in. For ninety minutes, the seven career military men sat at a long table in alphabetical order, dressed in conservative business suits and smiling nervously at the crowd of two hundred, while photographers pushed and shoved for better shots and reporters bombarded them with questions. Most of their responses were short and uninformative, but then the redheaded Glenn, who had recently set a transcontinental speed record and thus had some experience in dealing with the press, spoke thoughtfully about his wife and family and their support of his flying career. A few of the others rose to the challenge and supplied more in-depth answers. Glenn waxed eloquent on the program, comparing it to Orville and Wilbur Wright at Kitty Hawk: "I think we stand on the verge of something as big and as expansive as that was fifty years ago." If the crowd's reaction was any indication, he was the odds-on favorite to be the first in space.

The next day, the classmates of Gus Grissom's second-grader, Scott, hoisted the boy on their shoulders and carried him around the schoolyard. Then they called for a speech, and Scott told them what he knew about his dad's new job. The deification had begun.

Within days, the seven "astronauts"—the name they were given, to distinguish them from the Russian cosmonauts—became American heroes, long in advance of any astronautical achievements. They had volunteered for something that no one—not even their new colleagues at NASA—expected all of them to survive.

Virtually every hero myth of America—Daniel Boone, Davy Crockett, and so forth—was invoked. Perhaps more often, though, an enormously popular comic-strip and movie-serial hero came to mind. The parallels with the present conflict were vivid; each of these seven was Buck Rogers, an American transported to the twenty-fifth century, battling, as a book based on Buck's adventures put it,

> the terrible Red Mongols, cruel, greedy, and unbelievably ruthless, who for a time, all too long, utterly crushed a large part of humanity in a slavery frightful to contemplate. In their great battle craft, sliding across the sky...like a scourge over all North America, with their terrible *disintegrator rays* blasting men and entire cities into nothingness.

But what truly amazed people—and what was invariably emphasized in the newspaper stories over the next few days—was that their wives backed them in their desire to strap themselves into a capsule atop a rocket originally built to deliver nuclear warheads. Each wife was asked a variation of "Aren't you worried that he'll be killed?" more than once. Publicly, each one dutifully supported her husband's choice and downplayed the worry and the danger, which was nothing new to a test pilot's wife. Most Americans, it was clear, expected some of the men to die—after all, about half of the rockets launched blew up. Even Lloyd's of London, renowned for insuring almost anything, from Marlene Dietrich's legs to a yo-yo champion's fingers, would not cover these seven men.

The Mercury Seven, for their part, seemed unfazed. The test-pilot mortality rate at the time was horrendously high; in 1952, at the air force's test-pilot school at Edwards Air Force Base, sixty-two pilots had died in just thirty-six weeks. Overall, about one in four test pilots perished. Eight men had been killed getting just one fighter jet, the F-104, operational. Test pilots were on close terms with death, and they were even blasé about it. A few of the Seven had told their examiners that they were onboard with the program as long as there was a good chance of survival—by which they meant at least 50 percent. Because of this, some of the psychologists suspected that these test pilots might even have a death wish. They did not; they had simply become professionally inured to the concept, believing that accidents could be avoided through knowledge and careful planning.

The astronauts appeared to be a remarkably homogeneous bunch. Each was from the Midwest and had an IQ over 130, well above average. Each was a "superb physical specimen," though not muscle-bound—one newspaper likened them to "a group of square-jawed, trim halfbacks recruited from an All-American football team." All were from small towns, all were middle-class, all were Protestant, all were white (in fact, there were no non-white test pilots at the time), and each an only or eldest son. Six of the seven were veterans of World War II, the Korean War, or both. They spoke of duty and country and faith in platitudes, but it was clear they meant it. The American public, and the press, ate it up.

The Soviets hadn't been satisfied with their satellites. On January 2, nine weeks before the Mercury Seven were introduced, the Soviets had accomplished another first, launching a man-made object to leave Earth orbit. The plan was for Luna 1, a four-foot-wide metal ball, to crash into the moon, but it missed by 3,600 miles. When it achieved orbit around the sun instead, they renamed it Mechta ("Dream"). And the Pentagon reported receiving word from behind the Iron Curtain that another Soviet space spectacular was coming up soon—maybe they would even send a man into space.

CHAPTER THREE

"THE HOWLING INFINITE"

Man's first trip into space will be a new human experience, to be highly desired by courageous and adventurous men, but fraught with hardships, difficulties and danger.

TIME, APRIL 20, 1959

BY THE TIME NASA'S Space Task Group was formed, in the fall of 1958, Max Faget had been thinking about the difficulties of manned spaceflight for a few years, often while standing on his head. One particular obsession of his was the danger of atmospheric reentry, in which the friction from plummeting into Earth's thick atmosphere at ten thousand miles an hour or more would result in temperatures of about three thousand degrees Fahrenheit. Meteorites were also on his mind.

Faget was a man whose brain worked somewhat differently than others'—though usually successfully—when fixed on a problem; it was a trait that apparently ran in the family. His father, a doctor, had helped develop the first practical treatment for leprosy; one of his great-grandfathers, a New Orleans physician, had discovered an accurate way to diagnose yellow fever. Young Max grew up in Louisiana building model airplanes and submarines with his older brother and reading science fiction novels and *Astounding* magazine, the first "hard SF" publication that insisted on stories with a solid grounding in science. A gymnast in college, wiry and about five six, Max had an elfin appearance (decades later, it would invite comparisons to the Yoda character from the Star Wars movies). After receiving his engineering degree from LSU in 1943, Faget spent almost three years as a junior naval officer aboard a submarine in the South Pacific, eventually serving as executive officer. He had a penchant for startling people in conference rooms, restaurants, almost anywhere, by leaping over chairs and sometimes standing on his head to improve blood circulation to his brain while continuing discussions with colleagues.

His roommate at LSU had been a chemical engineering major named Guy Thibodaux, a Cajun also from New Orleans. Neither ever made the honor roll or took to rote learning, and while other students were pulling all-nighters for exams, they played pool and watched movies. Before they went off to war—Faget with the navy, Thibodaux with the army—they made a vow that if they survived, they'd reunite and look for jobs together.

In the spring of 1946, Guy got a call from Faget, who was following up on that promise and had an idea about where they should apply. Max's father had a 1941 Ford coupe that they could borrow. It had airplane tires—all that was available, since there was still a shortage of rubber—but it ran, so in June the two headed to the NACA's Langley Memorial Aeronautical Laboratory in Hampton, Virginia, where they walked in and applied for jobs wearing Hawaiian shirts, work pants, and sandals. The agency being what it was—somewhat eccentric in both hiring and methods, often finding employees at model-airplane contests—both were hired immediately. (It didn't hurt that the man in charge of making the decision was also an LSU graduate.) Thibodaux was put to work in rocket propulsion and Faget in ramjets for Robert Gilruth's newly created Pilotless Aircraft Research Division (PARD). At a starting annual salary of $2,644, they were now working with the world's leading experts on aerodynamics in the most exciting venture they could imagine—trying to break the sound barrier. Not bad for a couple of

Louisiana Aggies without graduate degrees. Both of them advanced quickly in the loose, merit-based NACA hierarchy—a "classless society where every member of the team was an equal contributor to the success of NACA's mission," Thibodaux observed later—and Faget was soon Gilruth's right-hand man in PARD. Only three years later, Thibodaux would be running his own section.

Faget had been working on supersonic aircraft, including the X-15, for years, and his unique ability to find the simplest solution to design problems often led to valuable breakthroughs. He understood, sooner than many of his colleagues, that there was no advantage to an aerodynamic shape in space, since there was no atmosphere—no air—to act on it or slow it down. With that in mind, he was excited about designing a craft that would operate in a vacuum, and he was even more excited when he and other Space Task Group members visited Huntsville just a few days after NASA started up. Faget, von Braun, and his engineers discussed working together to launch a manned capsule into space.

Faget and several others at the NACA—a small group that some at headquarters called the Space Cadets—had been wrestling with the thorny problem of reentry for a while. They'd been "bootlegging" a manned space program for at least a year before NASA was formed; they figured that since no one had told them *not* to do it, they might as well, even without official approval. They all agreed that space was where they were heading. The big question was: What shape should the spacecraft be? Initially, they chose a needle-nosed, streamlined spaceship—like the ones they'd read about in science fiction magazines and seen in space artist extraordinaire Chesley Bonestell's illustrations—to offer as little air resistance as possible. Someone even suggested using the X-15 experimental plane, designed to fly up to the fringe of space—maybe they could send it to the moon. Faget knew that wouldn't work. For one thing, after reaching Mach 6, the X-15 began suffering serious heat damage. Besides, its aerodynamic shape wouldn't dissipate that enormous heat, and with no heat-resistant external surface, the plane would completely disintegrate when it reentered the atmosphere at extreme velocity.

So when two of Faget's colleagues, Harvey Allen and Alfred Eggers, pointed out that meteors with rounded noses were aerodynamically stable and survived the searing heat of the plunge—they had been studying the concept for years—Faget and designer Caldwell Johnson came up with a blunt-nosed shape like a shuttlecock that would slow down the craft on reentry and create a shock wave that would deflect much of the blast-furnace heat away from and around it. A paper Faget presented in March 1958, "Preliminary Studies of Manned Satellites—Wingless Configuration, Non-Lifting," introduced key features of a simple but workable spaceship. His coworkers at Langley weren't convinced that a blunt design was a good idea, and neither were many others in NASA. The staff at Ames Research Center in California believed a craft with some lift would be better, and military flight surgeons argued a man would black out from the eight g's expected in this craft during reentry. But the unassailable simplicity and logic of Faget's arguments—and the fact that a ballistic craft could take only one path and thus its splashdown point could be easily predicted—eventually won them over, and his blunt body design was officially adopted.

After much refinement, endless wind-tunnel, spin-tunnel, heat, and drop tests, and trajectory work on primitive computers, the cone-shaped, blunt-bottomed Mercury capsule was finished. Faget added a thick ablative heat shield (the concept of which had been described in 1920 by rocket pioneer Robert Goddard and later fine-tuned by Eggers) made of an aluminum honeycomb and several layers of fiberglass. The outer layer of the shield would absorb some of the heat and burn away, or ablate, protecting the capsule itself during reentry. A pack of small rockets strapped to the bottom of the craft would also decrease its speed and massive g-forces during reentry. Faget and a couple of other aerodynamicists had determined the capsule would decelerate at eight g's or so, which would be bearable if a man was on his back on a surface designed to help him withstand that force. So Faget and his colleagues fashioned (and patented) a fiberglass contoured survival couch that would do the job.

Now, if they could keep the weight down and secure the Mercury capsule to the nose of a rocket powerful enough to launch it into space (first just a simple ballistic arc beyond Earth's atmosphere and then a larger booster that would reach the 17,500 miles per hour necessary to balance the Earth's gravitational pull and maintain a stable orbit), it just might work. And, of course, if they could find a way to keep its occupant alive.

The craft that would convey the first American into space was nothing like any spaceship in the comic strips or the movies—or in the previous history of manned flight. Some dubbed it the "Flying Ashcan." Since the underpowered Redstone's payload capacity was limited, the capsule wouldn't be very large—eleven feet long and six feet across at the wide end of the cone. And it would weigh only three thousand pounds; its shell would be constructed of thin but strong titanium covered with hundreds of heat-radiating shingles of equally strong alloys to resist the expected thirty-five-hundred-degree temperature of reentry. The crew compartment would be just big enough for one person. "You don't climb into it. You put it on," said John Glenn. The astronaut would sit on his personally shaped contour couch with his back to the heat shield, facing about a hundred and twenty switches, levers, buttons, and fuses a couple of feet in front of him. There was no computer; any trajectory or reentry calculations would be made by computers on the ground and transmitted by radio. The capsule's path could not be changed, although its attitude—the direction it was pointed—could, both from the ground and by the astronaut, with eighteen small thruster jets powered by hydrogen peroxide that altered the three axes of up-down pitch, right-left yaw, and side-to-side roll.

After the basic plans were set, McDonnell Aircraft, producer of many of the country's finest fighter planes, was chosen to build it. Its bid was far from the lowest, but the Mercury program had been placed on the Master Urgency List, meaning its administrators did not have to choose the lowest bidder.

It wasn't much of a spacecraft, this hollow meteor, but it would get the job done—the job of putting a man into space and returning him to Earth alive—if the rocket it perched on did *its* job. Because NASA was required to use only rockets already in production, von Braun's Redstone, designed for the battlefield, would be employed for the first flights, the suborbital ones that would be quick up-and-down trips. The kerosene-fueled Atlas, the new ICBM the air force was developing, would launch the Mercury into orbit on later missions, since it was the only one in the nation's arsenal with the thrust capable of doing the job. But the recent history of the Redstone and Atlas boosters wasn't encouraging. The Atlas, especially, had a nasty habit of exploding or malfunctioning in some other way. It had been designed to carry H-bombs, not humans, and its skin was so thin that a steel belly ring, like a large, jury-rigged hose clamp, would need to be fitted around its girth as a brace. But the rocket engineers and launchpad technicians, aided by newly hired safety and quality-assurance inspectors, committed themselves to doing all they could to keep their passengers from being blown to bits.

Some of the dangers of space travel were known. Many more were not.

A fragile human in the vacuum of space would die almost instantly. Even if he held his breath, it would take only seconds; the absence of external pressure would cause his lungs to rupture and send air into his bloodstream, resulting in a quick death when air bubbles lodged in his heart and brain. Even if his lungs didn't rupture, the deoxygenation of the blood would result in the loss of consciousness in fifteen seconds or less. As the water in his body vaporized and his oxygen disappeared, the moisture on his tongue, in his eyes, and elsewhere would begin to boil and bubble; his skin and the tissue beneath it would start to swell and turn bluish purple; and the gases in—and possibly the contents of—his stomach, bowels, sinuses, and other body cavities would release rapidly. His heart would continue to beat for a minute and a half or so. If pressure and oxygen were restored before then, the astronaut might survive.

And what effects would weightlessness have on a man? No one knew exactly, but several possibilities were postulated. Gravity, many experts asserted, was necessary for some body organs to function. Without it, eyeballs might explode or vision might blur, the heart might stop beating, esophageal muscles might constrict, the digestive system might shut down, the vestibular system of the inner ear might malfunction and cause extreme dizziness and nausea, or sleepiness might occur. The brain might simply cease to function.

And could an astronaut survive the fierce gravitational forces of acceleration and deceleration during liftoff and reentry and remain conscious without suffering any lasting damage? What would be the effects of space radiation unfiltered by the Earth's atmosphere? Perhaps it would burn retinas and skin, mutate DNA, sterilize gonads. A burst of deadly radiation from a solar flare might kill an astronaut. And heaven help him if he had a bout of space-sickness while wearing a pressurized spacesuit and helmet. Without gravity, the vomit would remain near his mouth and nose; there would be no way to wipe it away, and with every breath he would inhale more until he drowned—hardly a heroic death.

These fears and others consumed the medical community. And if the physiological dangers weren't enough, there were the psychological ones as well. An astronaut in a confined space for an extended period might become depressed and take his life. He might even succumb to what some psychologists called the breakaway phenomenon (a sense of being completely cut off from everyone on the planet) and decide not to return to Earth. Others suggested that the spaceman might faint or even die of fright during the flight. Some thought that he might go berserk.

There were other worries: meteors large enough to puncture a spacecraft's hull, extreme noise or vibration strong enough to rip a man's organs loose. All possibilities were carefully considered and researched. The prime concern of each mission would always be the safety of the astronaut, and toward that end, each system was refined to a point never before seen in any machine, vehicular or otherwise, from the planning and designing stages to production, training, and monitoring. The engineers gave the old term *redundancy* new meaning. Virtually every system—electrical, environmental, navigation, and so on—was backed up two and sometimes three times. The oxygen system, for example: If an astronaut's suit failed him in flight, he could open his faceplate and breathe the cabin's pressurized air, which was 100 percent oxygen (unlike the Earth's atmosphere, which is 78 percent nitrogen, 21 percent oxygen, and 1 percent other gases). If that failed, an emergency supply of oxygen—about eighty minutes of breathing time—was available; that would keep him alive long enough to finish an orbit and make an emergency reentry. If all three systems failed, he would be dead within minutes.

And though a mission might last only fifteen minutes, wherever possible, every unit was tested ten to a hundred times longer than that, until its reliability could be statistically measured and predicted to the nth degree. As von Braun described it:

A methodology was created to assess each part with a demonstrated reliability figure, such as 0.9999998. Total rocket reliability would then be the product of all these parts' reliabilities and had to remain above the figure of 0.990, or 99 percent.

Reliability of each part and redundancy in each system became central to NASA's culture. But not every system could be backed up, and no matter how well components were made or how rigorously they were tested, they occasionally failed, and 99 percent is not complete reliability.

The astronauts were supposed to be little more than passengers in a fully automated system; in this grand experiment, they were glorified guinea pigs or, at most, the final backup in emergencies. But after the press conference, the seven test pilots became national celebrities, and they realized they could use their celebrity to effect change in the program's hardware and in their role in the mission. Since the men couldn't be replaced without national embarrassment, their popularity gave them the power to make demands. The capsule was designed to fly without a man—at least at first—but the seven test pilots did their damnedest to change that.

In May 1959, the astronauts visited the McDonnell plant in St. Louis to inspect a mock-up Mercury capsule that was radically different from any craft they had ever flown. Deke Slayton stated the obvious: "The thing ain't got no wings!" They were surprised to find that there was no front window, just two small portholes that were too far away from the astronaut to be useful. Pilots needed a front window—at least, they had since Charles Lindbergh crossed the Atlantic without one and had to use a periscope to see around his extra-large fuel tank—and after they insisted, one was added, though it wouldn't be available for the first mission. Another problem was the door—there wasn't one, or not one that could be opened from the inside, since the hatch would be welded shut after the astronaut wedged himself into the capsule. That was fine for the structural integrity of the spacecraft but not for the man inside, especially if he wanted to egress quickly. So the hatch was redesigned with explosive bolts that allowed the astronaut to open it should that become necessary. There were other design problems the astronauts noted—the instrument panel and switch accessibility, for instance—that led to major changes. But the Seven didn't agree on everything. For example, Deke Slayton didn't like the three-axis attitude-control stick, which enabled a pilot to manage yaw, pitch, and roll with one hand. Slayton lobbied for an aircraft's stick-and-rudder foot pedals, but the other astronauts didn't join him, so the hand controller stayed.

Weight issues, and the brevity of the first few flights, dictated a meager reserve-fuel capacity. No one expected that to be a problem.

One factor dominated all spacecraft-design decisions and always would: the unforgiving equation involving weight, gravity, and thrust. The Atlas, which would be used for orbital missions, was over four times more powerful than the Redstone, but it would not be available for manned use for some time. The Redstone's limited thrust of eighty thousand pounds dictated that the capsule be small and relatively light. Every pound, every ounce, was important.

The Mercury Seven began training, and so did another group of American spacefarers. They underwent similar tests, and they would face the same mortal dangers, though they hadn't volunteered.

The job of astronaut was hazardous, but it was safer than the job of astrochimp. In the earliest days of space exploration, other life-forms had been hurled into the void atop primitive boosters: rabbits, mice, fruit flies, and so on. But as it became increasingly clear that *Homo sapiens* would at some point journey into space, nonhuman simians quickly became the standard passengers. (In Russia, dogs preceded humans, since they were frequent medical-research subjects and easier to work with than chimps.) Many nonhuman spacefarers had to give their lives for their hominid cousins before a spacecraft was officially "man-rated"—that is, deemed safe to transport a human into space and back to Earth.

The army had been investigating the biological effects of space travel on primates since April 1948, when a small rhesus named Albert was rocketed thirty-nine miles into the atmosphere in a capsule aboard one of von Braun's V-2s; his oxygen supply failed, and he suffocated. The next year, in June 1949, another rhesus, Albert II, rode a rocket eighty-three miles into the stratosphere—technically becoming the first Earthling in space—and survived, at least long enough to die on impact after a parachute failure. In September 1949, Albert III, a cynomolgus monkey, died when his V-2 exploded at thirty-five thousand feet, and in December, Albert IV, another rhesus, perished on impact after another parachute failure, which was also what happened to Albert V in April 1951.

Not until September 1951 did a primate launched into space survive the impact. A rhesus named Albert VI and eleven mice shipmates reached an altitude of forty-five miles and returned safely, though Albert VI died two hours later, as did two of the mice, from overheating in the sealed capsule while waiting for their recovery team. Eight months later, two cynomolgus monkeys named Patricia and Mike survived a quick jaunt fifteen miles into the atmosphere.

By that time, two Russian dogs had survived suborbital flights and returned to Earth safely. The Soviets flew so many animals that one orbital launch, in August 1960, carried two dogs, a rabbit, forty-two mice, two rats, many flies, and several plants and fungi. (One NASA engineer called it "the herd shot 'round the world.") They all survived the journey.

After the Mercury project became official, monkeys continued to be sent into space in capsules atop various rockets, and most survived; in May 1959, aboard a Jupiter, Able, a rhesus monkey, and Baker, a squirrel monkey, penetrated three hundred miles into space and survived thirty-eight g's during deceleration. The Mercury capsule and its planned booster, von Braun's Redstone, would be next.

A few days after their April 1959 press conference, America's human astronauts reported for training to Langley Research Center—a large, ordinary-looking World War I–era building they shared with the Space Task Group. They were given a single large office to use; seven steel desks and chairs were wedged into it in a U shape. Their nineteen-year-old secretary sat at a desk outside. Most of her job consisted of handling their mail, a task that would soon become overwhelming due to the sheer volume of it.

For the rest of 1959, the men took crash courses in everything space-related. These were airplane pilots who knew little about rockets and missiles. They received graduate-level instruction from scientists and engineers on astronomy, meteorology, geography, aviation biology, physiology, rocketry, and more. None of their teachers had ever taught astronauts before, so the lectures were wide-ranging, though the basic format drew heavily from traditional flight training and test-pilot methods. They spent endless hours becoming familiar with every phase of their spacecraft's planned flight and every one of its systems. Because the actual vehicles had not yet been built, the astronauts had to settle for reviewing design drawings and blueprints. Eventually the Mercury Seven would employ simulators, but since no one knew what the controls of their spacecraft would look like, they couldn't practice flying it, so most of their early training was environmental rather than procedural.

Scientists attempted to familiarize the men with various conditions of spaceflight, such as weightlessness, heat, pressure, acceleration and deceleration forces—in short, every aspect of what an astronaut might experience during a ride on a rocket into space. Training sessions involved heat and pressure chambers and many hours riding the unforgiving centrifuge, in which they endured as many as sixteen g's and learned techniques to build up their tolerance. Another training rig was a gimbaled, caged whirligig with the ominous acronym MASTIF (Multiple-Axis Space Test Inertia Facility). It consisted of three cages made of aluminum tubes, one inside the other and each hinged to the next, that rotated independently and on different axes. This fiendish device could spin a man at thirty revolutions a minute in three axes, and testing had determined a human tolerance of about thirty seconds, beyond which even the most experienced pilot would toss his cookies. A red "chicken switch" button set off a loud klaxon that told technicians to kill the machine. One ride on the MASTIF was enough for the Seven—"You even felt like getting sick if you just stood there and watched another astronaut take his turn," said Gus Grissom. Then there was the Slow Rotating Room, designed to accustom its occupants to a spinning spacecraft. Finally, they spent valuable time learning to deal with weightlessness while flying in a zero-gravity-inducing ballistic parabola in an F-100 trainer. C-130 and C-135 cargo planes could do the trick for thirty seconds, allowing them to float and do flips in the much larger cabin cleared out for just that purpose. Those sessions, at least, were pain-free and enjoyable.

The men were expected to keep themselves in top physical shape but were not told how to do it, so each one decided on his own regimen. John Glenn ran every morning—he felt he had to, to combat a weight problem. Scott Carpenter lifted barbells and exercised on a trampoline. Wally Schirra and Gus Grissom played a lot of handball, and the others joined in now and then, with Alan Shepard emerging as the best player. A quick, strenuous game of handball would become the favorite workout of the astronaut corps.

Although NASA employed a team of physicians, a twenty-three-year-old air force nurse named Dee O'Hara was assigned to monitor their health, at first only when they visited Cape Canaveral for a mission, then later on a regular basis. Initially she was intimidated by them—they had become major celebrities, and there were few women in the world of spaceflight—but the astronauts put her at ease right away, and they soon came to trust the always cheerful and very capable O'Hara. Before a flight, she was the only one they would allow to draw their blood. Pilots as a rule avoided doctors—physicians had the power to ground a pilot for any one of a seemingly endless list of reasons—but O'Hara made a deal with the astronauts: if one of them came to her with a medical problem, she would keep it private unless it compromised a mission.

Since there was no guarantee that a spacecraft would end up where it was supposed to when it returned, they also took water-, jungle-, and desert-survival courses in far-flung parts of the world. The survival trips made for great photo ops for *Life* magazine, which had negotiated a five-hundred-thousand-dollar deal with NASA for exclusive magazine and book rights to the stories of the seven men and their families. The money would be split among them equally over the next three or four years—the length of time the program was estimated to last. This was a boon to the men, who were each being paid only a standard officer's salary, and would enable them to buy or build houses later.

The Mercury Seven began to spend increasingly longer periods at the production plants contracted to build various components of the Mercury project, not only to attend design briefings and inspect and critique production for improvement, but also to inspire craftsmen and technicians there to the highest levels of workmanship. Grissom, perhaps the most introverted of the Mercury Seven and certainly the most tight-lipped, was present at a gathering of eighteen thousand employees at the Convair plant in San Diego, California, where the Atlas rocket was under construction, and he was asked to say something to the crowd. He walked up to the microphone and said, "Well—do good work!" and then turned and sat down. The workers roared their approval and adopted the phrase as their mission statement. Posters of Grissom captioned with his brief statement were produced, and a huge banner with the words was hung above the plant's work bay.

Each astronaut was also assigned a specific area of responsibility. Glenn was given cockpit layout, Schirra the pressure suit, and so on. Each would regularly brief the other six on developments in his specialty.

The Seven had barely begun training when they were flown down to Cape Canaveral on May 18 to see their first missile launch—an Atlas, similar to the one that would eventually boost one of them into orbit. A minute after the rocket lifted off and just after it began to nose over toward the horizon, it exploded into a million pieces. It was the fifth straight Atlas failure. The astronauts looked at one another, and Shepard said to Glenn, "Well, I'm glad they got that one out of the way." In December, only one-third of U.S. satellite-launch attempts reached orbit. That was a fine average in baseball but not encouraging in the field of manned spaceflight.

As a result, the astronauts were less than optimistic about their chances. At a December press conference they were asked about the odds of their coming through alive. Oklahoma-born Gordon Cooper, the youngest one at thirty-two, answered first. "Well," he drawled, "as the engineers say, barring any unforeseen circumstances, I'd say we've got one hundred percent chance of success." Alan Shepard added, "We might lose the first guy, but the second, third, or fourth would make it."

In July 1960, when another Atlas exploded soon after liftoff, that booster's failure rate reached 45 percent. Since the rocket had not met its mission objectives, an exhaustive review of the entire Mercury-Atlas program was undertaken. The Mercury-Redstone program continued, though that too was experiencing delays, which made for bad press. Despite the criticism and the fact that not a single mission had been carried out yet, Gilruth began wondering what would follow Mercury. No one knew for sure. He thought the program might be a dead end, phased out after its three-year goal was achieved. His Space Task Group had some grandiose plans, but nothing had been approved. For now, they all threw themselves into the formidable job ahead of them.

On the other side of the globe, the Russians hadn't sent a man into space yet, but they continued to earn headlines with various firsts. In September 1959, the space probe Luna 2 was deliberately crashed into the moon, becoming the first man-made object placed there, however violently. Three weeks later, on October 4, another probe, Luna 3, flew around the moon and sent back the first photographs of its far side. The United States was still a distant second in the prestige department, even though outside the Soviet Union, little information about the Russian space program was available. But its director—known only as the Chief Designer to the Russian public and the West—took note of the announcement of the Mercury Seven astronauts and began preparations for his own corps of spacemen.

In mid-November of 1959, von Braun's old mentor Hermann Oberth, who was now in the United States working as a technical consultant on the Atlas rocket, claimed to have intelligence reports that the Soviets had launched a manned spacecraft in 1958 that crashed, killing the pilot. In December, an Italian news agency announced unconfirmed reports from "most reliable sources" that four Russian cosmonauts, including one woman, had died in spaceflight. A few Soviet academicians promised a manned flight to the moon in the not-too-distant future. That seemed unlikely, but what would come next was anyone's guess.

Early in January 1961, six chimpanzees and their medical teams and handlers were moved from Holloman Aerospace Medical Center in New Mexico to Cape Canaveral. These Mercury Six primates—four females and two males—were ready, though perhaps not willing, to risk their lives for the sake of American prestige. Over the next several weeks, for hours each day, they were strapped onto small contour couches in mock-ups of the Mercury capsule, and they became accustomed to it and to the timed tasks involving a panel of red, white, and blue lights and two levers that they had been training with at Holloman. Two of them, Chang and Enos, even experienced brief spells of weightlessness on the cargo planes and trained on the centrifuge.

The astronauts had to put up with a lot from their fellow test pilots. Part of the problem was that the men would be flying a capsule requiring little actual piloting, since it would be controlled almost entirely by automatic electronic signals. "Backing up his onboard systems and taking over in the event of malfunction"—that was how one NASA official described the job of astronaut. It sounded far removed from the stick-and-rudder work they prided themselves on, and they had to endure ribbing from other test pilots, who pointed out that they would ride the rocket, not pilot it. "Man in a can," a popular phrase bandied about, became "Spam in a can."

The most vociferous critic of the program was America's best-known test pilot, Chuck Yeager, the man who had broken the sound barrier in an X-1 experimental plane in 1947 and would soon take command of the air force's test-pilot school at Edwards AFB. He was heard to say that the astronauts were going to have to sweep the monkey shit out of the capsule before they rode it into space.

Yeager hadn't attended college, thus making him ineligible for the program, which might have colored his opinion, and there very well might have been a degree of envy on the part of Yeager and other test pilots over the attention—and the *Life* magazine money, a hefty twenty-four thousand dollars a year each—the Mercury Seven were receiving. But even the folks at MIT, selected by NASA to develop the capsule's guidance, navigation, and control systems, cracked jokes about the monkeys: "After the chimp, the chump." The astronaut trainees tried to shrug off the ridicule, but it rankled. In October 1959, at the annual meeting of the exclusive Society of Experimental Test Pilots, Deke Slayton gave a speech specifically intended to, in his words, "defuse some of this Spam bullshit." He made the case that experimental test pilots were necessary in spaceflight, since the likelihood of a failure or an emergency would require their experience, knowledge, and quick reactions. The audience gave him a standing ovation.

But the apes made everything worse. Other test pilots sneered at the primates preceding the astronauts in the capsule. If a monkey could do it, they opined, it couldn't be much of a challenge. Did Charles Lindbergh have a monkey fly the *Spirit of St. Louis* first? The Mercury Seven were used to flying experimental planes before the kinks had been worked out. The increased hazards involved in a space venture made no difference to them—in their minds, they believed they had risked far greater dangers, not only during test flights but also in wartime combat missions. They understood that the simian experiments were necessary, but they didn't like it. They would gladly have flown the risky monkey missions themselves, the dangers be damned.

But given the complexities and inherent dangers of spaceflight, the complex calculations necessary for navigation and trajectory, the massive rocket thrust needed for liftoff, the precise in-flight adjustments required, and the perils of reentry, the creators of

Mercury had known from the beginning that the spacecraft would be controlled by a group of engineers on the ground who could monitor its many systems. The pilot would be secondary to the engineer, at least for the time being. And the chimp flights would continue.

In the summer of 1960, the astronauts started using a more sophisticated simulator, the Mercury Procedures Trainer, which provided a reasonable facsimile of actual flight in the spacecraft. The trainer featured an exact replica of the cockpit and instrumentation that used state-of-the-art computers to simulate every conceivable in-flight emergency situation—275 separate systems failures, to be exact. To further mimic an actual flight, the astronauts trained in their pressure suits until they could do entire missions with their eyes closed and still not miss flicking the right switch or pushing the right button. Eventually they began practicing in the capsule itself, each spending endless hours on his back strapped onto the contour couch Faget's team had designed specifically for his body to help him withstand the expected fearsome pressures of liftoff and then reentry into the Earth's atmosphere. And since there was no urine-collection device in the pressure suit—the first few missions weren't expected to last more than a quarter of an hour or so, and it hadn't been deemed necessary—they learned to just let go if necessary and allow their thick underclothing to absorb it.

As the space race became increasingly public and competitive, so did Gilruth's Space Task Group's search for engineers, the people who would take the calculations and theories of scientists and turn them into reality. In the summer of 1960, the Space Task Group had seven hundred employees. Two years later, that number would grow to over two thousand. At the same time, NASA was setting up recruitment offices in major cities that would help swell the ranks to sixteen thousand.

Such aggressive efforts were necessary if the Americans were to have any hope of competing with the Russians. At the time, there was such an engineer shortage in America that on one Sunday in 1958, the *New York Times* ran 728 want ads for engineers and scientists, and most of the ads were for multiple openings. Across the board, these missile, aircraft, and electronics companies lured new grads with good salaries, bonuses, and various amenities.

But it was hard to beat the appeal of NASA, whose message, sometimes in these exact words, was "We're going to the moon. Want to come along?" NASA recruited on college campuses, by word of mouth, and through referrals and ads in trade journals like *Aviation Week* and *Missiles and Rockets.* "Destination Moon!" shouted one such ad, and another began "You can be sure to play an important part in the exploration of space when you join NASA." Few engineers could resist this siren call. For a generation of young men raised on Buck Rogers and Flash Gordon; the smart, hard science fiction that began appearing in the forties; the sci-fi films of the fifties; and, of course, the writings and appearances of von Braun in books, magazines, and TV, it was nearly impossible to walk away from NASA. But even without the lure of romance, in the old-fashioned sense of adventure, the attraction was immense. The pay there was decent, around five thousand dollars for a young man with an aeronautical engineering degree. That was less than the private industry paid but still a good starting salary for a young, single college grad, especially a small-town boy from a midwestern school, as so many of them were, kids whose families didn't have much money and who had had to work hard to afford college. But more important, the program promised to be the largest engineering project since the Panama Canal, and a successful flight to the moon would be the greatest technological achievement in history.

It didn't take the kids fresh out of college long to figure out the culture of NASA, which was sink-or-swim—the ones who didn't blend in or who didn't learn fast enough were there one day and gone the next.

Early on, the engineers ran into the astronauts quite often. One young man had been miserably pursuing his master's degree when he noticed a NASA recruiting booth in the library. He signed up, quit school, and, four months later, after background checks, was told to report to Cape Canaveral. At the end of his first day, one of his co-workers drove him over to see a simulator, and the new hire was invited to take a spin. He did. When he climbed out, everyone was gone except Gus Grissom, who offered him a ride back to the main building. They jumped in Grissom's new blue Corvette and roared off half a mile up a gravel road at eighty-five miles an hour. Then the astronaut veered onto a two-lane paved road and floored it to a hundred and twenty. He turned to the engineer with a grin and said, "Are you having a good time?" Then he entered the freeway and pushed it to a hundred and forty. Just having fun with the new guy.

As America's seven instant heroes took a crash course in the brand-new job of astronaut, other areas of the program worked through various problems and difficulties. After more than a year of political infighting between the air force and the army—the latter reluctant to give up its Germans, who had been anointed great patriots after the successful Explorer launch—and of lobbying by NASA administrator T. Keith Glennan, the genteel former president of the highly regarded Case Institute of Technology, Eisenhower finally approved the Redstone Arsenal's transfer. Von Braun would be director, of course, and though his administrative heads were American-born, all of his sixteen technical department leaders had worked with him since Peenemünde except for one who had been a Luftwaffe pilot during the war. Many of the German specialists had been wooed by private industries, most of them defense-oriented, but they were happy at Huntsville and intensely loyal to von Braun, their savior. He had rescued them from almost certain death and brought them to the promised land to prosper and devote their lives to their truest love—rocketry. They all felt the same way he did about spaceflight. "We'd really like to go to the moon instead of aiming at puny targets two hundred miles away," one said, off the record.

Von Braun had initially considered NASA a "baby agency" that wouldn't be around for long, and he seemed to be favoring the air force as his next employer. "All I really want is a rich uncle," he told a colleague. The air force was not yielding the high ground easily. Why shouldn't it handle the exploration of space, which was after all just an extension of the airspace it already had dominion over? Von Braun's army team at Huntsville had been working on design studies for a super-booster since the spring of 1957, six months before Sputnik was launched. Von Braun was sure of America's future among the stars and confident there would be a need for a rocket powerful enough to launch heavy payloads—astronauts, probes, space-station components, and so forth—into space. Sputnik's launch had gotten those plans funded in August of 1958. But when Saturn's rising costs led to rumors of the project's cancellation and the likely loss of 75 percent of the Redstone Arsenal's jobs, NASA suddenly became a much more attractive home. Von Braun had come around, and when Eisenhower finally approved the transfer of his team and its facilities, he was convinced it was the right decision and that NASA would be well financed. Von Braun breathed easier in January of 1960, when his Saturn became a national priority and Eisenhower reluctantly agreed to a large increase in the booster's budget.

On July 1, 1960, the Redstone Arsenal was renamed the George C. Marshall Space Flight Center, after the chief of staff of the army under Roosevelt and Truman. Finally, von Braun and his rocketeers would be free of military supervision—in his words, they'd be working on "spaceflight for spaceflight's sake." Although some at NASA were cool to von Braun—Chris Kraft despised his "Teutonic arrogance" and his celebrity, and the two had almost come to blows at a party—Gilruth was glad to have him, or at least have his rocketry expertise. Privately, he told Kraft, "He doesn't care what flag he fights for." (After a few clashes early on, as each man jockeyed for more control within NASA, von Braun and Kraft worked well together and came to respect each other. But Gilruth never quite accepted the former SS member; after a few drinks, he would complain about "our damned Nazi.")

The astronauts, for their part, overlooked their former enemy's equivocal allegiance. Impressed with his passion for manned spaceflight and won over by his charisma, they had bonded with von Braun on a visit to Huntsville just a few months after their selection. He had only recently given up a longtime desire to voyage into space himself, and he took a liking to these men who would go in his place.

As the space program evolved, the astronauts endured long workdays, and much of their time was spent on the road. They frequently traveled and trained together, except for the week or so each month that they devoted to keeping up with their specific areas of responsibility. It was a busy schedule, and it would get even busier, for politicians, contractors, businessmen, and just about everyone else wanted to be seen with them, so the astronauts tried to oblige. Sometimes that meant accommodating a congressman or business leader supportive of the program; at the request of a Missouri senator, Scott Carpenter once appeared at a supermarket opening in St. Louis. The astronauts quickly became popular guest speakers for civic groups, and NASA's public relations department was eager to get the word out, so they developed a schedule: they would take turns, each astronaut spending a week at a time giving the same old speech and answering the same questions.

But "wine, women, and song" was part of the test-pilot credo, so the men managed to make time for after-hours carousing and, often, companionship, and since they were household names and oozed testosterone, they had no problem finding companions. John Glenn, married since 1943 to pretty, dark-haired Annie Castor, whom he had known since they were toddlers in the same playpen, resisted those temptations, but most of the others didn't. And when a local car dealer offered them all new Corvettes at ridiculously low lease prices, Glenn declined and opted for a station wagon instead. Over the next few years, stories of Shepard, Grissom, and Cooper racing their sports cars on the two-lane roads around Cape Canaveral, up and down Highway A1A, and through the small burg of Cocoa Beach, became legendary. (So did tales of their romantic entanglements.) When people saw a square-jawed guy in his mid-thirties wearing a short-sleeved Ban-Lon knit shirt and aviator sunglasses zoom past them going way above the speed limit, they knew they'd just gotten a glimpse of Cape royalty.

Before the air force began using the Cape as a missile-testing range in 1950, Cocoa Beach had been a sleepy little town with a few bars and restaurants. Even by 1959, the first year of the Mercury program, there were only a few decent-size motels in Cocoa Beach, and the Seven would often stay at the ninety-nine-room Starlite. It featured a coffee shop, a restaurant, and the space-themed Starlite Lounge, a far cry from the austere crew quarters and bunk beds on the second floor of the Cape's Hangar S, the structure originally built for Vanguard that had been transferred to the Space Task Group. The Starlite's manager, a gregarious Auschwitz survivor named Henri Landwirth, befriended the astronauts and went out of his way to indulge them and their friends. (When Bob Gilruth stayed there for the early launches, Landwirth made sure there was a pitcher of Beefeater martinis in his room at the end of the day.)

The motel quickly became Astronaut Central, attracting reporters, women, and anyone who wanted to get a glimpse of genuine American heroes before one of them got blown to bits atop a military rocket. Landwirth would later refer to the festive goings-on as "a giant fraternity party." The fact that he hired the most attractive waitresses he could find and that women hung around the large pool by day and the lounge by night hoping to meet an astronaut didn't hurt the Starlite's popularity. But late in 1959, when new owners began making big changes at the motel, Landwirth resigned and helped open a Holiday Inn nearby. All of the astronauts said they'd follow him to his new place of business, on one condition—that he would guarantee them rooms if they were in town. He agreed, so Astronaut Central, along with just about everyone associated with the Mercury project, moved to the Holiday Inn.

Glenn was older and perhaps more mature than the rest of the Mercury Seven, and he was worried that one big slipup would endanger the entire program, which still had more than its share of doubters in the government. Some scientists and congressmen insisted that human space exploration was too expensive and that using robots and machines would be safer, cheaper, and more effective. Any ammunition against the astronauts—the sullying of their public image, perhaps—might be used to cancel the Mercury project, which would sound the death knell for the advancement of manned flights. But most of the astronauts dismissed Glenn's worries. Only Carpenter sided with Glenn, as he did on most subjects. It wasn't long before the Mercury Seven became, at least away from the public eye, the Mercury Five and Two.

Things came to a head in December 1960, when Glenn was awakened by a two a.m. phone call in San Diego. It was a NASA public affairs spokesman begging for his help in convincing a West Coast paper not to publish compromising photos involving another astronaut. Glenn succeeded in doing that by appealing to the newspaper staff's patriotism. Back at Langley, Glenn called a closed-door meeting of the Mercury Seven and read his comrades the riot act. He said the next screwup could blow it for all of them. The program meant too much to the country, he told them, "to see it jeopardized by anyone who couldn't keep his pants zipped."

All the men except Carpenter told Glenn to mind his own business.

Dwight Eisenhower had his reasons for never warming to the idea of manned space exploration. In addition to the enormous costs projected, which would prevent him from balancing the budget, he was leery of the potential military involvement. Till the end of his administration, he was adamant that there was no need for a space race or something as far-fetched as a moon program. He hadn't expected much out of NASA, and he wanted to spend as little on it as possible. Toward that end, he cut the agency's budget; the money available would barely support Mercury, and nothing beyond. Only a last-minute intercession by Glennan and his deputy administrator Hugh Dryden kept Eisenhower from announcing in his final State of the Union address that the nation would be abandoning manned space exploration after the Mercury program's first orbital flight. In the closing months of 1960, Eisenhower withheld funds needed for feasibility studies on a Mercury follow-up program and for further work on the upper stages of von Braun's Saturn booster. The future of manned spaceflight, so dependent on funding, looked bleak, particularly if the Republicans won the upcoming election.

In November 1960, Eisenhower's vice president, Richard M. Nixon, lost a closely contested presidential race to a young, charismatic senator from Massachusetts, John F. Kennedy. A World War II hero, Kennedy, among his many other accomplishments, had written (albeit with significant help from his speechwriter Theodore Sorensen) a bestseller entitled *Profiles in Courage*, a book about decisive moments in American history.

One of the campaign issues that redounded to John F. Kennedy's advantage and may have decided the election was the ginned-up "missile gap." In truth, there was no such thing—or if there was, it was actually in America's favor—but Kennedy's repeated use of the term persuaded much of the electorate that the Eisenhower administration was weak on defense and that Nixon would be too. American launch failures after Sputnik, exaggerated public claims by the Soviets of their missile capabilities, and inflated U.S. Air Force estimates of the number of Russian weapons seemed to strengthen the assertion. After Kennedy's election, the term *missile gap*, and the concept, faded away.

At the time, Kennedy wasn't much interested in space, but his vice president, Lyndon Johnson—who had been instrumental in the creation of NASA—was, and Kennedy put him in charge of the space program.

NASA's aim was to get a man into orbit as soon as possible. After that, as far as the public knew, plans were vague. But by January 1960, the higher-ups at NASA had concluded that the long-term goal after Mercury should be getting a man on the moon—or at least in an orbit around it. Even before the agency's formation, Gilruth's Space Task Group had focused on what should follow Mercury, and all of NASA's field centers had begun research toward that goal. But they had been unable to proceed without executive approval, and Eisenhower's aim to balance the budget precluded any such frivolous expenses. As a result, the plans were limited to feasibility and design studies. Then, on July 29, 1960, NASA announced an ambitious program involving a three-man spaceship very different from Mercury. This new spacecraft was to be larger, more powerful, and maneuverable, capable of circling the Earth and perhaps flying around the moon; it was seen as an intermediate step toward the establishment of a permanent manned space station above the Earth that "should lead ultimately toward manned landings on the moon and the planets." When it was made public, the idea wasn't much more than a vague concept, without capital or contracts. And if the president didn't release the funds needed to develop something of substance, it would remain that way.

Still, a name had been chosen for the as-yet-undesigned spacecraft's project. Abe Silverstein, head of the Office of Space Flight Programs, had come up with Mercury a year earlier, and he took it upon himself to name this project also. Like most of the names for American space projects, it would come from the world of mythology, and after scouring lists of ancient deities, Silverstein settled on one. At lunch with Gilruth, Faget, and Charles Donlan, Gilruth's deputy director, he tried it out on them. They all liked it, so the program was officially named for the Greek god of music, medicine, and knowledge, a deity often identified with Helios, whose horse-drawn chariot transported the sun across the sky: Apollo.

CHAPTER FOUR

MAN ON A MISSILE

With Mercury we are using a device that has to work nearly perfectly the first time or somebody is taking a one-way trip.

WALT WILLIAMS, MERCURY OPERATIONS DIRECTOR

BESIDES BEING OLDER THAN the other six astronaut trainees, John Glenn had more combat experience. He had flown fifty-nine ground-support missions in the South Pacific during World War II and another sixty-three in Korea, where he shot down three MiGs in the last nine days of the war. Glenn was one hell of a pilot. He had been known to fly his plane up alongside another Marine's, slip his wing under the other plane's, and tap it gently. Ted Williams, the legendary baseball player who flew as Glenn's wingman in Korea, once said, "The man is crazy." The "Clean Marine," as the press dubbed him, had settled down since then, and he wasn't the engineer that Gus Grissom or Wally Schirra was, but he could still fly anything with wings better than almost anyone else.

He was also a master of the art of sniveling. The term was used among pilots to describe "maneuvering"—working yourself into a program or flight whether it was your job or not. In a sense, Glenn had sniveled his way into the Mercury program without even knowing he was doing it. One of the requirements was a college degree, which he didn't have—after Pearl Harbor was hit, he had quit college to enlist—and Glenn was dropped from the pool of candidates. But a former Marine Corps officer showed NASA's selection board Glenn's academic records, which included a surplus of credits from night school and his technical flight-test reports. Convinced that Glenn had more than the equivalent of a college degree, the administrators put him back on the list.

But his latest attempt at sniveling had been a failure.

Since their selection, all seven of the astronauts had vied to pilot the first flight into space. They worked well as a team, but each man thought he was the best and should be the first one to go up. That confidence—not arrogance, but a hard-earned faith in his own abilities—was hardwired into every test pilot, especially these seven: "Maybe just a little arrogance," Glenn would write later.

After the April 9, 1959, press conference to introduce the astronauts to the world, Glenn had decided to move to Langley AFB, where Mercury was located, and live there Monday through Friday so he could better focus on the program. While the other six commuted from their nearby homes, he returned to his family in Arlington, a hundred and eighty miles away, only on the weekends. To make up for his lack of engineering experience, he worked diligently to master the Mercury systems. He spent more time on the testing programs than the others, and his scores in most cases were slightly higher. He ran two miles before breakfast every day to lose twenty-five pounds and get himself into his best shape ever. As the oldest astronaut—thirty-seven at selection—Glenn felt he had to train harder than the others, and he did.

Although all of the Mercury Seven lived for competition, only one was as competitive as Glenn—Alan Shepard. As a kid, he'd been slow and scrawny, and he'd had to fight to keep up with others. He hadn't lost any of that scrappiness. When he saw how Glenn charmed the press and the public and how his hard work impressed the NASA brass, he decided to beat Glenn at his own game. If Glenn ran two miles every morning, so would he. Shepard started lifting weights, and he even quit smoking for a while. Glenn may have been the smoothest at that first press conference, and his open, smiling freckled face was certainly more camera-friendly than Shepard's snaggletoothed, slightly bug-eyed visage. But Shepard, despite a personal loathing for the media, strove to improve his relations with journalists, and soon he could work a press conference as well as Glenn. That competitive spirit carried into the classroom and every training exercise. No one studied more or worked harder than Shepard, and he was focused. "You tell him one time, and that was it," recalled one engineer. "He was really sharp." And he was not shy about his ambition. When a reporter asked him why he wanted to be the first man in space, his answer—"I want to be first because I want to be first"—was a marvel of tautology. Glenn was a fierce competitor, but Shepard was cutthroat. He would do anything to fly the first mission.

Shepard's hard work paid off. Early in 1961, Bob Gilruth called the seven astronauts together in their shared Langley office and announced that Al Shepard—later nicknamed the Ice Commander for the colder side of his Janus-like personality—was to pilot the first flight. The second would go to Grissom and the third to Glenn, who would also back up the first two missions. Six men were disappointed, four of them more than the other two. But each of them walked over to Shepard, shook his hand, and congratulated him. Shepard struggled to keep a huge grin off his face.

The choices would be kept secret from the press and the public until just before the first mission. The media was told that one of these three would be picked for the first flight, which created some awkwardness between them and the other astronauts, whom the press dubbed "the Forgotten Four." But the Clean Marine refused to accept his third-place finish. The next day, he wrote Gilruth a letter lobbying for a change—he felt he was the better choice. The Space Task Group director was unmoved and unpersuaded, and he ignored Glenn's plea. Inconsolable, Glenn began to withdraw into himself. "He was real, real shook," remembered an acquaintance. "It was the only thing Johnny ever lost in his life." It wasn't until his next-door neighbor, a close friend, told him his funk was hurting his family that he pulled out of it—somewhat. He swallowed his pride, shook off his disappointment or at least managed to hide it, and did the best job he could of backing up Shepard.

But the Mercury-Redstone was still experiencing delays, and its plodding progress elicited criticism from several quarters. At the program's first unmanned launch on November 21, 1960, Faget and the Seven had watched from an outside viewing area at the Cape. In the blockhouse, with its thick concrete walls and slot windows, von Braun joined several members of his German team and a mix of former army technicians and booster contractors as they prepared to launch. Save for a few brief holds to fix minor problems, the countdown went smoothly. When the clock reached zero, the booster lifted off amid heavy clouds of smoke. Everyone in the viewing area was impressed at how quickly the Redstone accelerated and disappeared. "My God, that was fast!" said Faget, but when the smoke cleared, the rocket was still there. It had risen about four inches from the launchpad, but then the engine had shut off, and the rocket had settled back unsteadily on its fins on the launch cradle. The only thing that had launched was the escape tower, which self-destructed four thousand feet in the air. Mechanics and technicians near the launchpad ran to hide under trucks and behind cars as debris rained to the ground as far away as a quarter of a mile from the viewing area.

In the blockhouse, the frantic Germans reverted to their native language as they tried to find out what happened. They communicated with a German booster engineer in the control center. Flight director Chris Kraft, a man with little patience for anything less than perfection, was not happy with how the first mission had gone; just about everyone had been working heavy overtime for months to get ready, and it certainly hadn't been his team's fault. He stormed over to the engineer, ripped his headphones out of the console jack, and yelled, "Talk to me, dammit!"

The cause of the misfire was soon found and fixed. But there was still plenty of work to do before the Mercury-Redstone was ready to carry a man into space. Amid growing concerns and pressure from government officials and the press, Gilruth insisted there wouldn't be a manned flight until it was safe, but privately, he was worried. "If we get those first three guys back alive, we're going to be damn lucky," he told one of his engineers. A month later, on December 20, they finally achieved a successful unmanned mission.

On January 31, 1961, the final Mercury-Redstone suborbital flight carrying a primate, the charismatic young Ham, was launched. (The thirty-seven-pound chimpanzee's original name was Chang; it was changed to Ham, an acronym for Holloman Aerospace Medical, and the name would be released only if he returned to Earth alive.) Ham was strapped into a Mercury capsule atop a Redstone rocket and launched into space. He survived despite several mishaps with the rocket and with Ham's protocol; he experienced sixteen g's, twice as many as expected, and instead of getting banana pellets when he pulled the correct levers of a psychomotor box in sequence with cueing lights, he received mild electrical shocks. This continued until splashdown, when water began leaking into the capsule. By the time the recovery copters fished the half-drowned chimp out of the sinking spacecraft, Ham was not in a good mood. But the life-support systems had worked fine, and the mission was deemed a success. Ham's sixteen-and-a-half-minute whirl showed that manned spaceflight was possible and that pulse and respiration rates and blood pressure were not adversely affected by weightlessness or under heavy g's.

The successful mission also meant that Ham, an immigrant from

Cameroon, was the first hominid—and the first American—in space. A newspaper cartoon at the time portrayed a pair of apes walking away from a successful spaceflight. One says to the other: "I think we're behind the Russians but slightly ahead of the Americans." Which they were, thanks to Ham.

Ham's human counterpart, Al Shepard, was also ready. He'd been ready for a while, spending many hours on the various procedure trainers and simulators. He'd flown a hundred and twenty simulated flights, working through every possible emergency and failure mode and learning to cope with them all. He'd also endured more time on the malevolent MASTIF, six hours and ten minutes, than anyone else. Glenn, too, was ready if need be—over the previous three months, he'd been Shepard's training partner and shadow, and the two had achieved a mutual respect and even friendship, though Shepard still liked to tease Glenn by calling him "my backup."

The two astronauts might have been prepared for the real thing, but Wernher von Braun's team at Huntsville was not. Ham's Mercury-Redstone had flown much higher and farther downrange than planned. Over the objections of Gilruth and most of his staff—including the astronauts—the conservative von Braun insisted on an additional unmanned flight. On March 24, 1961, it launched successfully, which officially made the Mercury-Redstone "man-rated." But it pushed Shepard's mission to early May.

By the end of February 1961, the Americans appeared to have drawn even with the Russians in the space race. After the program had repeatedly failed to attain its mission objectives, the schedule had slipped by a year, and the Mercury-Atlas program had undergone an exhaustive review over the past six months. Finally, three weeks after Ham's Redstone flight, a successful Atlas flight on February 21 boosted not only the Mercury capsule but the spirits of everyone in the Space Task Group. Since Kennedy's first days in office the previous month, rumors had swirled that the Mercury project—still considered a stunt by some in Washington—would be canceled or somehow handed over to the Pentagon. An in-depth examination of the program in April by a presidential committee had not improved morale. But the Atlas flight, combined with the Russian failures of one Sputnik in December 1960 and two more in February 1961, had NASA flying high and looking forward to putting the first man in space in just a few months.

On April 12, 1961, from a top secret cosmodrome on the desolate steppes of southern Kazakhstan, thirteen hundred miles southeast of Moscow, a young Soviet air force lieutenant named Yuri Gagarin was boosted into space. The cosmodrome's name, Baikonur, was deliberately misleading for security reasons. A real mining town of Baikonur was two hundred miles to the northeast; the rocket complex was actually near Tyuratam, a small village with a convenient rail station. Despite the ruse, the United States had known of the complex since 1957, when an American U-2 spy plane had discovered it.

In a type of spherical spacecraft called Vostok ("East"), Gagarin not only traveled beyond Earth's atmosphere but orbited the planet. The sphere shape was chosen for its inherent dynamic stability, though that required it to be completely covered with an ablative heat shield. The stocky Gagarin was the son of a peasant farmer and of pure Russian stock—the better to trumpet the superiority of the ordinary socialist worker. Another reason he was picked was his height; the five-foot-two pilot could eject safely through the hatch, the cover of which would be blown clear of the spaceship just two seconds before ejection.

Gagarin had been chosen from a half a dozen cosmonaut candidates. He was less than four years out of flight school and had only two hundred and thirty flight hours under his belt, and he was a passenger in every sense of the word during his 108-minute ride; the entire mission was controlled from the ground and automatically. There were manual controls, but the numeric code to unlock them was placed onboard in an envelope to be opened only in an emergency, since Gagarin's superiors weren't sure how a human would react to extended weightlessness. He didn't need the instructions, but because his capsule came down over land, Gagarin ejected at about twenty thousand feet and parachuted the rest of the way, a detail that would remain hidden until 1971. (The French organization that judged and maintained world aeronautical records required a pilot to land with his craft for the flight to be considered official, and the group did certify the flight.) And as with all previous Soviet rocket launches, the flight was kept secret until after the fact, when it was announced triumphantly. Major Gagarin—he was promoted even before his capsule landed—would soon embark on a worldwide publicity tour of non-aligned nations to help persuade them to hitch their wagons to the superior Soviet star.

America's reputation was further damaged by the response of Colonel John "Shorty" Powers, Mercury's press officer, when he was awakened in the middle of the night by a reporter calling for a response to the Gagarin flight. "We don't know anything about it," Powers snapped. "We're all asleep down here!" Everyone at NASA had been working long hours, most of them without a day off, but Powers's groggy response gave the wrong idea when it was paraphrased in a headline the next day: "Soviets Send Man into Space. Spokesman Says U.S. Asleep."

The manned mission shouldn't have come as a surprise. Over the previous eleven months, the USSR had released details of five orbital flights, all carrying some kind of biological specimens ranging from mice to guinea pigs to dogs—and, on the final two, mannequins. To anyone paying attention, the Soviets' goal was obvious.

The United States officially congratulated the Soviet Union, but Shepard and the other astronauts were furious. Not only had they been beaten into space, but Gagarin's flight was an extraordinary propaganda coup, producing roughly the same effect on the world and on the American public as Sputnik had. It was another momentous first for the USSR—not only the first man in space, but the first to orbit the Earth. NASA was several flights away from that—at least a year. But the Russians had already set their sights much higher. One of their leading scientists was quoted as saying that the flight "completes halfway the effort of sending man to the moon," and Gagarin echoed that. Who knew what Soviet space "first" would occur next?

No one in the free world did. The Soviet program, completely military, operated behind a veil of secrecy. Information about it was maddeningly meager, since little of it was conducted out in the open, and few details were released—and when they were, it was only after a successful flight. Occasional rumors made their way through the Iron Curtain, sometimes through so many informants that the final intelligence was highly questionable. Not even the Russian people knew much about their space program or could say where it was located, and few outside the program knew who headed it. His name was Sergei Korolev, and his position corresponded roughly with von Braun's. The Soviets feared assassination attempts from the West, and so Korolev and his most valuable assistants were kept anonymous. He was referred to only as Chief Designer in press stories. The U.S. intelligence community knew little more; a few fragmentary conversations intercepted when he called his office from his car had yielded nothing substantive.

Korolev's route to becoming Chief Designer had been a rough one. Born in 1907, he had been a brilliant young aviation and rocket engineer; he had founded the premier Soviet rocketry group in 1931 and seen it taken over by the military in 1933. He continued his work with rockets, specializing in design, but in 1938 he was found guilty of trumped-up charges of treason and disruptive activities. After severe beatings and torture, he was persuaded to confess and sentenced to ten years of hard labor—a victim of Joseph Stalin's Great Purge. He spent six years in the Soviet prison system, including a stint in a Siberian labor camp, where starvation almost killed him and scurvy resulted in most of his teeth falling out. But when World War II began and missile development became a national priority, he was released with the rest of the unexecuted rocket specialists and commissioned a lieutenant colonel in the Soviet army; his country needed his expertise. In 1945 he was sent to Germany to evaluate the V-2 work done by von Braun's rocket team. The Soviets co-opted those rocketeers who hadn't aligned themselves with the United States. The Germans moved to the USSR, but after five years, the Soviets had learned all they could from them, and they sent them packing. A few years later, Korolev was appointed to lead the Soviet space program. Like von Braun, he had a genius for management and strategy and a knack for inspiring his people to believe they could do anything if they worked hard enough.

In 1953, soon after Nikita Khrushchev came to power after Joseph Stalin's fatal stroke in March, Korolev met him to discuss his space plans. For all Khrushchev's outward rusticity, he was intellectually curious, and Korolev secured his support—as long as it didn't interfere with priority number one: the safety of the Soviet people. The success of Sputnik had further solidified Khrushchev's power.

The Sputniks and Lunas might not have put bread on a single Russian table or provided a car to any Russian family. But the Soviet people gloried in their space triumphs and, in the words of one Russian rocket designer, "felt proud and were thrilled to be citizens of the country that was blazing the trail for the human race into the cosmos."

* * *

At a press conference right after Gagarin's flight, President Kennedy told the nation that America would not try to match the Soviets in space but would instead choose "other areas where we can be first and which will bring more long-range benefits to mankind." But this second, much-ballyhooed defeat in the space race did not sit well with the president—or his military advisers. Beyond the question of prestige, it meant that the enemy might soon be able to intercept the increasing numbers of American spy satellites that had begun flyovers of the USSR the previous year.

Glenn, the press favorite, spoke candidly about the flight and put the best face possible on it. "They just beat the pants off us, that's all, and there's no use kidding ourselves about that," he told reporters. "But now that the space age has begun, there's going to be plenty of work for everybody." He and the other astronauts extended their personal congratulations to the Soviet program. In private, they were disappointed, none more so than Shepard. His suborbital flight had originally been scheduled for three weeks before Gagarin's, but von Braun's cautious approach had prevailed. Now he would be second, and that was a place the ultracompetitive Al Shepard hated. He consoled himself with the fact that he'd still be the first American in space, if you didn't count Ham. As the May 2 launch drew near and hundreds of newspeople invaded the Cape Canaveral area, spirits at NASA were buoyed—a cancellation at this point would be a fiasco.

Shepard was expected to experience six g's during launch and between twelve and fourteen upon reentry. A man sitting in a can atop a powerful rocket built to carry a warhead to the battlefield—despite Gagarin's survival, those rigors, combined with the unknown hazards of weightlessness, convinced some physicians consulted by NASA that "subjecting a human body to such stresses is practically equivalent to sending the astronaut on a suicide mission," wrote Walt Williams, Mercury's operations director, and that opinion was shared by many Americans. A tragedy would shatter national morale and might jeopardize the entire program, but Bob Gilruth and his Space Task Group were guardedly optimistic.

Bad weather scrubbed the mission, and it was postponed to May 4, then to May 5. The night before, Shepard and Glenn slept in bunk beds in the Cape's crew quarters on the second floor of Hangar S, three miles from the launch site. While they were sleeping, technicians began loading the propellants—kerosene in the lower tank, the pale blue, cryogenic liquid oxygen, kept below its boiling point of negative 297 degrees F, in the upper. When the tanks were full, to prevent them from bursting, the Redstone began venting plumes of condensed water vapor that swirled around the rocket. Other members of the pad crew filled the capsule's thruster jets with hydrogen peroxide.

In the spartan crew quarters, flight surgeon Bill Douglas woke the astronauts at about one a.m. to shower and shave. Then a cursory medical examination was made by nurse Dee O'Hara, who helped attach six biomedical sensors to Shepard. At 3:30 a.m., after a breakfast of bacon-wrapped steak and eggs with orange juice and coffee, Shepard was helped into his aluminum-coated pressure suit (a modified version of the high-altitude suit worn by navy pilots) by suit technician Joe Schmitt, a tall, quiet former aircraft mechanic who had been with the NACA since World War II. At 5:20 a.m., Glenn—who for almost two hours had been checking over every one of the capsule's 165 switches, dials, and meters—helped shoehorn Shepard into the tight confines of *Freedom* 7, the name Shepard had chosen for his Mercury capsule. Glenn wished him luck, reached in, shook his gloved hand, said, "Happy landings, Commander," and watched as the hatch was closed at 6:10 a.m. The launch was scheduled for 7:20, but it was delayed due to cloudy weather, then later because of a faulty computer.

The capsule would be blasted about a hundred miles into space, reach five thousand miles per hour, then fall back to Earth in a curving ballistic trajectory several hundred miles downrange. The flight was expected to last fifteen minutes, so no one thought a urine-collection system was necessary. But after Shepard had been sitting on the launchpad for more than three hours, the orange juice and coffee he had consumed made its presence felt. On the radio he asked Gordon Cooper, assigned as his voice contact in the nearby blockhouse, to check if he could get out and urinate. Cooper got back to him a few minutes later. "No," Cooper said, and then, imitating von Braun's clipped German accent: "Ze astronaut shall stay in ze nose cone." Shepard warned them that he would go in his suit if he could not get out for a minute—since he was on his back, the liquid would follow gravity and seep into his long cotton underwear. But that might mean a short circuit in his suit's biomedical sensors, and Mercury Control refused again. Shepard suggested they turn off the power to his suit. They did. He relieved himself. The liquid was eventually absorbed and mostly evaporated in the 100 percent pure oxygen atmosphere.

A short while later, there was another delay—the pressure in the liquid-oxygen propellant system was too high. While technicians tried to turn some of the valves by remote control, an impatient Shepard snapped, to no one in particular, "Why don't you fix your little problem and light this candle?" For some reason, that seemed to do the trick. The countdown soon resumed, and it was not interrupted again. As it hit the two-minute mark, flight director Chris Kraft asked each position in Mercury Control for a go/no-go opinion on whether to proceed. This would become standard operating procedure on every mission for every important decision, and only when all systems were go would the flight continue. Each flight controller and doctor asked replied, "Go, Flight," using the shortened version of Kraft's job title. He gave the okay to continue. The atmosphere in the room was tense, and Kraft himself, sporting a small Mercury lapel pin he would wear during every flight, was shaking so hard that his microphone fell off. The Redstone was nicknamed "Ol' Reliable" for its dependability, but anything could happen with a rocket. Fortunately, Max Faget had designed a fourteen-foot rocket-powered escape tower that was painted bright red and sat atop the black capsule, and it would lift the capsule far enough away, it was hoped, to protect the astronaut from a fireballing rocket explosion.

The pad rescue team—amphibious vehicles, armored tanks, helicopters, asbestos-suited firemen, divers, and boats—were all at full alert and ready to rush to Shepard's aid if need be. Douglas, the astronauts' personal physician, sat in an idling helicopter a safe distance from the launchpad with a medical pack strapped to his back. Next to the chopper was a small one-patient hospital. Farther afield, the navy's recovery fleet—an aircraft carrier, eight destroyers, a radar tracking ship, Marine copters, and pararescue and frogmen teams—was deployed in the anticipated recovery zone, in the Atlantic Ocean five hundred miles southeast of Cape Canaveral.

As a last line of security, just in case the main and reserve parachutes failed, Bob Gilruth had insisted that a personal-chute chest pack be placed on a shelf inside the crowded cockpit, the idea being that before he crashed into the Earth at an obscene speed, Shepard might somehow disentangle himself from several harnesses, hook on the chest pack, open the hatch, and maneuver himself out of the capsule while avoiding the dysfunctional parachutes above him. It was a situation that a circus contortionist might be better able to handle.

Forty-five million Americans—about 25 percent of the country's population—watched the launch of the black-and-white Redstone booster on a black-and-white live television broadcast, carried on each of the three nationwide networks. The president's chief science adviser, Jerome Wiesner, had suggested that the launch be held in secret, like the Soviets', to prevent a potential national embarrassment occurring in front of the entire world, but Kennedy had nixed the idea. America's space program would operate, at least for the most part, out in the open. At the White House, Kennedy and his wife, the vice president, and a few advisers followed the launch on a small TV. At 9:32 a.m., more than three hours after the capsule's hatch had been sealed, the countdown reached zero and liftoff commenced.

"*Freedom Seven* is still go!" Shepard said a few seconds later, and viewers across the nation shouted and screamed, "Go! Go! Go!" as one of their countrymen sat in a small capsule at the tip of a slender rocket that spewed out white-hot flame that then changed to a blinding yellow, and the Redstone slowly and loudly roared straight up into the sky. It shot higher and higher and vanished behind a large cloud, then reappeared, a white contrail behind it, and finally disappeared from sight completely. A thin white trail arced down behind the rocket, an indication that the red escape tower had jettisoned, meaning enough speed and altitude had been achieved by *Freedom* 7 that it wouldn't be needed. In the reinforced-concrete blockhouse some two hundred and fifty yards from the launchpad, where the rocket's launch and functioning were controlled, the only guest permitted that day allowed himself a small sigh of relief. His Redstone had done its job, and Wernher von Braun knew the most critical part of the flight was over.

Two miles away, in the Mercury Mission Control building, Kraft and fifteen other men sat at three banks of consoles in a square space not much larger than an average college classroom, about sixty feet by sixty feet. Each console position was supplemented by a group of experts on that particular system, all of them sitting in another room, and they would be consulted if a problem arose. On a large animated world map on the wall of the main room, the location and status of every tracking station and navy recovery ship were displayed, and the path of the spacecraft was plotted by a toylike plastic replica of the capsule that would move on wire tracks, powered by servos emitting a steady whine. Above the map, several clocks showed various times: Greenwich mean time, countdown, elapsed time, time to retrofire, and so on. Data derived from telemetry was displayed on boards on each side of the map. The individual consoles featured black-and-white TV monitors, analog meters and displays, and the occasional rotary phone. Kraft and his team—each wearing the uniform he had decided on, a short-sleeved white shirt with a thin tie—followed every facet of the flight, including Shepard's biomedical readings and the status of the life-support systems. A doctor in the first row could call for an abort, at least in the first few seconds, if the astronaut's life became endangered. Near him sat Deke Slayton, capsule communicator (CapCom), the primary contact between Mercury Control and Shepard, with Glenn and Grissom on either side. The idea was that another astronaut—and the CapCom was always an astronaut, usually a backup to that particular mission—would best understand the situation in both the capsule and on the ground and would be able to pass on information clearly. The Mercury Seven also felt that one of their own could argue with anyone who wanted to abort the mission—for instance, if a doctor decided a random biomedical reading looked suspicious and pronounced the astronaut in medical distress.

In Mercury Control, Shepard's calm voice could be heard loud and clear as he reported at every phase of the flight and detailed his body's responses to acceleration, weightlessness, and deceleration and his craft's responses to the forces acting on it. A couple of minutes after liftoff, the spacecraft separated from the spent Redstone booster. Shepard took over manual control of the small thruster jets and changed, one axis at a time, yaw, pitch, and roll—the direction the capsule was pointing, its attitude. And at the top of his trajectory, he peered into the viewer in the center of his instrument panel—a periscope extended several inches out into space—and marveled at the breathtaking sixteen-hundred-mile-wide panorama of the Earth below him. "What a beautiful view," he said, and he went on to describe the cloud cover over the Florida coast.

As *Freedom* 7 began its ballistic arc back to Earth, he tipped his craft into position for reentry—bell-side down, so the ablative shield could burn away as it protected the capsule from the deadly heat caused by hurtling through the thickening atmosphere at more than four thousand miles an hour. The three retro-rockets strapped onto the heat shield fired and reduced the capsule's velocity enough to allow gravity to take over. In just one minute, the capsule slowed to 341 miles per hour, and as the massive deceleration of eleven g's slammed Shepard into his contour couch, his voice became a strained grunt: "Okay…okay…okay…"

The small drogue parachute opened up at twenty-one thousand feet to stabilize the capsule; a few seconds later, the red-and-white main chute unfurled at ten thousand feet; and fifteen minutes and twenty-two seconds after liftoff, the gently swaying *Freedom* 7 and its hardy human splashed down safely in the Atlantic Ocean, 302 miles downrange from Cape Canaveral.

The spacecraft had risen to an altitude of only 116 miles and descended immediately after, and Shepard was weightless for just five minutes, but he had performed flawlessly. After a perfect recovery, he stepped out of a copter onto the deck of the carrier *Lake Champlain,* unharmed and in high spirits. A few minutes later, Shepard was brought to the flag bridge for an unexpected phone call. It was President Kennedy, another navy man, congratulating Shepard on his flight. Kennedy's pleasure at this success would help create a sea change in the American public's formerly apathetic attitude toward the Mercury program. Now, it seemed, they got it. Bob Gilruth and everyone in his Space Task Group breathed easier. The fact that Shepard hadn't died was not insignificant. No one knew what would come after it, but Mercury would continue.

Though the disdainful Khrushchev compared it to a "flea jump," Shepard's successful flight was a soothing balm to the nation's injured pride. And most of NASA's medical community, previously unsure whether a human could survive the known and unknown stresses of spaceflight, relaxed a little, although the apparent safety of Gagarin's flight had also allayed some of their fears. Perhaps man could function in space after all.

At a press conference soon after the mission, before hundreds of reporters, Shepard was smooth as silk. And he admitted to nothing more than some "apprehension" during the flight. No one would ever get Alan Shepard to say he was scared.

An American had penetrated the darkness of space in a rocket. And he had done it in full view of the world, live on TV, unlike the Russians. Surely that counted for something in the propaganda war. Outside the United States, it apparently didn't count for much. A poll taken after Shepard's flight showed that 41 percent of Western Europeans believed the Soviet Union was the stronger military power, compared to only 19 percent who believed it was the United States, and more of them thought the USSR was significantly ahead of the U.S. in overall scientific achievement.

Unlike Gagarin, Shepard had been able to adjust his craft's attitude by using the small thruster jets on its exterior, and the press made much of the fact that he had "driven" the spaceship, the first man to do so. Not everyone considered that flying, since there was no way to power *Freedom* 7. (Some pointed out that the astronaut exerted about the same amount of control over his craft as a glider pilot, though that wasn't quite true—a pilot could change a glider's trajectory.) Still, to the American public, it was enough, at least for now.

The president needed some good news. He'd been in office for two months, and he and the U.S. had just experienced an embarrassment as profound as any in the country's history.

The island of Cuba lies ninety miles off the southern tip of Florida. There, in 1959, a young, inspirational revolutionary named Fidel Castro had finally succeeded in leading a group of disaffected countrymen to oust dictator (and U.S. ally) Fulgencio Batista. As Castro quietly became a dictator himself and assumed military and political control, his country became increasingly communistic and drifted toward Russia's sphere of influence. As a result, Cuba's relations with the United States quickly deteriorated.

President Eisenhower had approved a covert plan for the CIA to train and arm a small army of Cuban exiles with the goal of overthrowing Castro's Communist government. The original plan called for guerrilla infiltration operations to gradually win over Cuban hearts and minds. But near the end of Eisenhower's administration, the plan grew larger, and eventually it called for them to land in the Bahía de Cochinos—the Bay of Pigs—and support the amphibious assault with heavy bomber and fighter cover. The Cuban people, the Americans were sure, would rise and join this Free Cuba cadre, and if the rebels could secure a beachhead for seventy-two hours, during which period a free Cuban government would be established, other nations in the hemisphere might recognize the new government and send aid. America's role in the invasion would be kept secret, to avoid the appearance of meddling in Latin American affairs. Despite Kennedy's personal misgivings, his aides advised him to approve the plan, and he gave the go-ahead less than two months after taking office.

On April 17—thirteen months after the plan's inception and only five days after Gagarin's orbital flight—fourteen hundred insurgents launched from Guatemala and Nicaragua on five small freighters and landed at the swampy Bay of Pigs. They waded ashore and fought courageously but after three days were overwhelmed by Castro's defensive forces. At the last moment, Kennedy canceled a vital early-morning air strike and any further support. The rebel brigade's fifteen B-26 bombers were effective at first but were soon disposed of by the small Cuban air force of two B-26s and five smaller fighter-bombers; these also disabled two invasion ships filled with much-needed supplies and ammo. Other supply ships were ordered to leave the scene.

The result was a fiasco of epic proportions. Hundreds of insurgents were killed or executed, and the remaining rebels, many of them without food and ammunition, surrendered three days after the landing. Some of those captured revealed America's complicity in the invasion, which was viewed as warmongering. The worldwide criticism of the United States and its young president was intense, as was the country's humiliation. When Khrushchev had telegrammed Kennedy expressing alarm at American involvement in Cuban politics, the president had told him that the U.S. was only supporting the one hundred thousand Cubans trying to resist the Castro regime. Few were convinced by his argument. Kennedy's standing with the American people, and his country's standing among the community of nations, would never be lower.

Shepard's successful flight, brief as it was, restored some of America's pride, and Kennedy took note of that and of the strong positive response to the mission. He decided he was not satisfied with a snail's-pace space race with the Russians, whose booster rockets were clearly more powerful than America's. That was a decidedly different attitude than his earlier one. Three weeks before Gagarin's triumph, Kennedy had heard arguments from NASA for a $308 million budget. The Bureau of the Budget agreed to only $50 million. The president's decision to increase that to $126 million—none of it for the Apollo program—hadn't exactly been a vote of confidence. It had seemed as if U.S. space exploration would come to an end once the U.S. caught up to the Soviets in the space race. The Mercury cost overruns and delays had soured a good many government officials and representatives of the American public on the space program, and there was nothing to indicate that enough of them would commit to a long-term, monstrously expensive project with little immediate and appreciable benefits to their constituents. Apollo, at this point little more than plans, schematics, and dreams, might be stillborn.

But on April 14, two days after Gagarin's triumphant flight, Kennedy had hosted a meeting at the White House that included his vice president, Lyndon Johnson, and a few advisers. Also present was the tough Washington insider that the president, with help from the vice president, had chosen as NASA's new administrator: a thick-necked, barrel-chested, square-jawed bulldog with a Southern drawl and a steel-trap mind.

James Webb, born in a small North Carolina town and the son of a school superintendent, was a former Marine aviator. He had begun working the halls and back rooms of government in 1932 as secretary to a U.S. congressman while earning his law degree. After World War II, he had served as director of the Bureau of the Budget, then undersecretary of state. Tough, savvy, and a resourceful negotiator—"the fastest mouth in the South," some called him, though he could also be an attentive listener—he had become a political operator nonpareil.

When Kennedy called Webb to the White House to offer him the post, he had been working in the private sector for almost a decade. Webb told the president that he wasn't the right man for the job: "You need a scientist or an engineer," he said. He also knew that many others had turned the job down (seventeen, Lyndon Johnson would remember later), since no one knew how long the nascent agency would be around or how much support it would receive from the new administration. But Kennedy would have none of it. He knew Webb's reputation, and he wanted someone who understood policy. Webb couldn't see a way to refuse his president. He agreed to take the job.

Within a few months, Webb stood up to the air force on a major matter of space policy, faced down the Bureau of the Budget director, and, in a House Space Committee hearing, became involved in a shouting match with a congressman while defending NASA's budget. Over the next eight years, the energetic Webb would use his considerable charm, experience, negotiation skills, and knack for translating complex space terms and concepts into understandable English to lobby for the fantastic amounts of money NASA needed from Congress—and the freedom to operate with minimal interference. And he was not a yes-man, as Kennedy would soon find out.

At the April 14 meeting, the president, exasperated and tired of being asked why the U.S. was second to Russia in space, asked the room, "What can we do? How can we catch up?"

While stumping for the presidency, Kennedy had used space only as an issue on which to criticize the Republicans. In his inaugural address, he had suggested that the two superpowers "explore the stars" together, and ten days later, he had asked the Soviet Union to join with the U.S. in several space ventures involving weather prediction, communications satellites, and space probes. Nothing would come of his peace overtures. But the success of Gagarin's flight and its extraordinary worldwide acclaim hadn't been lost on him. He had also noted the enthusiasm with which the country had embraced the astronauts and the jubilation with which Americans had greeted Shepard's suborbital flight. Maybe there was political hay to be made from space after all. Certainly, after his disastrous first hundred days, Kennedy needed a boost. And despite *Freedom* 7's success, the Mercury program continued to fall behind schedule. More space triumphs might be long in coming and trivial compared to the Soviets' accomplishments.

Lyndon Johnson, the farm boy from the Texas Hill Country who, through massive ambition and keen politicking skills, had become the most powerful man in the Senate before assuming the vice presidency, was also Washington's foremost advocate of the space program and had been for some time. He had called for research into a space program as early as 1949, to no avail. Upon assuming office and to give Johnson something to do, Kennedy had named him chairman of the National Space Council, a body created a few years earlier to help coordinate the nation's space efforts. Now Johnson was in the middle of everything, making phone calls, prodding Webb and others at NASA, marshaling all the forces he could with his well-known and none-too-subtle powers of persuasion. When Kennedy asked him what they could do to get ahead of the Russians, a few possibilities were mentioned. One was landing a man on the moon. Johnson told Kennedy that yes, it was possible, and he asked the president for a memorandum containing his thoughts and questions. He got it the next day, April 20, 1961, a one-pager asking Johnson to be in charge of "making an overall survey of where we stand in space." There were several questions about the program. The first was the most important:

> Do we have a chance of beating the Soviets by putting a laboratory in space, or by a trip around the moon, or by a rocket to land on the moon, or by a rocket to go to the moon and back with a man. Is there any other space program which promises dramatic results in which we could win?

Johnson and Webb spent the next several days consulting with NASA officials, scientists, top military brass, politicians, and even a few prominent businessmen on the feasibility of putting a man on the moon. When some of them expressed reservations, Johnson said, "Would you rather have us be a second-rate nation or should we spend a little money?" As the vice president polled these leaders, he also developed support for—and commitment to—the project. Webb was initially resistant to the idea but slowly came around as he watched Johnson work his contacts.

Von Braun, who had dreamed of personally exploring the heavens since childhood and had been jailed by the Gestapo for just mentioning space exploration, jumped on this idea like a Doberman on a burglar. In a nine-page detailed memo, he told Johnson that they had an excellent chance of beating the USSR to the moon.

Johnson delivered his evaluation a week after receiving the president's memo. The gist of it was that it was conceivable that the United States could circumnavigate the moon and possibly land a man on it by 1966 or 1967. The report set forth broad guidelines on how to get there and stressed the importance of manned spaceflight to national prestige.

Kennedy began talking to some of the men Johnson had consulted. Bob Gilruth told him it could be done—as a matter of fact, for two years he, Max Faget, and a few others at NASA had been researching how to put a man on the moon. The advantage of a moon-landing goal, he explained, was that it leveled the playing field for both countries, since neither had begun serious development of the massive boosters needed to lift a large craft into space. They'd both be starting from scratch, and he felt confident the United States could win that particular race. And no one needed to point out to the president the prestige that the winner would earn.

On May 8, 1961, at a ceremony held in the sun-dappled Rose Garden, the president presented Shepard with the NASA Distinguished Service Medal. Afterward, Kennedy and the Mercury Seven retired to the Oval Office. Sitting in his rocking chair, Kennedy listened as Shepard talked of his flight, and he asked several questions about the mission, the men's training, and the program. He even hinted at plans for a moon landing. The seven astronauts looked at one another, and Shepard said, "I'm ready."

Kennedy, it appeared, had made his decision.

About three weeks later, Kennedy gave a speech to a joint session of Congress on "urgent national needs," and he made a bold statement: "I believe that this nation should commit itself to achieving the goal, before this decade is out, of landing a man on the moon and returning him safely to the earth." He emphasized the necessity of "a major national commitment of scientific and technical manpower, materiel, and facilities…a degree of dedication, organization, and discipline which have not always characterized our research and development efforts."

Seven hundred miles away, in the main conference room at the Marshall Space Flight Center, von Braun and his board listened raptly to the president's address. When they heard his moon-landing directive, they cheered, and some yelled *"Ja!"* and "Let's go!"

At the time, the USSR had already landed Luna 2 on the moon, sent Luna 3 around it, and orbited a man around the Earth. The United States had a total of fifteen minutes and twenty-two seconds of manned-spaceflight experience. What the space program didn't have was the massive rocket needed to get to the moon, the spacecraft to convey the astronauts, or even a definite idea of how to navigate there and back. "Here we were struggling to get a 2,500-pound capsule up, and this thing [that Kennedy] just assigned us was going to require getting 250,000 pounds into earth orbit," remembered one core member of the Space Task Group.

A London bookmaker immediately set the odds at a thousand to one against a moon landing on Kennedy's ambitious schedule.

Gilruth and Webb were on a plane over the Midwest at the time of Kennedy's speech, and they asked the pilot to patch the broadcast through the radio. Though Gilruth had discussed the possibility of a moon landing with the president and was impressed with his quick grasp of spaceflight fundamentals, he was aghast when he heard the challenge—only then did he fully realize the scope of what was proposed and understand the task ahead of him as director of the Space Task Group. He couldn't believe Kennedy had actually gone through with it. "The concepts of manned spaceflight were only three years old, and voyaging in space over such vast distances was still a dream," he would write later:

> Rendezvous, docking, prolonged weightlessness, radiation, and the meteoroid hazard all involved problems of unknown dimensions. We would need giant new rockets burning high-energy hydrogen; a breakthrough in reliability; new methods of staging and handling; and the ability to launch on time, since going to the moon required the accurate hitting of launch windows.

It could be done. But the president's challenge was a formidable one (and it could have been even tougher; an early draft of Kennedy's speech had given a target date of 1967, prompting Webb to ask the White House to change the time frame to the end of the decade). As the normally taciturn Gus Grissom put it, "It's as if somebody had said, 'Let's build New York City overnight.'"

It would be an undertaking of enormous dimensions, and it required a new organization mode: a strong headquarters supervising several centers that would mobilize American industry and know-how to tackle a multitude of don't-know-how matters. A massive rocket engine was already in development by von Braun's Huntsville team, but a complex spacecraft, one with all the necessary features, would take years to design, develop, build, and test before it was man-rated. The sheer scale of the project would mean new factories, new testing and training facilities, new transport methods, and new systems of all kinds, from communications to environmental to others barely imagined.

The gauntlet was picked up with enthusiasm by the rest of Gilruth's Space Task Group. Faget and others had long been touting a moon landing as the follow-up program to Mercury. Both houses of Congress quickly approved massive increases in funding for the ambitious program—$1.67 billion for the 1962 budget alone, enough to get a strong start on Apollo—so NASA now had the money, the official approval, and a specific goal. Thus began one of the largest peacetime projects in U.S. history.

But first there were important—and dangerous—baby steps to be taken.

II.

AROUND

CHAPTER FIVE

IN ORBIT

Space is a risky business. I always considered every launch a barely controlled explosion.

AARON COHEN, NASA MANAGER FOR
APOLLO COMMAND AND SERVICE MODULES

IT DIDN'T TAKE LONG for Jim Webb to prove his worth.

NASA had received little cooperation from the air force in its first couple of years. It needed launch boosters and facilities—chiefly, the rocket range at Cape Canaveral—and the air force, jealous of the turf taken from it, had responded with only grudging assistance on that issue and others. In May 1961, Webb and Secretary of Defense Robert S. McNamara negotiated a document of agreement—former lawyer Webb argued about almost every line—that provided a mandate for NASA's peaceful, noncommercial exploration of space and clarified and strengthened its position vis-à-vis the military branches. The space agency would not have to kowtow to them and would receive ready support when needed—and it would be needed often.

Webb had experience working a budget. His time on Capitol Hill came in handy later that year when the president asked him how much it would cost to put a man on the moon. Webb's staff had given him an estimate of thirteen billion dollars, but Webb decided on a different amount, one that provided some flexibility to counter the optimism of NASA's technical experts. Webb didn't want to have to go back in a few years and ask for more money; who knew how supportive Congress would be then? So he told Kennedy that the lunar program would cost upward of twenty billion dollars—an outrageous sum. When Webb's people heard about it, they were shocked. "I put an administrator's discount on it," he told them. Though that figure would rise modestly through the years (the final price tag would end up a few billion more), it reflected well on NASA that the agency stayed close to its initial budget. Webb's decision to add his "administrator's discount" would pay dividends a few years later when he requested annual budgets of five billion dollars or more.

Nine and a half weeks after Kennedy's May 25 call to action, Gus Grissom—who took great pains to keep his classically Russian middle name, Ivan, on the down-low—was shoehorned into his own Mercury capsule. He'd dubbed it *Liberty Bell* 7, and it even had a crack painted on it. The fit wasn't quite as tight for Grissom as it had been for the first astronaut to fly; at five foot six, Grissom was five inches shorter than Shepard. The flight was planned to be a virtual repeat of the previous one, and if successful, it would be the last on a Redstone booster (provided the air force's more powerful Atlas ever became reliable and fully man-rated).

Glenn, still smarting from his third-place showing, was Grissom's backup. The two had been training almost constantly since Shepard's mission, spending plenty of time on the centrifuge and the trainers; each had done more than a hundred different simulations. Grissom was prepared for one more situation, since neither the doctors nor the engineers had addressed the problem Shepard faced. He wore a makeshift urine-collection device that he and Glenn patched together using a few condoms, some tubing, rubber cement, and a plastic bag.

At 7:20 a.m. on July 21, 1961, the Redstone lifted off. The flight went off without a hitch, and after six minutes of weightlessness, Grissom splashed down close to where Shepard had.

With the recovery copter hovering above, Grissom armed the side hatch—new in this model Mercury and held in place with seventy explosive bolts—loosened his helmet, unbuckled the several harness straps, and lay back on his contour couch, waiting for the chopper to hook the capsule and carry it onto the recovery ship. Once the capsule was safely aboard, Grissom would hit the plunger that opened the hatch, emerge as the conquering hero, and stride across the deck as men cheered and cameras recorded the moment for posterity.

A moment after Grissom lay back, he heard a dull thud and watched the hatch cover blow off and skip across the waves. Salt water began to pour into the capsule. Grissom threw his helmet off and quickly followed the hatch cover, swimming away from the foundering *Liberty Bell* 7. When he looked back at the capsule and saw it sinking, he realized he was entangled in a line attached to a dye marker.

He fought free of it. But as the chopper attempted to secure a line to the capsule and Grissom attempted to help, he realized his pressure suit was losing air through his neck dam, which wasn't tight enough, and through his oxygen inlet port, which he hadn't sealed, and water was entering the port. Along with some small models of his spacecraft, two sets of pilot wings, and a string of pearls for his mother, in the left leg he was carrying two rolls of dimes he planned to give to the children in his neighborhood. He was slowly sinking in a heavy suit that was becoming heavier by the second, and when another recovery copter approached, its rotor wash made floating even more difficult, and he began bobbing beneath the surface and swallowing seawater. He feared he was going to drown.

The second copter lowered a looped horse-collar lifeline. His head barely above water, Grissom grabbed the collar and hauled himself into it. He had been in the water no more than five minutes, but it had seemed like an eternity. As one copter reeled him to safety, the other cast the capsule loose—filled with water, it was now more than a thousand pounds over the chopper's lifting capacity. The *Liberty Bell* 7 sank to the bottom of the sea, almost three miles below, where it would remain for thirty-eight years, until it was salvaged in July 1999. The first thing Grissom said after reaching safety was "Give me something to blow my nose. My head is full of sea water."

Grissom had flown a hundred missions during the Korean War and never lost a plane, so the loss of *Liberty Bell* 7 pained him. He insisted that he hadn't hit the detonator plunger (though in his first debriefing, he said, "I don't see how I could have hit it, but possibly I did"), and he had no bruise or cut on his hand that would indicate he had. In private, he admitted to his best friend, Deke Slayton, that he might have accidentally banged into it. NASA investigated and found that that would have been unlikely but could not determine how the hatch could have opened by itself. Despite Grissom's protestations of innocence, a cloud hung over him, and he was acutely aware of it. It also didn't help that in the press conference following the mission, he had used the S-word; he'd admitted that he had been "scared a good portion of the time." Surprised, the reporter who had asked the question said, "You were what?" "Scared, okay?" was Grissom's reply. Slayton and other NASA managers defended Grissom, but of course, heroes weren't supposed to admit to feeling fear.

Despite NASA's embarrassment at losing the capsule after splashdown, the flight itself had gone well—it had been a nominal flight, to use the aerospace term for "as expected; according to plan." And newspapers grasped at every positive they could. The program "could now boast two successful space travelers to one for the Soviet Union," reported the *New York Times*, which went on to point out that "both American astronauts maneuvered their craft. The Soviet major [Gagarin] had left the controls to automatic devices."

The next mission would be an orbital one, the capsule launched atop the more powerful Atlas booster, fatter than the slender Redstone. But on August 6, 1961, just seventeen days after the *Liberty Bell* 7 mission, a twenty-six-year-old rookie cosmonaut, air force major Gherman Titov, was launched into space in a Vostok capsule. His flight lasted seventeen orbits and took twenty-three hours and eighteen minutes. He became the first human to eat (from tubes), sleep (for five orbits), and take photos of Earth (with a handheld camera) from space. He was also the first to experience spacesickness, which was like seasickness and included nausea, vomiting, and intense discomfort. His Soviet superiors were so concerned that they decided to organize a crash program of physiological and psychological research before launching another manned mission, though no one outside the Soviet Union knew that; even a year later, Titov would admit publicly only to some "unpleasant sensations" that soon "disappeared almost entirely."

The Space Task Group originally planned to have each of the seven astronauts make a suborbital flight before tackling any orbital missions. But after Titov's long flight and the lack of physiological problems during the first two U.S. flights, NASA officials concluded that more suborbitals were a waste of time. Since the oft-delayed Atlas was finally ready, the decision was made to go straight to orbital. That meant Glenn, next in line, had lucked into the first attempt at an Earth orbit—if and when the rocket was man-rated.

The Atlas—a very light rocket whose dime-thin, stainless-steel skin was supported only by internal pressure of the propellants—had proven a tough horse to break. Thirteen Atlas launches had resulted in blowups, some on the launchpad, some soon after the booster had climbed into the sky. To make matters worse, von

Braun had been angry when NASA chose the booster without consulting him, and relations between the Space Task Group and Marshall cooled even more. Von Braun and his team were developing a superbooster called the Saturn, but they could proceed only so far without a contract from a government agency—and, increasingly, NASA looked to be the only customer with any need for such monsters. The agency still hadn't decided how it would get to the moon, though it was clear a massive booster would be required. Marshall's future depended on it, despite the March 1960 success of Pioneer 4, a small, gold-plated probe that achieved lunar flyby after four failures and became the first American spacecraft to go into orbit around the sun.

On September 13, 1961, a Mercury capsule carrying a dummy was launched into space atop the ninety-five-foot Atlas and made one orbit—another nominal flight. Many weeks later, on November 29, one of the space chimps, thirty-nine-pound Enos—nicknamed "the Penis" for his habit of fondling himself—rode an Atlas into space and around the Earth twice. The mission and the performance of Enos in his lever-pulling duties went well enough. The next Mercury-Atlas flight would carry a man, and Bob Gilruth announced publicly that that man would be Glenn, with his fellow astronaut and good friend Scott Carpenter as his backup.

Glenn's orbital flight would utilize a worldwide tracking network, something new in this pre-communications-satellite era—an attempt to follow, as close as possible to real time, an object speeding above the Earth. It consisted of a global system of thirteen manned stations and ships around the world on the Mercury orbit path, each with a sixty-four-foot dish to receive signals. Each one would be in contact for up to eight minutes with the capsule every time it passed overhead, from horizon to horizon, in the vicinity; the craft would be entirely out of touch for only a few minutes between stations. The stations would receive telemetry and voice, track it by radar, and communicate by radio and landline phone. They were also in contact with Mercury Control and would relay messages from one to the other via teletype machines. These were kept in a small room just off the control center, and their sewing-machine-like clatter could be heard by the flight controllers when the door was opened by a runner rushing a message in.

The mission was scheduled for January 27, 1962, but near-constant weather problems scratched that flight, and the next, and the next. NASA psychologists worried about what all the delays were doing to Glenn's psyche. Even Yuri Gagarin sent him best wishes for a launch soon. Whatever his private feelings, Glenn never expressed impatience and continued training. He spent up to nine hours a day lying on his back in the Mercury Procedures Trainer, a replica of the actual capsule, simulating every conceivable mission failure, and he repeatedly practiced maneuvering the capsule's attitude with the manual controls.

After weeks of Atlantic storms and ten delayed launches, the flight was once again set, this time for February 20, 1962. On that morning, more than eleven hundred TV, radio, and newspaper correspondents from around the world convened at Cape Canaveral. A little after six a.m. at pad 14, as the moon floated among clouds in the west, the silver-suited astronaut was helped into the tight quarters of the spacecraft he had named *Friendship* 7 at the suggestion of his two children. Glenn, who would soon turn forty-one (an age that would prompt some to say that he was too old for spaceflight), was in good spirits, buoyed by Carpenter and frequent helpings of his favorite music, Puccini's *Madame Butterfly*.

This time the weather cooperated, though it remained slightly overcast. In Mercury Control, Al Shepard had the job of CapCom. Seven hundred and fifty feet from the pad, in the domed launch-center blockhouse fortified by ten-foot-thick walls and forty more feet of sand, engineers prepared to view the launch through periscopes. Carpenter apprised Glenn of the countdown progress, then patched through a call to Glenn's wife and family in Arlington. Over the course of two wars, Glenn and his wife, Annie, had developed a ritual they performed whenever he left home to go into combat, and they continued it now:

"I'm just going down to the corner store to get a pack of gum," he said.

"Don't be long," she said, fighting back tears.

After dawn came two holds, one due to a fueling problem with the liquid-oxygen propellant, the other to fix a broken bolt on the capsule's hatch. The Atlas's thin skin popped and crackled like a bowl of Rice Krispies as it expanded and contracted, and vapor whistled through the liquid-oxygen release valve. The countdown was resumed. Over in the blockhouse, Carpenter said, "Godspeed, John Glenn," seconds before liftoff, and at 9:47 a.m., the sixty million Americans following the launch on TV or radio held their breath as the white Atlas with the small black capsule and escape tower atop it slowly rose off the ground amid massive clouds of exhaust. This was the sixth launch of the Mercury-Atlas, and two of the previous five had exploded soon after liftoff. In the main concourse of New York City's Grand Central Station, thousands of commuters stopped to silently watch the flight's progress on a giant TV screen above the ticket windows, and throughout the city's subway system, train conductors asked their passengers to say a prayer for Glenn—the first time the transit authority had used its communications system for anything besides subway operations. The call to prayer would be repeated every ten minutes for almost five hours.

Five minutes after liftoff, when *Friendship* 7 achieved an orbital speed of 17,545 miles per hour in the thinning upper reaches of the atmosphere, a huge roar erupted in Grand Central Station. As the spacecraft began to follow the curve of the Earth, Glenn's vital signs were taken—blood pressure, pulse, respiration, temperature—and he read eye charts to make sure his eyeballs hadn't changed shape and impaired his vision (one of the doctors' worries). Then he relaxed between more experiments and measurements and tried to appreciate the wonder of his situation and the pleasures of weightlessness. Sunsets, sunrises, the Earth below him with its continents and oceans and swirling clouds, all in the most brilliant colors.

Over Mexico on his second orbit, Glenn began to experience problems with his automatic attitude thrusters and then his gyroscopes. He used the one-stick controller that determined his craft's attitude semiautomatically—a system known as fly-by-wire—then switched to full manual control, which used more fuel; he could hear the pop of the rocket thrusters emitting their bursts of hydrogen peroxide. Glenn prepared for reentry. He would hit the atmosphere at Mach 24.

But another issue had arisen. In Mercury Control, a yellow warning light began to flash, and an engineer told Chris Kraft of the problem: telemetry received from *Friendship* 7 indicated that the ablative heat shield and landing bag were no longer locked onto the craft, and the heat shield was now held in place only by the titanium straps of the retro-rockets on top of it.

Kraft was busy, but Al Shepard heard the technician. This was serious news if accurate—as bad as it could get. If the heat shield jettisoned when the retro-rocket package (retropack) that sat on it did, Glenn and his craft would vaporize upon reentry into the atmosphere.

The tension in Mercury Control increased as word spread from console to console—though not to the congressmen and other VIP visitors in the glassed-in viewing gallery in back or to the TV and radio audiences. Shepard and Kraft discussed the problem. Soon there were more than a hundred people in back rooms at Cape Canaveral, Houston, and St. Louis trying to decide what to do. Three miles away at Hangar S, twenty McDonnell engineers pored over blueprints of the Mercury engineering systems and wiring diagrams looking for a solution. And they couldn't leave Glenn up in orbit until they found one—he didn't have the fuel or oxygen to remain in space for much more than the planned three orbits.

The problem appeared insoluble. Time was running out, and Kraft couldn't get an answer out of his consultants and experts. "Either you guys are going to give a decision in five minutes, or I'm going to make one myself," he thundered. They finally gave him an answer, and Kraft rendered his own decision. The retropack of small rocket thrusters designed to slow the craft during reentry was still clamped onto the capsule, and hopefully its three titanium straps would keep the heat shield from being torn off. Sooner or later, the retropack, about the size of a bushel basket, would burn up and melt away, since it wasn't designed to remain on the craft and hit the Earth's atmosphere at 17,000 miles per hour, but if it could stay on long enough and if the heat shield wasn't ripped off or damaged by the burning thrusters...When they got hold of Max Faget, he told them that the shield could survive reentry with the retropack still fastened—theoretically.

Glenn was told not to jettison the retropack, though he wasn't told why—Kraft thought it would be better not to worry Glenn with another problem, since he was busy with the faulty automatic control system and with trying to control the capsule. But from questions and commands he received, Glenn knew that his craft was probably not shipshape.

"We are recommending that you leave the retro-package on," a tracking-station technician told him, "through the entire reentry."

Glenn said, "What is the reason for this? Do you have any reason?"

"It's the judgment of the Cape. Cape Flight will give you the reason for this action."

As he approached the outer limits of the atmosphere, Glenn assumed full manual control of *Friendship* 7's attitude, then used the semiautomatic fly-by-wire system to assume the proper reentry angle. If the approach was too sharp, the capsule would burn up; too shallow, and the orbit might not decay fast enough before the capsule's oxygen ran out in a few hours. Either mistake would, in all probability, result in Glenn's death. Few in NASA believed they'd get through the program without a death. Glenn might be the first.

Shepard coached him down. Moments before reentry, he said, "We are not sure whether or not your landing bag has deployed. We feel that it is far safer to reenter with the retro-package on. We see no difficulty at this time in that type of reentry. Over."

"Roger," Glenn said. "Understand." There was no fear in his voice, but one NASA official thought he sensed a note of resignation, as if Glenn believed death was only moments away.

Friendship 7 entered the atmosphere and a layer of ionized air surrounded it, preventing communication. The blackout would last about four and a half minutes. It felt like an eternity to those in the Mercury Control room, its air a blue haze from the many cigarettes and pipes. One doctor bowed his head and prayed. Somewhere out there about a hundred miles above the Earth, Glenn was "falling through space like a shooting star," one NASA official would remember thinking.

His reentry occurred at the proper angle. As he tore through the atmosphere, the capsule rocked from side to side, and Glenn tried to steady it. He felt and heard a bang and then watched through the porthole as large pieces of the retropack ripped off in flaming chunks and fell away; one of the metal straps smacked against the window before it disappeared. Then burning remnants of the heat shield did the same, and Glenn didn't know if it was secured or if he was about to burst into flame, and he could almost feel the incinerating heat at his back. With the hand controller, he continued to help steady the bucking capsule as he strained against eight g's of deceleration.

Through the window, he could see a bright orange glow sheathing the craft. The ionized air around it was about ninety-five hundred degrees, only slightly less hot than the sun.

Glenn radioed, "*Friendship* 7. I think the pack just let go. *Friendship* 7. A real fireball outside. Hello, Cape. *Friendship* 7, over... Hello, Cape, do you receive? Do you receive? Over."

No one could hear him in Mercury Control. Its radar was tracking the capsule, but no one knew if Glenn was alive. A group of men got up from their chairs and walked over to gather behind Shepard's console. One of them urged him to keep talking.

"*Friendship* 7," said Shepard, "this is Mercury Control. How do you read? Over."

There was silence. Shepard said, "*Friendship* 7, this is Cape. How do you read? Over."

More silence. Then they heard Glenn's voice: "Loud and clear. How me?"

As cheers sounded in Mercury Control, Glenn's deceleration force eased, the rocking abated, and the orange glow faded. His small drogue parachute opened a bit early, at 23,000 feet, the orange-and-white main chute opened at 10,800 feet, and *Friendship* 7 smacked into the Atlantic Ocean within six miles of the USS *Noa*, the destroyer assigned to recover him. After four hours and fifty-six minutes, Glenn had returned to Earth safely. An entire nation following Glenn's mission on TV and radio could finally relax. Aside from losing five pounds from dehydration due to the heat, Glenn was fine. The *Noa* reached him seventeen minutes after he splashed down and hoisted the capsule aboard. Moments later, he hit the hatch release plunger and got two skinned knuckles when it snapped back into place (as Grissom would have if he'd opened his hatch deliberately). Glenn stepped out onto the destroyer's deck bathed in sweat. The now-traditional phone call from the president awaited.

Glenn was not happy that he hadn't been told about the heat-shield problem (although it turned out that the issue was a defective indicator light; the heat shield had been secured the whole time). He made his feelings clear to his superiors, particularly Kraft—the pilot was entitled to any and all information concerning his ship.

Upon reuniting with his wife and children, he burst into tears.

For his day's work, he received $245 (before taxes) in bonus flight pay.

Public reaction to Glenn's flight and his stoic, calm bearing in the face of his potential death by fire was overwhelmingly positive. And his successful mission—he had made three orbits—pulled the Americans closer to the Soviets. It also whetted America's appetite for more space adventures and made the president's lofty goal seem more possible. Why *not* go to the moon?

Glenn was hailed as the greatest American hero since Lindbergh and he received even more adulation and attention than Shepard had. Later that year, toy *Friendship* 7 space capsules were popular Christmas gifts; unlike the original, they were made in Japan. The flight of *Freedom* 7 hadn't inspired any toys. And NASA rushed out a thirty-minute film entitled *The John Glenn Story* with an introduction by President Kennedy.

Parades and appearances, a White House reception, and an address to a joint session of Congress followed. On a cold March morning in New York, four million people turned out on what was officially declared John Glenn Day for a parade for all the astronauts—though it was Glenn they were really there for. The ticker tape and shredded paper fell thick as snow, and the New York sanitation department later measured the paper garbage at thirty-five hundred tons—more than Charles Lindbergh had received after his 1927 flight from New York to Paris. Glenn was the third American in space, but his fame had quickly eclipsed that of his predecessors.

There would be longer Mercury flights and more complicated and more challenging ones. But Glenn would be the gold standard for astronauts. Americans and countless others around the globe responded to something in him—an earnestness, a likability, a goodness. If these seven men were knights, he was Sir Galahad the pure, undergoing a trial by fire and emerging triumphant. That aura would follow him for the rest of his days.

CHAPTER SIX

UNDER PRESSURE

We knew that human beings are never perfect.

ROBERT VOAS, ASTRONAUT TRAINING OFFICER

THE ORIGINAL IDEA OF the Mercury Seven was that any one of them could fly any mission—they were all equally capable. But as it turned out, that wasn't true. Some astronauts were better at certain tasks than others. The program's remaining flights made that clear, as each man's strengths and weaknesses were revealed under the harsh public spotlight.

The next flight was supposed to be a virtual repeat of John Glenn's three-orbit mission, this time with salty-tongued Deke Slayton the man in the can. But plans, like rockets, have a way of going awry.

For almost two decades, the Wisconsin farm boy had been flying planes on dangerous missions, including sixty-three combat sorties over Europe and Japan in World War II. Slayton was probably the best stick-and-rudder man of the Mercury Seven, and they were all damn fine pilots. He was also as tough as they come, so when NASA doctors told him a few months after he'd been selected that they'd detected a minor heart arrhythmia called idiopathic atrial fibrillation, Slayton just shrugged. He was in superb physical shape. To him, it was no big deal—about every two weeks, his pulse would, as he put it, "act up," and eventually, especially if he went for a run, it would return to normal. But just to make sure his condition wasn't an issue, he quit smoking, stopped drinking coffee, and started jogging every morning.

Though air force regulations stated that anyone with atrial fibrillation could not fly high-performance aircraft, the physicians assigned to the Mercury program had decided that the condition didn't affect Slayton's flying, and they recommended that he be accepted into the spaceflight operations. That lasted until a few months before Slayton's mission, when Jerome Wiesner, the president's top science adviser—who had opposed manned space exploration from the start—got involved and advised Jim Webb not to let Slayton fly. What if something happened, anything, and then it got out that the astronaut had a heart condition? Atrial fibrillation could cause fainting or a stroke, and if that happened while a pilot was flying, it could lead to catastrophe. In addition, the condition could affect his heart rate and blood pressure, making it impossible to accurately assess the effects of spaceflight on humans—a main purpose of the Mercury program. They had plenty of men in perfect physical shape ready to go. Why take a chance if they didn't have to?

Webb convened a panel of flight surgeons; they found Slayton fully qualified. But two days later, three eminent civilian cardiologists examined him and disagreed. On March 15, 1962, the NASA higher-ups told the astronaut he was grounded indefinitely. When they insisted that the devastated Slayton attend a news conference about his grounding the next day, he didn't think he could—"I could have killed everyone in that room," he would remember later. But somehow he got through it, and publicly and privately he vowed he would support the program in any way possible. Most of the other astronauts were on his side and personally asked President Kennedy to reinstate him, but without success.

Sputnik 1, launched on October 4, 1957, by the USSR, weighed a mere 184 pounds. The beach-ball-size sphere was the first artificial satellite, and its launch triggered the space race. *(All photos courtesy NASA unless otherwise indicated)*

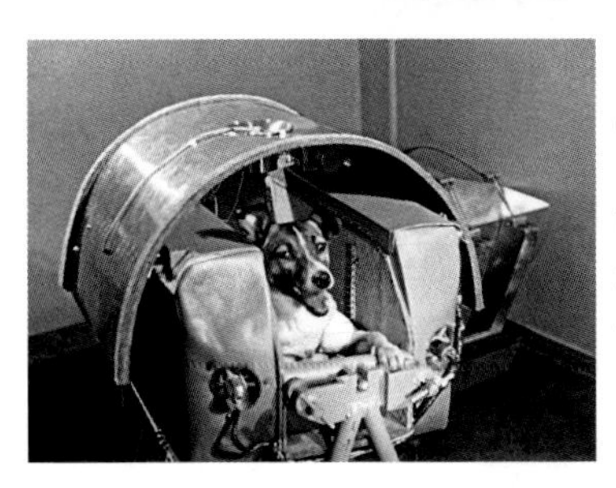

On November 3, 1957, a month after Sputnik 1 went up, the first living being to orbit the Earth was launched into space aboard Sputnik 2. Laika died within hours from overheating and stress.

The V-2 rocket developed by Wernher von Braun's team at Peenemünde was the first long-range guided ballistic missile. *(Author's collection)*

The V-2 was capable of inflicting horrific damage—this photo shows the results of the last one to hit London, on March 27, 1945. *(Author's collection)*

Von Braun (center, with cast) surrendered to the U.S. Army on May 2, 1945. At left is Charles Stewart, CIC agent; Magnus von Braun is at right in leather jacket. The others are members of von Braun's rocket team.

When von Braun and his rocketeers were brought into the U.S. as part of Operation Paperclip (above, with von Braun in the first row, seventh from the right), they spent years assembling and launching V-2s, built from parts shipped from Germany, and improving their rocket expertise. The Redstone and Jupiter missiles were essentially larger V-2s with extra stages.

On January 31, 1958, von Braun's Jupiter-C launched the first American satellite, Explorer 1. At a celebratory press conference, von Braun (right) raises a model of the rocket with Jet Propulsion Laboratory director William Pickering (left) and scientist James Van Allen.

Bob Gilruth (right), an inspiring manager as well as a brilliant aeronautical engineer, was picked to head the fledgling Space Task Group; Chris Kraft (left), who did more than anyone to create the space-age Mission Control Center, was among its first members.

The idiosyncratic designer Max Faget, seen here in his navy whites during World War II, when he served as executive officer on a submarine. *(Courtesy Carol Faget)*

The Mercury Seven, the test pilots chosen to battle the Red Menace (from left to right): Scott Carpenter, Gordon Cooper, John Glenn, Gus Grissom, Wally Schirra, Alan Shepard, and Deke Slayton.

Once the astronauts began staying at the Cape's Starlite Motel, with its space-themed lounge, it quickly became a Cocoa Beach hot spot. *(Author's collection)*

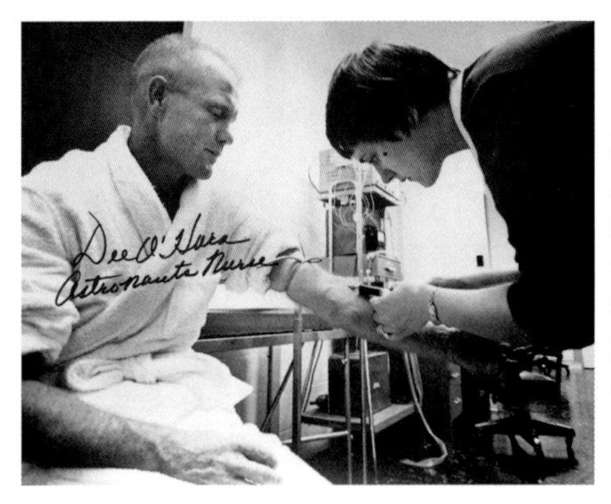

Nurse Dee O'Hara, seen here with John Glenn, was hired to attend to the Mercury Seven's medical needs. She stayed with the manned spaceflight program through the early seventies.

Sergei Korolev (left), shown here in 1956 with academician Mstislav Keldysh, was the lead rocket engineer and spacecraft designer for the Soviet space program until his untimely death in January 1966. Keldysh would later play a part in the Apollo 8 mission. *(Alamy)*

The diminutive (five-two) Yuri Gagarin (left), shown here with Korolev, was the first human in space, and the first to orbit the Earth, on April 12, 1961, in Vostok 1. *(Author's collection)*

Headlines like this one blared from every newspaper in the U.S. the next morning. *(Author's collection)*

Feature Index

The Huntsville Times

Where Progress...

Covers The Valley!

Man Enters Space

'So Close, Yet So Far,' Sighs Cape

U.S. Had Hoped For Own Launch

Hobbs Admits 1944 Slaying

Soviet Officer Orbits Globe In 5-Ton Ship

Maximum Height Reached Reported As 188 Miles

VON BRAUN'S REACTION:

'To Keep Up, U.S.A. Must Run Like Hell'

Praise Is Heaped On Major Gagarin

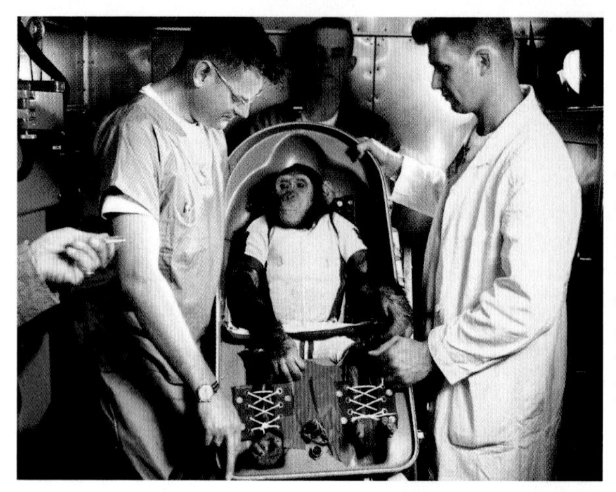

Ham was one of several chimps who flew into space before their astronaut counterparts. His January 31, 1961, flight on a Mercury-Redstone assured NASA officials that it would be safe for a human to fly.

On May 5, 1961—twenty-five days after Gagarin's flight—Alan Shepard became the first American in space. His Mercury craft *Freedom* 7 was boosted into space by one of von Braun's Redstone rockets. Shepard reached an altitude of 116 miles in a flight that lasted fifteen minutes.

Three days later, President John F. Kennedy presented Shepard with NASA's Distinguished Service Medal in a Rose Garden ceremony at the White House. NASA administrator Jim Webb, at Kennedy's left shoulder, looks on.

On May 25, 1961, three weeks after Shepard's successful mission, President Kennedy stood before Congress and threw down a massive challenge: "This nation should commit itself to achieving the goal, before this decade is out, of landing a man on the moon and returning him safely to the earth."

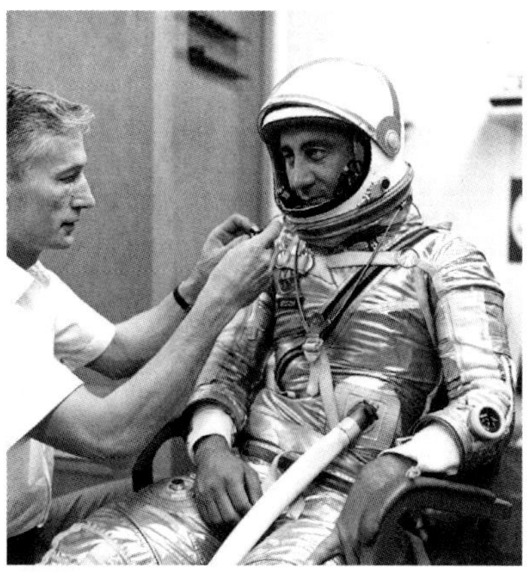

Suit technician Joe Schmitt (left) prepares Gus Grissom for his July 21, 1961, *Liberty Bell* 7 mission, a virtual repeat of Shepard's fifteen-minute flight.

Mercury Control Center at Cape Canaveral, here seen during John Glenn's February 20, 1962, flight, was a relatively simple setup. The mission's three orbits are plotted on the screen.

Scott Carpenter's May 24, 1962, mission was essentially the same as Glenn's, but Carpenter ran afoul of Chris Kraft, who vowed he would never fly again. Carpenter never did.

Wally Schirra, seen here with pad leader Guenter Wendt, "Der Pad Führer," flew his *Sigma 7,* a textbook engineering flight, on October 3, 1962.

Only Gary Cooper's superb piloting allowed *Faith 7* to splash down safely on May 16, 1963, after urine short-circuited several systems. The worn-out astronaut was all grins when the hatch was opened on the deck of the recovery ship *Kearsarge.*

Valentina Tereshkova, a former textile-factory worker and amateur skydiver, was chosen from five "cosmonettes" to be the first woman in space on Vostok 6, launched June 16, 1963. Before the flight, she was inducted as a lieutenant into the Soviet air force and later promoted to the rank of major (as shown here, in 1969). *(Courtesy RIA Novosti)*

To make things more complicated, Slayton's backup pilot, wavy-haired Wally Schirra, wouldn't be taking his place. Bob Gilruth decided that since Scott Carpenter, who had backed up Glenn on his flight, had more simulator time than Schirra, he would be a better fit for the mission. Now the Irish-Italian Schirra got his Irish up—and he became even angrier when he was named Carpenter's backup. Schirra, a fighter pilot, being replaced by a bomber pilot—he didn't think that was right. (Carpenter had actually flown patrol planes in Korea.) But Gilruth calmed him down with the news that the next mission, more of an engineering flight with even more experiments, would be his.

Slayton, meanwhile, continued making lifestyle changes, even giving up alcohol and, in his words, "everything that was fun." But the periodic arrhythmia continued, so when NASA began looking for someone to run the astronaut office—it had been decided by this time that more astronauts would be needed for post-Mercury programs—Al Shepard suggested his buddy Deke, and the rest of the Mercury Seven approved. If they had to have a boss, it might as well be one of them, not an outsider with little understanding of the job.

In September 1962, Gilruth made it official: Slayton would become coordinator of astronaut activities. His first assignment would be to pick a new group of spacemen. Though he would never completely get over his failure to fly into space on a Mercury mission, Slayton would take to his new career with relish. And when his job description came to include choosing crews, he swore to himself that he would be scrupulously fair in his selections. And he stuck to that vow—for the most part.

If there was a dreamer among the seven astronauts, it was Scott Carpenter, the self-described indifferent student, former shoplifter, and "no-good" youth from Boulder, Colorado, whose parents had separated when he was two. His mother was diagnosed with tuberculosis and placed in a sanatorium, and Carpenter's maternal grandparents had largely raised him. At the age of twenty-one, he had a near-fatal auto accident—he fell asleep at the wheel after a night of partying and drove off a Colorado mountain road. He fractured ribs, broke a leg, and spent two weeks in a hospital; after that, he decided to settle down. While enrolled at the University of Colorado, he married a smart, attractive, green-eyed blonde named Rene (pronounced "Reen"). In the fall of 1949, with a pregnant wife and lacking one heat-transfer course for an aeronautical engineering degree, he dropped out of college to join the navy. This fact had somehow been overlooked when the NASA administrators were looking for astronaut candidates. They'd just assumed he'd graduated before he joined up.

He came into his own in the military, eventually flying a patrol plane in the Pacific during the Korean War, then jumping to test-pilot school when it was over. Though he had the weakest test-pilot credentials of the Mercury Seven and only three hundred hours of flying time in jets (the others had from seventeen hundred to twenty-one hundred), he had excelled at many of the physical tests at Albuquerque's Lovelace Clinic. He had also flown a photo-reconnaissance fighter whose advanced camera was similar to a device planned for Mercury, and he had experience in celestial navigation and communications. All of these assets—and his keen, curious intelligence—had outweighed his relatively meager jet time.

By the time he became an astronaut, the wiry, movie-star-handsome Carpenter and his wife had four children. He was the most introspective of the Mercury Seven, though some NASA officials considered him "flaky" or "vague and detached."

The new guy in Mercury Control was anything but flaky. Sitting next to Kraft in the control room for the first time, in the new position of assistant flight director, was twenty-eight-year-old Gene

Kranz, a former air force fighter pilot. After a couple of years as a flight-test engineer with McDonnell, in October of 1960 the crew-cut Kranz had hired on with the Space Task Group. Two weeks after Kranz started, and shortly before the first Mercury-Redstone launch, Chris Kraft needed someone to do an important job and found only Kranz available. He sent him down to the Cape to talk to the test conductors and write up a countdown procedure and some mission rules—what to do if a given thing went wrong, covering every situation. Kraft knew that in a world where rockets flew fifty to sixty times the speed of airplanes, there would be little time for the controllers to discuss the next steps if something went awry. They had to be prepared in advance.

Kranz spent weeks on that job, talking to virtually everyone involved in a launch. He combined information from the control center, the blockhouse, and the tracking stations into an effective and thorough countdown procedure and wrote a thirty-page summary of mission rules. Kraft, impressed, had made him procedures controller, in charge of coordinating communications between the tracking network and Mercury Control, for the first three Mercury flights. Kranz's good work had been rewarded by his promotion to assistant flight director for Carpenter's flight.

Carpenter would be boosted into space on another Atlas rocket, make three orbits, and dive back into the atmosphere at almost 18,000 miles per hour—easy enough, even if Carpenter had only a couple of months to prepare for the flight and his job would be complicated by a host of science experiments.

At least one man in Mercury Control had doubts. Kraft had had problems with Carpenter while getting him up to speed as CapCom during Shepard's launch; he'd become convinced that Carpenter wouldn't do a good job as CapCom, and when he shared his concerns with Slayton, Deke replaced Carpenter. Kraft had also objected to Carpenter being named to pilot this flight, claiming he might jeopardize himself and the mission. He knew Carpenter had little experience as a test pilot and felt that he was of no help as an engineer; Kraft believed he'd somehow slipped through the astronaut selection process. (If true, Carpenter's superior athletic ability and superb conditioning had probably swayed the administrators, since the NASA doctors were obsessed with the potential physiological hazards the astronauts might endure.) On the morning of May 24, 1962, after a smooth liftoff and orbit insertion, Carpenter made his first questionable decision. When he put his capsule into automatic control mode, he found that the pitch control wouldn't function properly. But he didn't tell Mercury Control for fear that they'd bring him down early.

As he circled a hundred miles above the Earth, Carpenter ran through the tasks on his checklist—releasing a balloon, photographing terrestrial features, eating solid food, and a few others—and began to admire the view. He stopped focusing on his flight plan and started experimenting with the craft's attitude thruster jets, which changed the direction it faced. He also began paying more attention to the view than to the spacecraft and its systems, ignoring his decreasing fuel levels, prompting several requests from Mercury Control for him to switch to unpowered drifting flight to conserve fuel; he didn't. After he'd maneuvered *Aurora* 7 every which way as if it were a bumper car for two orbits, Mercury Control again ordered him to go into drifting mode. He did, but when he became fascinated by sparkling "fireflies" that were produced every time he smacked the spacecraft's side with his hand, he began using the jets again to swing around and see where they came from. (The fireflies, which Glenn had noticed also, were really just vaporized drops of moisture vented from the capsule that instantly froze into ice particles.) And when Carpenter was asked to help with recordings of his blood pressure, he declined; "I've got the sunrise to worry about," he said.

During his final orbit, he fell behind in making the necessary preparations for reentry, such as storing equipment and aligning the craft to enter the atmosphere at the correct angle—critical to help *Aurora* 7 slow down. Then he realized his automatic control wasn't working properly, so he took over the controls manually to reposition the craft. Time was becoming tight. He soon switched to the semiautomatic fly-by-wire but forgot to turn off the manual control, which wasted precious fuel. Some of the tracking stations reported that Carpenter sounded "tired" and "confused"; Kraft thought he sounded delirious.

Another component of the automatic control system failed, and Carpenter was forced to manually turn on the retro-rockets. He did it, but he was three seconds late—enough to change his reentry attitude and thus his trajectory and landing point. Mercury Control didn't know if he had reentered at the correct attitude. If the blunt end of the capsule was pointed at too shallow an angle, *Aurora* 7 might bounce off the thick atmosphere and circle the Earth in a slowly deteriorating orbit until its power and oxygen gave out—a few hours, tops. Most of the men working in the control room expected to lose one or two astronauts on the Mercury project—after all, that was the usual number of fatalities when pilots were testing a new airplane. The crisis in Glenn's flight, the supposedly loose heat shield, had turned out not to be a crisis at all, but this mission looked like it might end badly.

The angle was off by about ten degrees. That was good enough for a successful reentry, albeit a sloppy and shaky one that necessitated the early release of the drogue and main chutes to steady the wildly oscillating capsule. By the time the capsule dropped softly into a gently swelling sea, Carpenter was two hundred and fifty miles farther downrange than expected and out of radio contact. The recovery forces in the area knew roughly where he was—two planes had received the capsule's beacon signal—although no one knew whether he was alive or dead. To make matters worse, the last telemetry from his heart monitor had shown an irregular EKG; Kraft thought he might have had a heart attack.

Word of his downrange splashdown leaked out to the press, and the nation held its collective breath. Millions returned to their TVs, and CBS newsman Walter Cronkite told his audience, "We may have...lost an astronaut." Finally, a navy copter spied *Aurora* 7's green dye marker in the water and the bright orange life raft attached to the capsule containing a very relaxed Scott Carpenter, contemplating his situation while he ate a Baby Ruth candy bar. He was oblivious to his rescuers until two frogmen parachuted into the sea and swam over. When one called to him, Carpenter nearly jumped out of the raft. Safely choppered to the recovery carrier USS *Intrepid*, he said the frogmen "broke up his tranquility."

Flight director Chris Kraft was livid. Despite his steely demeanor during a mission, he had developed an ulcer he self-treated with a pint or two of milk before every launch, and during this one he had barely controlled his fury. Anxiety had been high in Mercury Control as they all waited for news of Carpenter, and John Glenn was on the phone with Rene Carpenter, trying to keep her updated and calm.

Kraft hadn't been the only one with reservations about Carpenter's readiness; Walt Williams had also been concerned. His worries were confirmed. Carpenter had joyrode, neglected to inform Mercury Control of a problem, paid scant attention to his instruments, and ignored instructions from the ground.

Kraft, who would one day declare that "Flight"—meaning the flight director—"is God," and who was called Jesus (pronounced "Hay-soos," in the Spanish way) by some of his flight controllers and the Prussian General by others, appeared to see this mission as the first test of his control, and he responded autocratically. He was further affronted when Carpenter was quoted as saying, "I didn't know where I was and they didn't either." The claim that Mercury

Control—his group—had lost an astronaut even briefly was likely the final straw for Kraft, who had once snapped during a botched Redstone launch, "The damn Germans still haven't learned who they work for. Everyone in this control room must work for me." Once a mission began, Kraft was a general commanding his troops, and he would brook no insubordination, real or perceived. He had once come close to blows with Slayton over whose man, Kraft's operations guy or Slayton's astronaut, was in charge at a remote tracking station.

Though Carpenter had made mistakes, Kraft ignored the fact that *Aurora* 7's navigational systems had failed, that finding the right balance among the three attitude-control modes was tricky, and that anything but perfection resulted in excessive fuel consumption. A slight flick of the wrist on the three-axis control stick would initiate the powerful thruster jets, and unless that was quickly corrected, the system would gobble fuel at an exorbitant rate. (This problem would be fixed on later Mercury flights.) None of this swayed Kraft, however. After the astronaut was located, in a voice loud enough for everyone in Mercury Control to hear, he said, "That son of a bitch is never gonna fly for me again."

Carpenter did, finally, receive his college degree. When he returned to Boulder for a hometown hero's welcome, the University of Colorado awarded him a bachelor's in science on the grounds that his "subsequent training as an astronaut has more than made up for the deficiency in the subject of heat transfer."

Despite Carpenter's errors in judgment, the mission was portrayed as a success in the press, and Americans began to think they were catching up to the Soviets in space. Yes, Titov's August 1961 flight had lasted seventeen orbits to just three each made by Glenn and Carpenter, but the U.S. had now matched the USSR's total of two men in orbit and had launched four men into space, twice the number of cosmonauts.

* * *

That postorbital glow lasted ten weeks, until August 12, 1962, when Vostok 3 carried Major Andrian Nikolayev into orbit. The next day, Lieutenant Colonel Pavel Popovich in Vostok 4 joined him on a similar trajectory, both of them circling the Earth. Nikolayev's sixty-four orbits and Popovich's forty-eight far surpassed their American counterparts' orbits. Their two capsules came within four miles of each other and landed safely in Kazakhstan six minutes apart. Though it was not a true rendezvous, the double mission proved the Soviets were still far ahead of the American program. The United States once again issued congratulations, but the feat initiated calls in the Senate and from the press for a military program to counter the Soviet capability in space—how would America respond to a nuclear attack from Soviet satellites? But President Kennedy resisted these demands, and NASA was allowed to stay true to its plan of slow, steady, and safe development.

Next up was Schirra, the acknowledged master of the "gotcha"—what the astronauts called the elaborate practical jokes that the seven of them were constantly pulling on one another and anyone in the vicinity to lighten up the almost unrelieved seriousness of their work. Despite his nonstop and painful punning, Schirra was all business when it came to flying, and he was the obvious choice for the next mission, which called for six orbits and a bevy of experiments. A man who thought of himself as the consummate engineering test pilot, Schirra was eager to show the positive difference an expert pilot could make, a pilot who wasn't distracted by the sights available out the window and the poetry of the moment. In short, he would be Carpenter's opposite.

Though Schirra hadn't given up his heavy smoking habit—the men who put together his survival kit included a pack of Tareyton cigarettes—he trained hard and without much help from Glenn, as Schirra made clear in a TV interview. According to him, Glenn was busy doing public appearances and had no time to talk to him. Schirra had never gotten along well with Glenn, since he thought of Glenn as a PR machine and himself as an engineer and a pilot—and certainly not as a "poet," he said, in a veiled reference to Carpenter. Schirra being a navy man, he began taking the "captain of his ship" concept to an extreme, developing an attitude about being told what to do.

The flight of Schirra's *Sigma* 7 on October 3, 1962, was nominal, as expected—in virtually every way, a textbook flight. Over nine hours, he used his thrusters sparingly, drifting uncontrolled for three full orbits, and ignored the spectacular view as much as possible. He was lauded for his professionalism and even forgiven for referring to flying in automatic control as being in "chimp mode." He splashed down just a half a mile from the navy's recovery ship.

After five Mercury flights, the decision to use test pilots as astronauts had been vindicated. Every mission so far had involved its share of errors and malfunctions; even Schirra's Atlas had made an unplanned clockwise roll after liftoff. Indeed, each astronaut had to deal with problems, some larger than others. The craft's attitude thruster jets were a constant headache, and NASA technicians had modified them yet again.

Before the next mission, though, a Cuba-related crisis occurred. Ironically, it was in part a response by Russia to the United States placing forty-five of von Braun's Jupiters—equipped with nuclear tips—within easy range of Soviet targets.

After the Bay of Pigs fiasco, Fidel Castro had requested nuclear missiles from the USSR to deter future invasions. Nikita Khrushchev had agreed, particularly in view of America's nuclear-tipped Jupiters in Turkey, on the Russian border, and in Italy, just a hundred miles across the Adriatic Sea. Construction of launch facilities in Cuba began as Soviet ships started transporting missiles. Rumors of their arrival began filtering into the U.S., and on October 14, 1962, photographs by an air force U-2 spy plane confirmed them. After several days of carefully weighing the available options, on the evening of October 22, Kennedy delivered a televised address to the nation. He announced the discovery of the missiles and their range of up to two thousand miles, which put most of the United States within reach, and blamed it all on the Soviet Union. He vowed that the U.S. would not hesitate to use military action. He invoked the Monroe Doctrine—which had come to mean that if any nation fired a missile in the Western Hemisphere, it would be considered an act of war against the United States—in his decision to blockade any and all offensive military equipment on any ship bound for Cuba. He euphemistically called the naval blockade a "quarantine."

Sixteen months before, in June 1961, Khrushchev and Kennedy had met in Vienna to discuss ways to solve some of the world's problems. Little progress was made, and the Soviet premier appeared to have gotten the better of the young president; he believed him to be weak and callow. But Khrushchev would have a different estimation of Kennedy after this crisis. Few countries had approved of the strategy behind the Bay of Pigs debacle, but most of the free world supported America's decision to blockade the Russian missiles. Several days of bluster and posturing by both countries followed, and the situation escalated dramatically. While the United Nations tried to broker a truce, Strategic Air Command bombers, many armed with nuclear weapons, were put on high alert.

Six days after Kennedy's announcement, the two countries reached an accord. The Soviets would remove the missiles from Cuba under UN watch, and the Americans would lift the blockade; in private, they also agreed to remove their Jupiters from Italy and Turkey. (The missiles had been declared obsolete and were scheduled to be taken down anyway, but the Russians didn't know that.) The Soviets also decided to remove 162 other nuclear warheads on

Cuba that the United States was unaware of—they didn't trust Castro not to start a nuclear war.

For thirteen days in October of 1962, the world had teetered on the brink of nuclear disaster—Kennedy would later estimate the odds of escalation to war as having been "between one in three and even." Though both leaders took great pains to avoid conflict, it appeared to the world that Khrushchev had backed down while Kennedy had refused to yield. Both Kennedy and the United States regained the stature and respect lost at the Bay of Pigs.

By January 1963, most of Gilruth's creative people had moved on to Gemini and Apollo, and Mercury was being run largely by operational personnel. The end of the program was in sight, but there was one last mission, and it would be a doozy.

The flight was to be twenty-two orbits—more than a full day—with plenty of experiments and procedures loaded into the schedule. Gordon Cooper, at thirty-six the youngest of the Mercury Seven and the only completely healthy one yet to fly, was Slayton's pick to pilot the spacecraft, which was beefed up with extra batteries, extra oxygen tanks, and extra water. No one questioned Gordo's piloting skills. He had been flying since he was six, when his father let him take over the controls—only once they were in the air—of the family's small plane. Those who dismissed him as a hick and mistook his soft-spoken, laid-back style for laziness were surprised at the sharp insights delivered in his Oklahoma twang. But he often seemed to be more interested in racing cars and speedboats and having off-hours fun than in training. And Cooper couldn't keep from infuriating management with his flying stunts and flat-hatting—that time-honored maneuver of flying close to the ground at dangerously high speeds. There had been doubts all along that he would be able to focus on the mission, which was why he was the last active astronaut to fly. But he'd finally settled down, and he'd done excellent work as Schirra's backup.

Al Shepard, however, saw a chance to get into space again. He began lobbying to be given Cooper's flight, insisting that he'd do a better job. But NASA, impressed with Cooper's newfound focus, decided to stay with him—and made Shepard his backup. They gave Shepard a sop: one more long, possibly three- or four-day Mercury mission was being considered, and if it happened, it would be his.

Even though he had spent enough time on the procedures trainer that he could go through an entire simulated mission with his eyes closed, Cooper almost lost the flight anyway. On the day before his launch, in a fit of pique over a minor pressure-suit alteration he hadn't been told about, he climbed into an F-106 fighter jet and proceeded to perform loops and rolls through Cape Canaveral's restricted airspace—and he finished off by flat-hatting the NASA administration building. An enraged NASA official told Shepard that the flight was his. Only after Slayton and the other astronauts objected to the change was Cooper, properly contrite, reinstated.

Mercury Control began making final preparations. This flight needed round-the-clock monitoring, meaning an additional shift—another team of controllers and another flight director. Kraft picked a young, capable, British-born Canadian named John Hodge, who had been hired in 1959 after a fighter jet program called Avro Arrow had been canceled. Each shift would work ten or twelve hours. Several tracking stations had also been added, since the Earth would spin appreciably during the day-and-a-half mission, meaning Cooper's path over it would change significantly on each orbit.

Early in the morning of May 15, 1963, Deke Slayton knocked on the door of the crew quarters in Hangar S to wake Cooper. As the chief astronaut, Slayton felt it was his duty to do so. At 8:04

a.m., a short while after a relaxed Cooper took a brief nap on the launchpad during a delay, his *Faith* 7 lifted off. He was scheduled for twenty-two orbits—at ninety minutes an orbit, almost a day and a half. The first eighteen orbits were nominal, and he took another short nap before his second orbit was even finished, though when he awakened, he was startled to find his arms floating before him with his hands dangerously close to the many switches on his control panel. Cooper remained cool as a cucumber, and his oxygen consumption was so low that CapCom Shepard told him, "You can stop holding your breath."

Despite his laid-back attitude, he was sharp and focused when he needed to be—as well as surprisingly eloquent. He used a recorder to tape his observations, and critics couldn't complain, as they had about other astronauts, that his descriptions were unimaginative:

> As the sun begins to get down toward the horizon, it is very well-defined...it is a very bright white, almost the bluish-white color of an arc lamp....The sky begins to get quite dark...the light spreading out from the sun is a bright orange color which moves out from under a narrow band of bright blue.

After checking on the eleven experiments he'd been tasked with doing, Cooper slept on and off over the next several revolutions, sometimes waking and marveling at the sight of Earth from a hundred miles above, often photographing what he saw. During the nineteenth orbit, recurrent surges of electricity began playing havoc with the spacecraft. Two orbits later, a short circuit shut down the automatic control system. Cooper would have to take complete control of the capsule and hold it steady in all three axes, and he would have to manually line up the capsule's angle of attack by hand and fire his retro-rockets at the exact right second.

While Mercury Control went into a frenzy of activity trying to find out what was happening, Cooper remained so calm, despite the increasing level of carbon dioxide in his suit—"We probably could not have made another revolution," said a NASA life-support systems engineer afterward—that they told him to take a go-pill (an amphetamine). That was just about the point when part of his automatic reentry system failed, and then another.

"Things are starting to stack up a little," Cooper radioed on his twenty-second and final orbit, sounding like an unconcerned commercial-airline pilot circling for a landing, then he drily proceeded to list the systems that had conked out. "Other than that," he said when he'd concluded, "everything's fine."

Everything wasn't. Cooper lined up a scribe mark etched on his window with the visible horizon to time the firing of each retrorocket. He performed a completely manual reentry—the first time in the Mercury program—and *Faith* 7 splashed down just four miles in front of the recovery ship *Kearsarge*. The flight was more proof that people could function well for long periods in space—a capability, some newspapers pointed out, that would be important when the air force developed a "space patrol," one of the many half-baked aerospace projects floated by the military.

Later, NASA engineers tore *Faith* 7 down to its skin, and they found the reason for the many failures experienced on the mission: Cooper's urine-collection system had leaked, penetrating and short-circuiting several systems. The Mercury capsule had only just survived thirty-four hours in space, limping home barely functional. "If it hadn't been for Cooper's superb piloting job," said Kraft, "the mission would have failed."

In hindsight, the decision to use test pilots made eminent sense. In emergencies, they did not panic or freeze—they acted. Besides their coolness under pressure and ability to deal with real-time problems under the most difficult conditions, they were familiar with the experience of evaluating an aircraft from the ground up, one step at a time, and expressing that evaluation in terms that an aeronautical engineer could understand. And their skill and knowledge were even more essential in these first days of manned space exploration, when NASA was inventing the spacecraft much like the Wright brothers had invented the airplane a half a century before—by repeated trial and error. That knowledge, experience, and skill had been invaluable on the Mercury flights. On every mission but one—Schirra's—the pilot found it necessary to take manual control of the attitude during reentry, and each mission saw its share of systems failures that required human interaction.

Even before Cooper's return, NASA officials had pitched the longer Mercury flight—the one promised to Shepard—to administrator Jim Webb. Now that they knew that there was less cosmic radiation than the space docs had expected and that it was tolerable because of the capsule walls and the spacesuits, they wanted to learn more about the effects of long-term weightlessness. A three-day flight would contribute significant knowledge and keep the U.S. in the game, so to speak, since there were no manned spaceflights planned for at least a year. At the time, Shepard and Cooper were the only astronauts left in the Mercury program—with the end of it in sight, the others had been assigned elsewhere, to Apollo and another program in the works—so Shepard was the obvious choice after Cooper's flight. He named his spacecraft *Freedom 7 II* and painted the name on its side.

At an astronaut event at the White House, Shepard told President Kennedy how much he and the others wanted to do the extended flight. Kennedy deferred to Webb, who wouldn't forget Shepard's temerity in going around him to the president. On June 12, 1963, four weeks after Cooper's flight, Webb announced that there would be no further missions. The Mercury capsule was a sophisticated machine, perhaps the most complicated ever built, but in space, its limitations and weaknesses had been revealed. To push farther into space would require a bigger, better, even more complex spacecraft.

The program had achieved its stated goal of putting a man into orbit and returning him safely to Earth—and it had accomplished much more. The gravest medical fears had been allayed. A human could function in space and endure heavy g-forces. The effects of long-term weightlessness were still not well known—it appeared that there was some loss of muscle mass and bone density and possibly more issues—but it was clear that in limited exposure, an astronaut could work without suffering incapacitating injury.

The astronauts had also proven the importance of the man to the machine—at least, this particular one. The man-in-a-can talk eventually died down as each successive mission demonstrated the value of a trained, experienced engineer-pilot. The Mercury Seven had captured the world's imagination with their competence in high-pressure situations. Even Chris Kraft, who tolerated no independent pilot initiative, said, "Man is the deciding element."

And they had done it the way Bob Gilruth wanted to do it—carefully, step by step, with *safety* and *reliability* the program's watchwords. Of all the program's accomplishments thus far, Gilruth was most proud of "six men up and six men back." The seven astronauts looked up to him, and he thought of them as his "boys." "He died a thousand deaths every time one of those things was launched," Kraft would remember.

It was time to move on, to another program already in progress that would entail longer flights—much longer—and greater dangers.

A month after Cooper's flight, the Soviets put on another space spectacle. On June 14, 1963, Vostok 5 blasted Lieutenant Valery

Bykovsky into orbit. Two days later, Vostok 6 launched another cosmonaut, or "cosmonette," as *Time* magazine put it—the first woman in space, and the first civilian. A former textile-factory worker and an amateur skydiver, Valentina Tereshkova had little flying experience; she was clearly picked for her "ordinary Russian" status, since one of her backup cosmonauts was a licensed pilot with a degree from the well-respected Moscow Aeronautical Institute, a kind of NACA university.

Tereshkova was twenty-six and almost five foot seven, taller than many of the male cosmonauts. She made forty-eight orbits during her three days in space, and Bykovsky made eighty-two in five days. The two capsules held parallel orbits, at one point coming within three miles of each other, and returned to Earth just three hours apart—yet another display of Soviet superiority in space. Tereshkova would never fly again, nor would any of the women who had trained with her; she had been space-sick and exhausted for much of the time and barely functioned, though this information was hidden from the Russian public and the West. That November, Tereshkova would marry cosmonaut Andrian Nikolayev and later give birth to a healthy baby, allaying fears that exposure to cosmic radiation might cause genetic damage to human reproductive cells.

And that was that. There would be no more Vostok flights. The Soviets, under Sergei Korolev's shrewd and resourceful direction and propelled by Premier Nikita Khrushchev's insatiable demand for more space spectaculars, also had something more ambitious in mind.

Over the next few years, details about the Apollo program and its massive Saturn V rocket emerged, and Soviet leaders who had been reluctant to underwrite Korolev's pet project, the N1—a booster even larger and more powerful than the Saturn—were finally persuaded of its necessity. Korolev sold it to them as an all-purpose launch vehicle, with thirty engines clustered on the first stage alone.

But its prime use, in his eyes, was as a booster to carry a cosmonaut or two to the moon—and maybe even to Mars and Venus. The Chief Designer spent more than a year lobbying for it and had a personal meeting at the Kremlin with Khrushchev in July 1964. Finally, after all concerned realized that the Americans were not just talking about a moon landing but proceeding to make it happen, Korolev's proposal for funding was approved by the party's Central Committee.

For the first time, the Soviet program to land a man on the moon was made a top priority. The goal was to do it by late 1967 or 1968—an ambitious timetable, especially given the fact that the Americans had a three-year head start and seemingly limitless funding. But the resourceful Chief Designer had performed miracles before. Maybe he could again.

CHAPTER SEVEN

THE GUSMOBILE

It was like the Blue Angels at 18,000 miles per hour.

WALLY SCHIRRA

THE ASTRONAUTS CALLED THE Gemini spacecraft the Gusmobile, and for good reason.

During Grissom's Mercury mission, the hatch had blown open without warning, allowing the capsule to fill with water and sink to the bottom of the sea, and Grissom had taken the loss of the capsule hard. He and the rest of the Mercury Seven, as well as his NASA bosses, knew it wasn't his fault, and an in-house investigation had concluded that he hadn't initiated the firing of the hatch. But he knew there were people out there who thought he'd panicked and popped it on purpose. More than anything, Grissom wanted—needed—another flight so he could redeem himself.

But his chances weren't looking good. It appeared that the Mercury program would be over before Grissom would make it into space again. With five others who had yet to fly, he was at the back of the line; not only that, but Al Shepard was already angling for another mission, and Gus wasn't going to beat *him* out. After Grissom's nominal flight—except for that last development—and the confirmation that a human could indeed survive weightlessness in space, NASA had canceled the two other suborbital flights originally planned. What would be the point? John Glenn, up next, would fly the first orbital flight. There would be more of those, but not six more.

Still, there might be another opportunity knocking. A new program was in the works.

In December 1961, Bob Gilruth had announced a new manned spaceflight program to bridge the gap between Mercury and Apollo—Mercury Mark II, a "two-man Mercury." A month later, it was renamed Gemini, after the constellation of the twins, Castor and Pollux, appropriate for a two-man capsule. The program was necessary because when Apollo was first announced, no one at NASA knew exactly how the spacecraft would land on the moon or even what kind of spaceship it would be. At the time Kennedy made the man-on-the-moon speech, two main methods, or modes, were under consideration. One was termed Earth-orbit rendezvous (EOR); the other, direct ascent.

If you asked someone in the street how the journey would happen, he'd probably describe a large streamlined rocket with tail fins—not unlike von Braun's V-2—that would blast off from Earth, fly the 239,000 miles to the moon, and, once it got close enough, turn around and use the rocket's engine to brake and settle on the surface. The spaceship would return to the Earth in the same manner. That's how it was done in movies like 1950's *Destination Moon* and *Rocketship X-M* (which used actual stock footage of V-2 rocket launches) and in countless comic books and science fiction stories. But launching a single, self-contained spacecraft carrying enough fuel for liftoff from the moon and the return trip as well as life-support systems, a heat shield, and the many other necessities for such a journey would require a huge booster.

This direct-ascent method was favored early on by the Space Task Group, and von Braun's rocket team was already designing the massive booster. The four-hundred-foot-tall Nova would cluster eight large engines in its first stage, and with two other stages, it would have a combined twelve million pounds of thrust to boost eighty tons into orbit. There would be other logistical problems—for instance, no one knew a surface that could bear that kind of weight or a pad that could survive the launch of such a behemoth. Nor could this beast be ready by the end of the decade.

The other mode, EOR, was just as complex. A large, spinning space station revolving around Earth would be the world's first space construction site, and there, a smaller spacecraft would be assembled. Its components would be launched separately from Earth...or maybe the complete spacecraft would be launched by another of the large boosters von Braun's team was developing, the Saturn V, and its propellants sent separately. No one was sure how it would be done, exactly, especially since any space construction would require—besides extremely large hard hats—techniques of rendezvous and docking that had never been attempted as well as the hazardous transfer of hypergolic liquid fuel, the kind that combusted when combined, from tanks to the spacecraft. And how would all these parts be connected? Screwed, bolted, nailed, clamped, welded, glued, soldered, or some other method? Max Faget was not a fan of EOR: "Every time you'd tell them what was wrong with one way of doing it, they'd tell you, well, they were going to do it the other way," he remembered later. "As far as I know, those problems never got solved." He and most of the Space Task Group were behind the direct-ascent mode. Von Braun supported EOR since it would involve his Saturn booster and was similar to plans he had laid out in his *Collier's* articles.

Neither of these choices addressed the other end of the voyage—descending to, and then ascending from, the moon's surface, which was probably solid...but might not be. After all, a respected Cornell scientist named Thomas Gold maintained that the moon's surface was just a deep layer of dust, into which a spacecraft would disappear. And even if it was solid, how would the crew guide it down to the ground safely, even in one-sixth gravity? For the Nova, the ladder from the rocket would need to be at least as long as a football field; would an astronaut be able to safely clamber up and down that wearing a spacesuit, since even a small puncture in that suit could mean death? The more closely one examined the direct-ascent mode, the more problems one found.

At least these two methods allowed the spaceship's occupants to return from their journey. One plan, conceived at Lockheed and advocated by Bell Aerospace as late as June 1962, didn't. Two engineers there—one with the title of head of human factors—claimed that America could beat the Russians to the moon only if an astronaut was sent there on a one-way trip. After all, it was already technically possible to get a man to the moon; getting him back safely to Earth was the issue. After landing in a modified Mercury spacecraft, the astronaut would stay there, with oxygen, food, and supplies, and live in a pressurized hut drop-shipped earlier. More supplies would continue to be delivered. It would probably take at least a year and twenty-two cargo rockets to get a one-man moon base up and running—and several more years before NASA figured out how to bring him back.

There is no evidence that NASA ever actually considered this suggestion.

But there was another option.

John Houbolt wasn't the first one to come up with the idea of lunar-orbit rendezvous (LOR). A self-educated Russian mechanic named Yuri Kondratyuk had suggested it in 1917 and so had an Englishman, Harry E. Ross, in 1948. Houbolt wasn't even the first engineer at Langley, NASA's revered research center, to suggest LOR; two others, Clint Brown and Bill Michael, had published a paper on the subject in May 1960 suggesting that a small "bug" with two astronauts leave a "mother ship" orbiting the moon, take them down to the lunar surface, and then return them. A team from Chance Vought Astronautics, a respected name in the aircraft business, also presented the idea of modular spacecraft at Langley around the same time. But this was before President Kennedy's May 1961 directive, when a lunar landing became a priority, and no one at NASA thought much of LOR—in fact, researchers laughed at the idea.

As a boy in Joliet, Illinois, Houbolt had entered his creations in model-airplane competitions like so many other future NACA engineers. Once he'd even jumped out of a hayloft with an umbrella. He had a pilot's license, and he had never wanted to be anything but an aeronautical engineer. Houbolt's job as associate chief of the dynamic-loads division had nothing to do with the manned space program. But he also chaired a committee assigned to study rendezvous as it pertained to space stations. At one meeting in August 1960, the subject of a moon landing came up. He began researching the various approaches that used rendezvous techniques and became fascinated by the concept.

He wasn't the only one. Bob Gilruth and others at NASA realized early on that orbital operation techniques such as rendezvous and docking would be needed at some point and thus should be developed. Toward that end, the Mercury follow-up program—initially called Mercury Mark II, since it would involve an improved Mercury capsule—began to take shape. Canadian James Chamberlin, the former chief designer of the canceled Arrow fighter jet, was assigned to work on it. The program was sold to Congress on the basis of its capability to intercept, inspect, and repair satellites and to support a space station, still a strong possibility at the time.

By the time Chamberlin's team got finished with its design plans, the Gemini would be a complete makeover of the Mercury in countless ways. First and foremost, it would be an operational spacecraft with enough power, through larger and more rocket thrusters, for its pilot to fly it in the vacuum and microgravity of space; he could not only alter its attitude through yaw, pitch, and roll but also change direction and speed in all three axes (up/down, forward/backward, and left/right). It would also be a much more easily serviced craft. Almost all its service components, previously crammed into the small Mercury cabin, would now be attached to the craft's outside or in the detachable adapter-module shroud that looked like nothing so much as a hoop skirt on Mercury's big sister.

Other researchers at Langley besides Houbolt had been studying rendezvous since 1959; by May 1960, there were eleven separate studies under way on the subject. It was becoming abundantly clear that rendezvous maneuvers would play an important part in any kind of space operations, whether it was LOR, EOR, or direct ascent. Within that cadre of researchers, the prematurely gray, forty-one-year-old Houbolt became known as "the rendezvous man." He was brilliant—he had taught the Mercury Seven course on navigation—and at some point in the summer of 1960, after much study of rendezvous as it applied to a lunar landing, he had an epiphany. He realized how much LOR would simplify all the parts of the process up and down the line, from development and testing to manufacturing and flight planning and operations. "I vowed to dedicate myself to the task" of pushing the concept, he said later.

Houbolt was by nature reserved and reticent, but from that moment, he took to advocating for LOR like a Baptist preacher spreading the Gospel. He began converting people, first a few NASA folks and then others, convincing them of its advantages—and the benefits of rendezvous in general—in countless briefings, lectures, presentations, and one-on-one talks. But the initial reaction from the Space Task Group was less than enthusiastic.

Houbolt gave a presentation in December 1960 at NASA's Washington, DC, headquarters, and most of the Space Task Group's top managers were there. When he pointed out the weight savings of LOR—a reduction by a factor of 2 to 2.5—the normally soft-spoken Max Faget stood up. "His figures lie," he said heatedly. "He doesn't know what he's talking about." There was some truth to Faget's claim—in his early calculations, Houbolt had underestimated the weight required, using an inadequate guidance system and a tiny, unpressurized, and unrealistically light lunar lander, not much more than an open platform and a rocket engine under the seat that "looked very much like a motorcycle," observed one NASA engineer. Nonetheless, in the gentlemanly world of science, it was a shocking display of vehemence. Von Braun shook his head and said, "No, that's no good." After the meeting concluded, Faget stood out in the hallway and told anyone who would listen about the flaws in Houbolt's claims. Houbolt merely suggested that Faget and others who were skeptical should first take a look at his study.

Arguments as to the most viable lunar-landing mode continued through 1961. For much of that time, LOR was the long shot; most favored was EOR. Many at NASA, Bob Gilruth included, agreed with Faget. Houbolt couldn't even get the Space Task Group to study his scheme. The year wore on, but Houbolt refused to give up. He continued to give rendezvous talks at NASA headquarters, usually to unreceptive audiences. Some of this was due to politics; von Braun's team in Huntsville preferred either direct ascent or EOR. The former would require their monster Nova booster, and the latter two or three big Saturns per launch—double the work for them, since LOR would utilize only one Saturn per mission. (And because no one knew what rockets would be needed or wanted after Apollo, that meant double the job security.)

One variation on EOR would employ von Braun's long-cherished space station idea. Even those at Houbolt's home, Langley, thought LOR too complex and risky, and *Time* magazine opined that "at first glance [it] seems like a bizarre product of far-out science fiction." If a crisis occurred in low Earth orbit during the EOR process, the spacecraft could most likely return home quickly—but what if there was a life-threatening problem near the moon, 239,000 miles away, during LOR? There would be no chance of rescue. The thought of astronaut corpses circling the moon indefinitely kept many people at NASA awake at night. Most at Langley preferred the direct-ascent method.

As various NASA study committees passed on LOR, a frustrated Houbolt decided to make a leap of faith. Risking his job, he went above the heads of his direct superiors and wrote a three-page letter pleading his case to NASA deputy administrator Robert Seamans. On May 25, soon after hearing Kennedy's moon speech, Seamans appointed another committee to assess the various lunar-landing modes. Houbolt was heartened; surely this group would give LOR proper consideration.

After examining several concepts, the committee's final rating placed LOR a distant third. The committee members thought it too risky, and even absurd, to send a lone astronaut (the plan at the time) down to the lunar surface in a small module and hope that he could successfully launch and rendezvous with a larger spacecraft orbiting the moon.

One more task force was immediately formed to focus on EOR. Its chair refused to even let Houbolt discuss LOR. At another meeting that included three hundred potential Apollo contractors and several Space Task Group members, Houbolt tried again. Faget and several others told him to forget LOR.

The lunar-orbit-rendezvous method finally gained some traction a month later, during another committee meeting, when direct ascent's Nova began losing steam as a viable option. Houbolt gave an impressive presentation to the group, then a well-received one to the Space Task Group, and its members started to come around. One of the first was Chamberlin, who spoke positively about LOR to Gilruth. By then he had been tapped by Faget to design the Mercury follow-up craft and was heavily involved in Gemini concepts, and he realized that the necessary rendezvous and docking experience were attainable in the new program.

It still wasn't enough. In November 1961, Houbolt sent another letter to Seamans—this one a nine-page epic in which he described himself as "a voice in the wilderness." Seamans's first reaction was less than sympathetic. "I'm sick of getting mail from this guy," he remembered thinking. "I thought of picking up the phone and calling Tommy Thompson, Houbolt's superior at Langley, and telling Tommy to turn him off. Then I thought, 'But he might be right.'" Instead of upbraiding Houbolt, he took the letter to Brainerd Holmes, who was directing the manned-spaceflight program at the time. Holmes read it and grimaced, but he said he'd give LOR renewed consideration.

He did. The Space Task Group finally began taking the "bug approach" seriously. Eventually, after several more reports and presentations and on closer inspection of the insurmountable difficulties of both direct ascent and EOR, everyone saw that the data provided undeniable evidence: LOR was the most sensible way to land on the moon. Even Faget became convinced, especially since direct ascent's problems of "eyeballing that thing down to the moon didn't have a satisfactory answer," he said later. By early 1962, the Space Task Group threw its weight behind Houbolt's vision.

The last holdout against LOR was the Marshall Space Flight Center. But on June 7, 1962, at the end of a daylong meeting at Huntsville that included a six-hour session on Marshall's recommendations for EOR, von Braun shocked everyone, including his Marshall associates, by announcing his support for LOR. After an earlier presentation by Houbolt, von Braun had asked him to send several papers on the mode; those had helped make up his mind. At the meeting, von Braun presented a detailed listing of the deficiencies of the other modes, and the advantages of LOR. He concluded, "We believe this program offers the highest confidence factor of successful accomplishment within this decade." Afterward, von Braun graciously sent Houbolt a personal copy of his remarks.

On July 11, Jim Webb—who had originally been a supporter of direct ascent—made the announcement: NASA had selected the lunar-orbit-rendezvous method for the job of landing men on the moon. For Houbolt, it was vindication, finally, for what would come to be seen as eminent common sense. As his division chief said to him, "Congratulations, John. They've adopted your scheme. I can safely say I'm shaking hands with the man who single-handedly saved the government twenty billion."

Webb and NASA would still have to defend their selection of LOR to the president. When Webb and Jerome Wiesner, Kennedy's chief science adviser, openly disagreed about it, the agency was forced to justify its choice in the public arena. In September 1962, during a presidential tour of Huntsville, Kennedy brought up the issue with Wiesner. Von Braun and Webb joined in, and the discussion quickly became a heated argument between Wiesner and Webb. Kennedy eventually sided with Webb, whom he trusted on the subject. The matter was finally settled in November, when NASA announced that Grumman had been selected to manufacture the lander portion of the Apollo spacecraft—the lunar excursion module, or LEM. (The name was soon shortened to lunar module, or LM, when *excursion* was deemed too frivolous-sounding, but it was still pronounced "lem.")

The LM would be one of the three modules, or self-contained units, constituting the spacecraft. The three modules would sit atop the massive three-stage Saturn V rocket that would launch Apollo into space. The conical command module would house the astronauts in a hospitable environment during their journey to and from the moon. It would be connected to the cylindrical service module, which would provide electricity, propulsion, and storage for various consumables. The two segments would operate as a single unit—the command-service module—for the entire trip until the service module was jettisoned just before the final reentry into Earth's atmosphere. The final piece of the puzzle was the spindly four-legged LM, designed to operate only in the airless vacuum of space and specifically in the moon's weaker gravity. During liftoff, it was housed directly beneath the service module with its legs folded up. Before the translunar voyage commenced, the command-service module would turn around and dock with it, nose-first. The LM consisted of a descent stage and an ascent stage. When a successful landing was made, the upper ascent stage with its own rocket engine would blast away from the lunar surface and reunite with the command-service module. After its two occupants had scrambled through the docking tunnel, the ascent stage would be discarded.

The Space Task Group was quickly outgrowing its facilities at Langley, Virginia, and a much larger center was needed. After a site-selection committee examined nearly two dozen areas that met all the requirements—moderate weather and proximity to water, a major airport, and a top university, among others—a decision was made. The Space Task Group's new home, and the place from which each mission would be controlled immediately after its launching, would be Houston, Texas. The Humble Oil Company had donated a thousand acres of land near Clear Lake, twenty-five miles southeast of downtown Houston, to Rice University, which in turn had offered it to NASA. (Humble owned a significant portion of the area surrounding the tract—the nicest part of Clear

Lake—and knew they would eventually make millions developing it, hence the roundabout donation.)

Some people—particularly several Florida politicians—thought the eighty-eight thousand acres of flatland on Merritt Island, immediately west of Cape Canaveral across the narrow Banana River, that NASA had recently purchased was tailor-made for the home of the Manned Spacecraft Center (MSC), as the Space Task Group would henceforth be known. But Merritt would be used only for NASA's new launch facilities, which included the world's largest hangar, the Vehicle Assembly Building (VAB), where multiple huge moon rockets could be produced simultaneously, and several capacious launchpads. Vice President Lyndon Johnson argued that the activities of the air force launching ground on Cape Canaveral might interfere with communications during long missions, and other arguments were made against the Cape. Site requirements would shift over time, and in the end, the steamy city of Houston won out.

It was lost on no one that Texas was LBJ's home state and that Houston was part of the district of Congressman Albert Thomas, chairman of the House Appropriations Committee, which controlled funding for NASA projects. Thomas hadn't been happy when another state had received a NASA field center when the agency opened in 1958. Now Thomas, Webb, and Kennedy engaged in a bit of quid pro quo that would not be made public until decades later—Thomas would help out on a few bills the president wanted passed if Houston got the MSC. (Webb would later insist that politics did not enter into the decision, despite the agreement between Kennedy and Thomas.) The plot got more byzantine: Thomas's roommate at Rice had been George R. Brown, and his Brown and Root construction firms would become one of the largest in the world. Brown was also a good friend, and a major financial backer, of Lyndon Johnson. His company would receive most of the contracts to develop the many large buildings needed at the MSC.

Some NASA employees comfortably domiciled near the Langley or DC facilities were loath to move to a city with stifling summer heat and a state looked down upon by most East Coast residents. "Texas was someplace out west that they saw in a movie with Gene Autry," recalled one engineer. Early visits reinforced their worries; the NASA site was nothing but cow pastures and mosquitoes, and the devastation left by Hurricane Carla was visible everywhere. Gilruth had some of the same concerns, but Webb said to him, "What has Senator Byrd ever done for you?"—the point being that Virginia Democrat Harry Byrd had provided little if any support for NASA. More than one person observed that the small body of water was neither clear nor a lake. Nonetheless, all but a few of Gilruth's people made the move in the spring and summer of 1962, and after the enthusiastic welcome and strong support, both personal and political, they received from the city and its residents, most were glad they did.

While new facilities—a dozen or so large concrete buildings, with more to come—were being built, MSC's various divisions were temporarily housed in offices spread out all over the city and up and down the Gulf Freeway, the main highway between downtown Houston and Clear Lake.

Most of the Mercury Seven and their families moved to Timber Cove, five minutes east of the MSC on the north side of Clear Lake, where they built modest houses—typically a three-bedroom, brick, midcentury-modern home with a spacious kitchen, if Mrs. Astronaut had a hand in designing it—and acclimated to the very different Texas culture and the more humid Houston weather.

The Glenns and the Carpenters split a lot on a tree-lined cul-de-sac, and the Grissoms and Schirras lived next door to each other just a block away. The Coopers settled across tiny Taylor Lake, a

Clear Lake estuary, in another recent development called El Lago. The Slaytons built a home five miles southwest, in the town of Friendswood. Al Shepard and his family decided on a luxury apartment near downtown Houston, twenty-five miles away from Clear Lake, which gave Al a chance to race up and down the Gulf Freeway in his Corvette. And a couple of the men's spouses had an idea. Almost every military post had its officers' wives club, so why not an astronauts' wives club? Their casual get-togethers would bloom into a full-fledged mutual-support system for both club members and their children, especially during the stressful flights. After the novelty of being astronauts' wives wore off and the women realized what they'd signed on for—the thankless job of raising children who hardly ever saw their fathers and the constant threat of their spouses' deaths always hanging over them—they would need the support, particularly when a husband died. That would happen all too frequently.

When Project Gemini was announced in December 1961, von Braun was strongly opposed to it. Nobody had consulted him about it, and NASA would use the air force's Titan II as a booster and not his Saturn (which was too large for a souped-up Mercury craft). Grissom became the chief astronaut assigned to the development of the new program. While the others trained for their upcoming Mercury flights—Shepard believed he could wangle one more out of his superiors and continued to remain heavily involved in Mercury—Grissom began spending much of his time at the McDonnell plant in St. Louis, working with Jim Chamberlin on the new spacecraft. Since Gemini would be more or less an expansion of Mercury, no other firm had been considered for the project. He sat in the mock-up for hours at a time, delivering to its designers his opinions on virtually every aspect of the spacecraft—"From the way the cockpit was laid out to what instruments went where," remembered astronaut John Young. Grissom was determined to ensure that the Gemini was a pilot's spacecraft.

On the morning of October 3, 1962, an Atlas booster launched Schirra's *Sigma* 7 capsule into space, and a group of clean-cut fellows stood close together on the Florida coastline a few miles away and watched intently. They were eight members of the New Nine, as they were known, NASA's new astronaut trainees: Neil Armstrong, Frank Borman, Pete Conrad, Jim Lovell, Jim McDivitt, Tom Stafford, Ed White, and John Young. (Missing was Elliot See, who was clearing up personal business and hadn't reported for duty yet.)

They had been presented at a press conference two weeks earlier, and their jobs were to help man the many Gemini and Apollo missions planned over the next several years; there were far too many for the original Mercury Seven, and each flight would also require a backup crew. In his new position as coordinator of astronaut activities, Deke Slayton had overseen the selection of the nine, culling them from 253 applications. They had endured the same battery of physical and psychological tests that the original seven had, from treadmills and steel eels to ink blots and cold water in the ears, though since NASA now knew that a human being could survive at least a day in space, a few of the wackier trials had been dropped. (These men were also test pilots, so someone had sensibly decided that they didn't need more time on the centrifuge.) Two of them were civilians, though that term was misleading—both See and Armstrong were former navy aviators.

The day before they were introduced at the press conference, the New Nine had gathered together for the first time at Ellington Air Force Base, near the new Manned Spacecraft Center in Houston, to be briefed on their jobs, their schedules, and their responsibilities. "There'll be plenty of missions for all of you," Bob Gilruth told them. With ten manned Gemini and more than a dozen Apollo missions planned—about sixty seats to fill, plus backup crews—there would be more than enough flights for everyone in the room.

Slayton got up and discussed the many pressures they would face, including business propositions and freebies. "With regard to gratuities," he said in typically blunt Deke-speak, "if there is any question, just follow the old test pilot's creed: Anything you can eat, drink, or screw within twenty-four hours is perfectly acceptable, but beyond that, take a pass!"

The men smiled nervously. Walt Williams held up his hand and said, "Within reason, within reason."

Unlike the Mercury Seven, this new group of astronaut trainees was, on average, slightly more educated—three of them had master's degrees—and since NASA had raised the maximum height to six feet and lowered the age limit to thirty-five, they were slightly younger and a tad taller and heavier. And like their predecessors, they were all married with children. Most of the Mercury Seven were too busy to offer much of a welcome, and the fact that nine new guys meant that each man's piece of the $500,000 *Life* pie was now only $16,000 didn't increase their hospitality. Nevertheless, soon after the nine's arrival, John Glenn invited all of them to dinner at his house in Timber Cove.

The new astronaut trainees would find that people believed their predecessors' after-hours reputation applied to them too. But the complexity of the post-Mercury spacecraft would mean more time devoted to simulation, training, and interacting with contractors, leaving them fewer off-hours—though many of them managed to squeeze in some playtime. In that era, celebrity infidelity, provided it was discreetly conducted, was ignored by the press. In the eyes of most of the public, the reputations of these American heroes remained lily-white. And though most of them were good family men, others found it hard to resist the companionship of the many women eager to meet America's space heroes and partake of their high-test testosterone. Like their predecessors, they moved their families to Houston, mostly settling in the Clear Lake area, and began adjusting to the frenetic schedule of an astronaut.

Also like the Mercury Seven before them, they underwent an intensive education program about space operations. Each one received a two-inch-thick flight manual that he became intimately familiar with. They were taught by experts in various fields—computers, guidance and navigation, communications, rocket flight, astronomy, orbital and reentry mechanics, meteorology, environmental control systems, and much more. With the Mercury Seven, the New Nine underwent survival training for contingency landings in jungle, desert, and water. And after their basic training, each received a technical assignment in a different area. Besides steady visits to the Cape, they made frequent field trips to the Marshall Space Flight Center in Huntsville; to St. Louis, where McDonnell was producing their Gemini spacecraft; and occasionally to Boston, to familiarize themselves with MIT's computer-guidance system, and Worcester, Massachusetts, to get fitted for spacesuits. And they began spending hundreds of hours in the Gemini simulators, running through every conceivable abort situation, each performance recorded and assessed, and making many thirty-second-long flights on the C-135 for experience in weightlessness.

With the help of their cut from the *Life* deal, the New Nine also moved into the Clear Lake area, most of them to Timber Cove or El Lago. (One moved into Nassau Bay, a new development just south of the MSC.) The astronaut candidates settled into their new routines, and their families settled into their new lives, new homes, new schools, new friends, and new surroundings.

By now, with the addition of the Gemini and Apollo flights, NASA was preparing more definite schedules. Some of the astronauts might fly more than once, but the increasingly complex missions required extensive and intensive training and preparation and were so tightly scheduled that each crewman would have his hands full for a while. And attrition was inevitable, due to injury, death, or some other reason, such as Scott Carpenter's blackballing. Even more astronauts would be needed. Based on a point system he had developed using academics, pilot performance, and character/motivation as criteria, Slayton got on the job.

On October 18, 1963, just a year after the New Nine's selection, NASA held a press conference to introduce a third group of astronauts: Edwin "Buzz" Aldrin, William Anders, Charles Bassett, Alan Bean, Eugene Cernan, Roger Chaffee, Michael Collins, Walter Cunningham, Donn Eisele, Ted Freeman, Richard Gordon, Russell Schweickart, David Scott, and Clifton Williams. The Final Fourteen, they called themselves. They were slightly younger and slightly more educated than their predecessors. The test-pilot requirement had been dropped, and new candidates were expected to have a background flying military jet fighters, so while they were less flight-experienced, they were even more engineering- and research-oriented than those in the previous class.

But even with intensive training and classroom instruction, frequent visits to the various plants scattered across the country that were manufacturing the myriad spacecraft parts, and countless other duties, they too found time for fun, both with their families and without them.

The space race had intensified, and Steve Bales, the boy from Fremont, Iowa, had not lost his childhood desire to be part of it. But his family didn't have much extra for college; his mother worked in a beauty salon, his father in a hardware store. From the age of twelve, Steve had mowed yards—at one time he had forty that he kept up—and saved his money. When he got older, he worked summers as a hired hand on a farm, cleaning out hog pens, walking fields cutting thistle and sour dock, and driving a tractor to mow hay. In the end, all his savings wouldn't be enough.

But Sputnik had sparked an encouragement of science and technology, and scholarships and low-interest loans were available to those who wanted to major in those subjects. In 1960, Bales applied for and received a scholarship to study aerospace engineering at Iowa State, 107 miles away from home. In February 1962, he watched John Glenn's *Friendship* 7 mission on TV, and when footage of Mercury Mission Control was shown, Bales wondered what it would be like to work there. He began reading all he could about NASA, and in the spring of 1964, during his senior year, he filled out a federal-government job application and sent it to Houston. Though he was a few credits short of his degree, he was hired as an intern that May. He decided to finish school the following semester, and soon he was in Houston giving tours of the new Mission Control Center. That summer he learned as much as possible about every job in the Mission Operations Control Room (MOCR—pronounced "moaker") by talking to every flight controller he could. He went back to college and finished his degree, then returned to Houston in December with a job.

The Gemini program was gearing up, and soon Bales was assigned to the complex area of Mission Control handled by the flight dynamics officer (FIDO), who was responsible for determining the location of the spacecraft and its trajectory. Mostly he observed and learned, helping to write mission rules. Eventually he switched to the guidance officer (GUIDO) console, where he helped monitor the guidance systems of the spacecraft, including its computer. He was not much younger than most of the other flight controllers in the room, and, like them, after a lot of self-instruction, classwork, and on-the-job training, he had gotten thrown onto the hot skillet of Mission Control. The ones who survived their trial by fire might someday become "steely-eyed missile men," the people who made informed life-or-death decisions: go, no-go, and, occasionally, something in between. Bales survived and thrived; at the age of twenty-three, he graduated to full-fledged controller—GUIDO—on Gemini 9.

About that time another new guy came aboard: Jack Garman, a big, friendly twenty-one-year-old whiz kid born in Oak Park, Illinois, and hired right out of the University of Michigan. His major in engineering and minor in computers had helped get him a job working on Apollo's onboard software system, then being developed by MIT. He'd interviewed with several companies, most of which were involved in some way with the space program, and he'd picked NASA even though it offered the lowest starting salary—he wanted to be on the inside of the program, not the outside.

As soon as he finished his last class, he drove to his parents' house in Chicago, then got back in his car and headed south—he didn't know exactly where Houston was, only that it was somewhere below Dallas, but he figured he'd be able to find it. Upon arriving there, he was given a choice: he could work with the big Mission Control Center ground computers or the onboard software, the Apollo Guidance Computer (AGC). He chose the one that would fly, the AGC, and though he felt intimidated at first, he found that almost no one there knew much about computers. After a month's worth of intensive computer classes in Houston, he returned to find that he knew more about the machines than anyone at Mission Control, and since he was now an "expert," they made him a group leader. When Mission Control asked for help with the AGC, he volunteered to man the GUIDO staff support room down the hall, where a group of experts were on hand to advise each of the MOCR positions during a mission. For the next few years, he lived and breathed the AGC, flying up to MIT frequently to confer with its inventors. He had little time for recreation, but he didn't care. He loved what he was doing. Besides, when he did go out, if people found out that he worked for NASA, he was treated like a king. Houstonites loved what the agency had brought to their city.

Sometimes the sense of magic that had lured him into the space program would disappear. But during a flight, Garman could sit at a console and read the cascading numbers on his screen and realize he was "looking at a computer that literally was out in space," and, as he would remember later, "it got to be awesome again."

Bales and Garman were only two of the many brilliant—and very young—flight controllers who thrived in the high-pressure environment of Mission Control. Managers like Chris Kraft rarely had to fire those who weren't working out, since the people who couldn't take the pressure usually left of their own accord. The ones who stayed helped define the role of flight controller, and they became legends to generations of their successors.

In late 1963, after his bid to fly one last Mercury flight was denied, Alan Shepard had been chosen to command the first Gemini mission, a short five-hour shakedown cruise designed to test the new spacecraft's maneuverability. Tom Stafford, a standout among the New Nine, would be his copilot—except he wouldn't be called that. Though Stafford's duties as Shepard's crewmate could accurately be described as those of a copilot, Deke Slayton had decided that no astronaut on a Gemini or an Apollo flight would ever be referred to by that term. Shepard and every other lead pilot on every mission would be the command pilot, or commander. His crewmate would be the mission's pilot.

Shepard was pleased, and unsurprised, at being selected. First to fly Mercury, first to fly Gemini—the first Apollo flight, he hoped, would be next, and perhaps he'd even be in on the first moon landing. He and Stafford began training, spending many hours in the Gemini flight simulator, which could run through a full mission with all its potential abort situations.

But Shepard had been keeping a secret. He'd been suffering from dizzy spells, many severe enough to incapacitate him. The first had occurred a few months after his Mercury selection. The next one hadn't happened until late 1963, years later; he began having them in the morning, soon after he rose. The attacks were so bad that they left him helpless on the floor, his head spinning and his stomach roiling. It was all he could do to drag himself to the bathroom to vomit. He'd tried everything—a private doctor, medication, vitamins—but nothing worked. And then the episodes began occurring more frequently and were joined by a ringing in his left ear. When he had an attack one day while giving a lecture and had to be helped off the stage, he had no choice but to tell Slayton, who convinced him to see the NASA flight surgeons. The diagnosis—Ménière's syndrome—was a serious one; doctors didn't know what caused it, and there was no known cure. Shepard was grounded. He couldn't even fly a NASA jet alone. He and Slayton, also still unable to fly due to his heart arrhythmia, shared that embarrassment.

Shepard, who had just turned forty, considered leaving NASA—it looked like his astronaut career was over. But in June 1964, Slayton, who had moved up to assistant director of flight crew operations, offered Shepard his old job as head of the astronaut office, and Shepard took it.

Shepard didn't handle being grounded nearly as well as Slayton had, and having to act as mother hen to the astronauts wasn't easy for him. He came down on them hard, earning himself the nickname of the Ice Commander. He believed he was just running a tight ship, but even Slayton thought he was excessively critical and told him to ease up. (During the interview part of the astronaut-selection process, Shepard seemed to delight in asking tough questions and intimidating interviewees. "His cold eyes seemed to look right through me," remembered one.) He started dabbling in investments, some of them requiring his time as well as his money. He became part owner and vice president of Baytown National Bank and got involved in wildcatting and a cattle ranch. Often, he'd go to the office for an hour or two and then spend the rest of his day handling his outside business interests; "He was never there," recalled one astronaut. But he was Alan Shepard, the first American in space, and rules seemed to be bent for him. No one else could have criticized one of his charges for not being an astronaut twenty-four hours a day and gotten away with it—but Shepard did.

He and the rest of the original seven didn't let the new guys forget that there was a pecking order. "Don't feel so smart," Grissom told a member of the New Nine one day. "You're just an astronaut trainee," Schirra told another. "You don't count for anything around here." In their view, no one was a true astronaut until he'd flown into space. Those who hadn't were just apprentices.

Grissom had been Shepard's backup command pilot for the first Gemini flight, so when Shepard was grounded, Grissom was named prime crew. And after Gus got through with it, the spacecraft—no longer just a capsule, since its more powerful translational rocket thrusters could change its course and orbit, unlike the primitive Mercury—was a pilot's dream, a smooth-handling sports car compared to the Model T Mercury. Even the myriad controls and displays were laid out in a way that made it clear a pilot had had a direct hand in the design.

The idea to make it a two-man vessel had come from Max Faget, who had been named the MSC's director of engineering and development after the move to Houston. It seemed like common sense to use two men, especially given the longer missions and more complex operations now involved. Faget's shop also favored a landing system that included a parachute that could be steered and had retro-rockets to cushion touchdowns on the ground. But problems in that system's development seriously delayed implementation, and it was finally scrapped.

Faget did not like the proposed landing system that was officially a part of the Gemini program and was planned for use in the last few flights: the Rogallo wing, an inflatable paraglider that would allow the craft to land on a runway using skids and would grant its pilot some maneuvering capability. Grissom and NASA civilian research pilot Neil Armstrong tested an early trainer version of the wing, but there were too many kinks to work out, and it was canceled after twenty-seven million dollars and years of development had gone into it. So the spacecraft would land as Mercury had—with an unguided splashdown into the water, which required a huge fleet of naval vessels to be ready in the primary, secondary, and contingent recovery areas.

The new, improved Mercury Mark II would also be twice as heavy as the original Mercury capsule, with 50 percent more space inside, so it would need a more powerful launcher. A new ICBM booster called the Titan II caught Chamberlin's eye. It was being developed for the air force to launch nuclear warheads, and its total thrust of 430,000 pounds would be the most powerful in the nation's arsenal. The first version of the Titan had an overly complicated engine, and early tests revealed its unreliability—"If the rocket got out of sight where you couldn't see it, it was classified a success," remembered one engineer. Another problem was its "pogo"—a liquid-fueled rocket's tendency to vibrate longitudinally, which could result in structural failure, blur its occupants' vision, and possibly inflict serious injury. These problems delayed the program by more than a year. But after some simplification and once the pogo and the other problems were fixed, Titan proved a reliable and powerful booster, and the final version of it received accolades from the astronauts. "A young fighter pilot's ride," one described its quicker acceleration at liftoff. But the bigger rocket was also louder at launch; one astronaut said it was "deafening…like a large freight train bearing down on you."

Its fuel system used hypergolic propellants that ignited on contact with each other and required neither an ignition system nor the super-cold storage facilities that liquid oxygen demanded. This greatly simplified launching, though the spontaneous combustion also made it more dangerous.

With that in mind, Chamberlin decided to use ejection seats instead of Mercury's escape tower, in case of an explosion during launch. Neither Faget nor Kraft supported that decision—the force of the ejection would subject the astronauts to twenty g's, which might mean serious injury or death, and it was so dangerous that the system was never actually tested with a human—but the two were overruled. Chamberlin's background was in jet aircraft, and jets used ejection seats. For easier access and egress, two large hatches would be installed. But the interior was still so cramped that an astronaut couldn't stretch his legs out straight. There was about as much space as you'd have in the front seat of a Volkswagen Bug.

Despite the cramped quarters, astronauts loved the Gemini. They would no longer be glorified passengers unable to fly the craft—this spacecraft was all "stick," with the pilot controlling virtually every movement. "My Gemini spacecraft was the orbital equivalent of a fighter aircraft…my favorite flying vehicle," said Schirra. The only thing the ground control could do was update the computer. And the astronauts would no longer have to put up with jokes about sweeping the monkey shit off the seat, since no chimp alive could learn to fly it. Each man would have his own set of controls, though they would share the joystick, which was mounted between them. And the Titan, for all its quirks, offered a smoother, quieter post-liftoff ride than the Atlas.

The McDonnell company had learned much from creating the Mercury capsule, and Gemini was an improvement on it in virtually every way. A primitive but helpful onboard computer, a rendezvous radar, and after the first few missions, instead of large batteries, compact fuel cells that produced more electricity, weighed much less, and yielded a useful by-product, water, came standard on the Gemini, like disc brakes on the new 1965 Corvette.

To the astronauts in the NASA bubble—working and training long days and nights at Cape Canaveral and the new Manned Spacecraft Center in Houston, visiting aerospace contractors' plants for weeks at a time, spending barely enough time at home to reacquaint themselves with their families—real-world events only occasionally seeped into their consciousness. Politics, sports, popular culture, entertainment, and even the violent race riots and the civil rights protests and opposition to the Vietnam War sparked by the emerging counterculture of the new Left—to the astronauts, this was mostly background noise. But not the events that occurred on November 22, 1963, in Dallas.

Earlier that year, President Kennedy's fervor for manned spaceflight had flagged; were the results really worth the money? In a September 20 speech to the United Nations, he had even suggested that the United States and the USSR combine their efforts toward a lunar landing, though the Soviet system was still too deeply entrenched in insularity to accept such an invitation. In congressional hearings, the space program and its massive budget had experienced its first solid opposition. Jim Webb had been forced to use all his powers of persuasion and to resort to scare tactics—*The Russians are still ahead in the space race*—to keep the Senate from cutting NASA's 1964 budget by half a billion dollars. On November 12, the president had issued a memorandum calling on Webb to begin developing a program of cooperation with the Soviet Union in the field of outer space—a program that would include lunar landings.

That order would not be carried out. On November 16, President Kennedy toured the new "moonport" facilities at Cape Canaveral. Wernher von Braun had flown down to give him a guided tour of pad 37B, which was being prepared for an unmanned Saturn I launch. In dark sunglasses, Kennedy posed for photos under the mammoth booster while Secret Service agents stood by nervously. He walked around the prototype of the rocket that would launch three men to the moon, hopefully before the end of the decade, per his directive. Gus Grissom and Gordon Cooper took him up in a navy copter to give him a bird's-eye view of the new moonport. Then he flew to Houston, the first stop on a quick jaunt through Texas to shore up political support for the '64 elections, and took a look at the fast-growing MSC. On November 21, at Brooks AFB in San Antonio, he gave a speech in which he told the story of a group of boys hiking across the Irish countryside who came to a high orchard wall and tossed their caps over it, forcing them to find a way over: "This nation has tossed its cap over the wall of space and we have no choice but to follow it. Whatever the difficulties, they will be overcome." His initial indifference to manned spaceflight and his previous waffling on its importance became enthusiastic approval in public.

From San Antonio he flew to Fort Worth and spent the night there. The next morning, the president and his entourage made the short flight to Dallas's Love Field. At a luncheon later that day, he was planning to give a speech defending the space program's cost. "This effort is expensive," it went, "but it pays for its own way, for freedom and for America." In Dallas, in an open limousine, he and the First Lady, along with Governor John Connally and his wife, were driven through the large friendly crowds lining the streets. At 12:30 p.m. in downtown's Dealey Plaza, the president was shot dead.

Americans were devastated, and millions around the world shared their sorrow. Everyone in NASA was grief-stricken; von Braun, in his office three days later doing paperwork while the funeral played on TV, broke down and cried, the only time his secretary ever saw him do so. Many in NASA worried about what would happen to the United States space program now that its biggest public supporter was gone. Lyndon Johnson, the new president, had been a strong advocate of it, but a vice president had the luxury to choose and nurture his pet projects. A president had far less leeway. And consuming an increasing amount of Kennedy's attention over the past year had been the growing conflict in Vietnam, where Communist forces from the north were threatening to take over U.S.-backed South Vietnam. The attempt to stem the "red tide of Communism" before it spread through the rest of Southeast Asia was demanding more American soldiers and dollars every day, and it showed no signs of slowing down.

Many of the astronauts had met President Kennedy, and the Mercury Seven had spent time with him at the White House and at NASA facilities. But his assassination hit John Glenn especially hard. Since his triumphant flight twenty months before, Glenn had become close to the Kennedy clan, much to the amusement—and, likely, jealousy—of some of the other Mercury astronauts. The president and his brother Robert had adopted him into their extended family, and the Glenns had spent time at various Kennedy houses and compounds. Glenn became especially close to attorney general Robert Kennedy; he and his brother had even suggested that Glenn challenge an aging incumbent in the Ohio Democratic primary for a Senate seat. Glenn was interested, but he thought he still had something to give the space program—and there remained that outside chance of a moon flight, the astronaut's holy grail. He had continued to train with the astronauts, and he had asked Slayton and even Bob Gilruth for a Gemini flight assignment.

But they hadn't given him a definitive answer, and Glenn began to suspect he'd never get a mission. Then a NASA official told him that Washington believed that he was too valuable to the program to risk losing him on another flight...and then he asked if Glenn was interested in becoming an administrator.

The death of the president helped Glenn make up his mind. Seven weeks after the assassination, on January 16, 1964, Glenn resigned from the space program. The next day he announced his candidacy for the U.S. Senate seat in Ohio and began campaigning. The initial reaction was promising.

Six weeks later, in a hotel while on the road, Glenn slipped while adjusting a bathroom mirror and smacked his head on the shower door's metal track, knocking himself out. He came to and found himself in a pool of blood. Worse than the concussion he sustained was the damage to his inner ear's vestibular system, which controls balance. His symptoms were much like those of his rival Shepard—extreme dizziness and nausea. After weeks in a hospital bed, he withdrew from the Senate race. His political career appeared over before it ever started.

The Russians, meanwhile, were not flying as often as they had been. Since the June 1963 Vostok 6 flight that had included Valentina Tereshkova, there had been no manned missions...or at least none made public. That hiatus ended with another Russian space first on October 12, 1964: three men were launched in Voskhod 1.

It was only much later that the West learned the men went up in a stripped-down, one-man Vostok capsule modified to carry three cosmonauts, a strategy insisted on by Soviet politicians to upstage the announced two-man Gemini program. There was so little space in the cabin that there were no ejection seats, and the three short men—a pilot, a physician, and one of Korolev's design engineers—could not wear spacesuits. They underwent just a few months of training, and for most of that time they had to diet to reduce the capsule's payload to a sustainable level. When they squeezed into their spacecraft, they wore nothing but lightweight clothing. Fortunately, there was neither a booster accident during the launch nor a cabin depressurization during the brief sixteen-orbit, single-day flight. The spherical capsule dropped to Earth safely with the help of a braking rocket added to the parachute lines, since the cosmonauts had to make a ground landing. Besides being the first multi-person space mission and the first with no one wearing a spacesuit, Voskhod 1 set an altitude record of 209 miles.

During the flight, the crew spoke by radiotelephone to Premier Nikita Khrushchev, who was at his villa on the Black Sea. This would be the last time he spoke publicly as the leader of his country. Though he had overseen a relaxation of the Stalinist terror tactics and introduced some academic and cultural openness, many of his policies were unsuccessful, and other Soviet politicians decided it was time for a change. Later that same day, he was summoned to Moscow, where he was removed from office on October 14. Two offices, actually—Leonid Brezhnev replaced him as first secretary, and Alexei Kosygin took his job as premier. The country's two new leaders greeted the cosmonauts upon their arrival in Moscow eight days later.

Five months after that, on March 18, 1965, cosmonaut Alexei Leonov accomplished the world's first EVA—extravehicular activity, or space walk—when he crawled through an inflatable air lock and floated out of Voskhod 2 at the end of an eighteen-foot umbilical cord. He spent twelve minutes floating and cavorting outside as the spacecraft orbited the Earth—the first human satellite. But when he tried to reenter the craft, he found that his spacesuit had expanded and stiffened so much that he couldn't bend his legs to fit through the air lock. As his oxygen supply dwindled down, he decided on an extreme measure. He partially depressurized his suit by opening a valve and letting air bleed out, then jammed himself inside headfirst; his crewmate, Pavel Belyayev, hauled him through the air lock.

Leonov was exhausted. He had worked so hard that his suit was filled to the knees in sweat. Years later, he would reveal that he had a suicide pill to swallow in case he was unable to reenter the craft and his crewmate was forced to abandon him. (This was an option no American astronaut would ever be given, despite persistent rumors to the contrary.) Leonov would later insist that he hadn't been scared; "There was only a sense of the infinite expanse and depth of the universe," he said.

Belyayev and Leonov's capsule landed a thousand miles east of its intended recovery point, wedged between two large fir trees in the Ural Mountains. The two cosmonauts spent a frigid night huddled together, trapped inside the craft while wolves howled nearby, and were found the next morning by a rescue team on skis. But those details came out later; at the time, the mission was deemed another Soviet space spectacular. The Russians appeared to have the inside track on a lunar orbit or landing, a belief they reinforced in various public comments. "But our immediate goal, the target before us, is the moon," said one spokesman, and under the headline "Sorry, Apollo!," a *Pravda* article bragged that "the gap is not closing, but increasing." Evidence of how close the Soviets were to that goal could also be seen in Leonov's environmental system: a self-sufficient backpack, far more complex—and troublesome in terms of getting in and out of an air lock—than a relatively simple umbilical. Such a backpack would be needed for a walk on the moon but was hardly necessary for an EVA in space.

But there would not be another Soviet manned mission for more than two years. It was America's turn.

CHAPTER EIGHT

THE WALK, AND A SKY GONE BERSERK

We try and plan for the unknowns. It's the unknown unknowns that you have concerns about.

BOB GILRUTH

IF THERE HAD BEEN a space equivalent of *Car and Driver* magazine, its editors would have voted Gemini the Spacecraft of the Year.

By the time the spacecraft was man-rated and ready to fly two astronauts, its total cost had ballooned to $1.35 billion, almost double its original budget of $700 million. But Gus Grissom had done his job well. It was a pilot's dream in every way that counted. And he had gotten his wish, thanks to Al Shepard's inner-ear problem—he would command the first Gemini mission. To prepare, he and his shipmate John Young, one of the New Nine astronauts, had spent many hours in the Gemini simulators, precise duplicates of the capsule's cabin in which a crew could approximate a complete mission, from liftoff to touchdown. Included were realistic visuals outside the windows and tilting and vibration that faithfully simulated the feel of launch and reentry. Nominal missions were flown at first, with various failure and abort situations added later.

At 4:40 a.m. on March 23, 1965, Deke Slayton knocked on the bedroom door of the crew quarters of Hangar S to wake up Gus Grissom and John Young. Slayton had first done this for Gordon Cooper on the final Mercury mission. He would do it for twenty-four flights altogether. He had also begun choosing the crew for each flight, subject to the approval of Bob Gilruth and a few senior NASA officials, though Slayton would be overruled only once.

Over the next twenty months, from March 1965 to November 1966, sixteen Americans would roar into space on ten different missions, roughly one every two months. Gemini's crews would comprise three of the Mercury Seven, every remaining member of the New Nine, and five astronauts from the 1963 selection. Each flight would incrementally increase the knowledge and experience needed to reach the ultimate goal of a man on the moon. Each would face its share of problems, some minor, some major. None would endanger the lives of its crew except one.

Two unmanned Gemini missions preceded Gemini 3, though, because of the piloting required in the new spacecraft, there were no more flights for the astrochimps. They had done their duty, and most of them were eased into retirement at various zoos to live long and happy lives.

On March 23, 1965, Gemini 3 blasted into space from Cape Kennedy's launch complex 19, as would every Gemini mission, to begin a three-orbit shakedown cruise. Grissom had tried naming the craft *Molly Brown*, after the popular Broadway musical *The Unsinkable Molly Brown*, a dig at the criticism of his Mercury flight. (Grissom had a sense of humor; his second name choice was *Titanic.*) NASA management was unamused, but allowed the name to stick. It would be the only Gemini spacecraft to have a name beyond the flight designation.

In the launch control center several hundred yards away—a blockhouse similar to its predecessors but even better fortified—was Wernher von Braun. It was his fifty-third birthday. His Marshall team had nothing to do with the Gemini program, and he and his center were busy enough; though costs and problems were rising on his Saturns and their huge engines, several test

Saturns had been launched into orbit successfully. He was already worrying about Marshall's post-Apollo fate, since there were no plans for more large boosters. But he wanted to see for himself the launch of a spacecraft capable of maneuvering and changing orbit.

Grissom's flight went as smoothly as expected. "Grissom Maneuvers the Gemini" ran the large headline on the front page of the *New York Times* the next day, emphasizing the superior piloting control available on the spacecraft. Grissom raved about its maneuverability and handling; unlike Mercury, Gemini had powerful thrusters that allowed it to change orbit and effectively go where its pilot wanted, short of leaving the Earth's gravity (which would require a larger rocket engine). Every astronaut who flew it agreed with Grissom. Passengers no more, they were finally flying in space.

An EVA had been planned for a later Gemini flight, but after Alexei Leonov's space walk five days before Gemini 3—a mission clearly timed to preempt Grissom's—NASA officials decided to move it up. Ed White, a world-class athlete who had barely missed qualifying for the 1952 Olympic track team, seemed a perfect choice for an activity that might be strenuous. In the weeks leading up to Gemini 4, some newspaper articles suggested the EVA might have been advanced to match the Soviets. Flight director Chris Kraft took offense when asked about it in one interview: "We're not playing Mickey Mouse with this thing," he replied testily. "I don't think it's very fair to suggest we're carrying out a propaganda stunt." But on a wall in the MOCR at the Manned Spacecraft Center was a neatly lettered poster that belied Kraft's denial: WE ARE 301 MAN-ORBITS AND 443 MAN-HOURS BEHIND THE RUSSIANS IN SPACE FLIGHT TIME.

For the first time, Kraft would lead his ground team from Houston. The MOCR would handle all future manned missions, taking control as soon as the rocket cleared the launch tower. The new

Mission Operations Control Room in Building 30 (there were two MOCRs, actually, one on the second floor, used for simulations and practice runs, and one on the third, used for all the Gemini and Apollo Saturn V missions) was larger and more up-to-date and would host not one or two but three shifts of flight controllers for around-the-clock operations. Gene Kranz would oversee one shift; he named his team White, in contrast to Kraft's Red team. John Hodge, the pipe-smoking engineer whose gray hair and British accent gave him a distinguished air, would handle the third shift, Blue. (Each flight director picked his team's color; when a flight director left, that color was retired.) He had been Kraft's assistant for a while but was promoted to flight director when it became clear that the last Mercury mission, Gordon Cooper's daylong *Faith* 7 flight, would require a third shift. And the tracking network had been increased to some twenty-odd stations around the world to improve communication. Data radioed from the spacecraft was received at NASA's Goddard Center near Washington, DC, digested, and then sent to Mission Control in Houston.

On June 3, 1965, ten weeks after Grissom's flight, astronauts Jim McDivitt and Ed White, El Lago neighbors and good friends since their time together studying aeronautical engineering at the University of Michigan, orbited the Earth for four days in Gemini 4. The mission had two main goals. One would end up being a perplexing failure, the other a smashing success.

A spacecraft had never tried to rendezvous with, or even approach closely, an object in a different orbital path. But rendezvous and docking would be required for a lunar landing using LOR, the chosen method. Command pilot McDivitt tried to maneuver his Gemini close to the spent upper stage of its Titan II booster, but as he pointed the nose of his craft toward the target and activated his thruster jets to close the distance, a curious thing happened—the booster moved away and downward. A few minutes later, he tried again, once more with no luck. After a few more attempts, Kraft finally told him to quit trying. Without an onboard rendezvous radar, planned for later Gemini missions, rendezvous would have to wait.

What this revealed was the complexity of orbital mechanics, which on the most basic level worked exactly the opposite of how it did with aircraft. Adding speed while in orbit raises a ship to a higher orbital path, where it will paradoxically slow down, since the craft's orbital speed is a direct function of its distance from the center of gravity of the object it's circling—in this case, Earth. The ship's target will now be traversing a lower, and shorter, orbit, and will consequently move faster around the Earth. To catch up to a target ahead of it or in a higher orbit, the ship needs to reduce its speed and drop into a lower orbit. At the correct moment, a burst of speed will lift the craft close enough to the target's orbit to eliminate all relative motion between them, at which point these paradoxical effects virtually disappear, and station-keeping, or flying in formation in space, is achieved. Only then can docking be attempted. McDivitt's failure was a lesson learned, and much of the subsequent Gemini missions would involve perfecting orbital mechanics and rendezvous maneuvers.

The second goal, however, was achieved. A few hours after the failed rendezvous, the cabin was depressurized, and after some difficulty, White's hatch was opened. A hundred miles above the Earth, he went drifting out into the void of space connected to the craft by a twenty-five-foot, gold-tape-wrapped umbilical cord supplying his oxygen. For twenty minutes, he floated around the capsule, maneuvering with bursts of compressed air from a small zip gun and shooting photos with a Hasselblad on his chest. McDivitt also had a camera, and his vivid images of his cavorting crewmate, the arc of the Earth and its clouds, continents, and oceans behind him, became iconic. Each man had decided to have an American flag patch sewn onto the left shoulder of his spacesuit, and every subsequent astronaut would wear the Stars and Stripes there.

White enjoyed himself so much that he needed a bit of coaxing to return to the craft. When he finally pulled himself into the cabin, he stood on his seat, drinking in the view. "This is the saddest moment of my life," he said. He needed McDivitt's help to close the hatch, and by the time he was securely fastened in his seat, he was exhausted. But his performance impressed his superiors; a few weeks later, he would be named backup commander for Gemini 7, and soon after that, he'd be assigned to the crew of the first Apollo flight. McDivitt was also rewarded; he was given command of another early Apollo mission.

The Gemini flights continued at such a rapid pace that eventually some of the American public lost interest. White's space walk, especially the striking color photographs of it taken by McDivitt that ran in magazines and newspapers worldwide, was a hit, but the novelty of a man in space, one blasted into the heavens atop a massive rocket, was wearing off. Though manned spaceflight was anything but routine—a hundred things could go wrong during launch or reentry, and a thousand in between—it began to appear that way to most Americans, especially since NASA avoided talk of a mission's danger and stressed its safety. No astronaut had died during Mercury, and Gemini appeared to be just as safe. Besides, everyone knew that the Apollo moon landing was the main attraction. To many, Gemini seemed a warm-up act, and like most warm-up acts, it attracted a smaller audience.

Near the end of August, Gemini 5 stayed in orbit eight days—the minimum length of time needed for a lunar landing and return to Earth—and easily exceeded the Russians' five-day flight of Vostok 5. But its mission, with command pilot Gordon Cooper and New Nine astronaut Pete Conrad at the controls, was plagued by problems, from issues with the fuel cells (on their maiden voyage) and the electrical systems to low oxygen levels and jammed thrusters. The complications didn't help Cooper's mood; occasionally, his attitude seemed peevish, which only worsened his reputation with his NASA bosses.

But the eight-day ordeal eased the fears of some doctors about the dangers of prolonged weightlessness and also about the human ability to travel to the moon and back. The next two missions would further test long-duration spaceflight and the perplexing problem of rendezvous.

The Soviets hadn't sent a man into space since Gemini had begun, and the growing accolades the American space program was receiving weren't sitting well with them. After Gemini 5, they accused the United States of conducting clandestine military activities. "The real purpose of the program is obvious," claimed a Russian newspaper, and the article insisted that the astronauts had brought spy cameras to photograph Soviet activities below, despite the fact that the spacecraft's orbital path had not carried it over the USSR once. To make matters more urgent for the Russians, midway through that flight, President Johnson announced the official approval of the Manned Orbiting Laboratory (MOL) program, in which Geminis would shuttle air force crews to a cylindrical laboratory for up to thirty days of reconnaissance experiments and defense research. The Soviets responded, predictably, with plans for their own military space stations. The Cold War had not thawed appreciably.

Publicly, Deke Slayton would claim that every astronaut could fly any seat in any flight, but privately, he didn't really believe that, nor did he put the theory into practice in his crew selection. "All astronauts are created equal, but some are more equal than others," he would write later. Some men in the 1962 group—John Young, Ed White, Tom Stafford, and Frank Borman, for instance—seemed to ooze that indefinable quality of leadership, and they became Slayton's early favorites, though the rest of the class wasn't far behind. It was loaded with well-educated engineers who just happened to be experienced test pilots as well. Slayton had told them that there would be plenty of flights for all of them, and as the Mercury ranks thinned and Gemini flights began to stack up, that seemed to be the case.

Not so for the October 1963 class of additional astronauts, the Final Fourteen. With the test-pilot requirement already dropped, the half a dozen without that experience wondered if it would work against them. None of them could figure out what Slayton based his selections on—if it wasn't test-pilot experience, what was it? Off-duty socializing? Sucking up to him and Shepard? How they fared in the many courses they took, or how they weathered the centrifuge and other machines?

Those picked to command a mission had a say in who their shipmates would be, and they often chose men who shared their service affiliations or other experiences. The navy aviators looked out for their own, as did the air force pilots. Men who had graduated in the same test-pilot class at Edwards or the navy test-pilot school at Patuxent took care of one another. Crew commanders who were West Point grads tried to get other former cadets as crewmates or at least recommend them for crews. Annapolis grads did the same. Some of this was successful, but not all—Deke had his own reasons for picking who he did. For instance, for the upcoming Apollo crews, he decided that in the first few flights, the command module pilot, who at one point would be orbiting the moon alone while his crewmates were in the LM, would not be a rookie astronaut.

Since no one knew Deke's criteria, everyone competed in any way possible to at least snag a backup role on a flight. About halfway through Gemini, a pattern began to emerge, though it wasn't a hard and fast rule: after a mission, a backup crew would skip two flights and be named the prime crew on the third.

The Gemini missions continued. On December 4, 1965, Frank Borman and Jim Lovell blasted into low Earth orbit on Gemini 7, beginning a marathon fourteen-day flight. Ten days later, Gemini 6, with Wally Schirra and Tom Stafford aboard, launched; their mission to dock with a radio-controlled Atlas-Agena target vehicle had been delayed when the Agena exploded six minutes after liftoff. After the craft reached orbit, Schirra skillfully maneuvered to within a foot of Gemini 7. During three revolutions of the Earth, the two vehicles kept within one hundred yards of each other in an impressive feat of station-keeping. Twenty-five hours and fifteen minutes after liftoff, Gemini 6 splashed down, having completed the world's first manned spaceflight rendezvous. Gemini 7 dropped into the Pacific two days later, its two occupants weary, sore, and extremely fragrant—but healthy. Two of the three frogmen who attached the flotation collar to the reentry module after splashdown vomited when the hatch opened and they got a direct blast of fourteen-day-old air and the men who had lived in it.

Gemini 7's two weeks in space reaffirmed that an eight-day lunar voyage could be made safely, and it further tested the Mission Control team. The flight was Kraft's final one as a flight director; after Gemini 7, he would turn full-time to his duties as director of flight operations. To join Kranz and Hodge, his two other Flights, he tapped a backup flight director named Glynn Lunney, the youngest of the original members of the Space Task Group. Lunney had been involved with Mercury from the start, both in mission planning and as a flight guidance officer controlling the trajectory of the spacecraft, before working backup on a couple of early Gemini missions. Kraft also started grooming a couple of others, including Cliff Charlesworth, formerly a civilian physicist with both the navy and the army.

But Kraft was still there for every mission's launch and much of the remainder of its flight, sitting in the back row watching over the operation he had created and the men he had hired to work it, biting his tongue occasionally but letting his new flight directors make the tough decisions. During every Mercury mission, but never at any other time, he had worn a Mercury lapel pin for luck. Now he did the same with a Gemini pin. And after every successful splashdown, he would light up a good-luck cigar. Soon many in Mission Control were doing the same thing, and the light haze from cigarette and pipe smoke became even thicker.

With Gemini 7, America had clearly surpassed the Soviet space program. All that was missing—besides more EVA experience—was successful docking. That would be the number-one objective of the ambitious Gemini 8 mission, a three-day flight that would also feature an extended space walk and several important experiments. Neil Armstrong, a quiet former navy aviator and X-15 pilot, would command the mission. His copilot—rather, his pilot, since Deke Slayton had decreed before Gemini began that no astronaut would ever be called a copilot—was Dave Scott, one of the Final Fourteen, on his first spaceflight. Scott had it all: good looks, confidence, a master's in astronautical engineering. He was a fighter pilot's son, a fighter pilot and test pilot himself, and married to the daughter of a retired air force general—clearly one of NASA's fair-haired boys, evidenced by the fact that he was the first in his astronaut class chosen to fly into space. He and Armstrong had been training for six solid months, and the upcoming mission featured an extended EVA for Scott.

Armstrong had been the backup command pilot on Gemini 5, so, per Deke Slayton's crew-rotation process, he was duly selected to command Gemini 8. His backup crewmate, Elliot See, the other civilian member of the New Nine, had also been a former navy pilot. It was a good match. The two had shared an office, and they became close—at least, as close as Armstrong ever got to anyone. See was serious and soft-spoken, even gentle, but a fine pilot. Though See was in line to fly with Armstrong on Gemini 8, Slayton had decided See wasn't physically capable of a potentially strenuous EVA, so he'd picked the muscular Scott, and See actually got a promotion: to the command pilot seat for Gemini 9, scheduled for June.

See and his Gemini 9 crewmate, Charlie Bassett—See's opposite, an athletic, friendly extrovert—had been training for several months. On the morning of Monday, February 28, 1966, they flew from Houston to St. Louis for a two-week session on McDonnell's rendezvous-docking simulator. According to Tom Stafford, who knew See well, he was "a capable pilot, if a little shaky on instruments." As they approached Lambert Field, he was at the controls of their T-38, the fast but fragile jet trainer astronauts used to fly all over the country. On his second approach, in rain and snow, heavy fog, and limited visibility, See descended quickly from low cloud cover to find he was too far left of the runway. He banked left and dropped altitude to keep the runway in sight, planning on another approach, but a building loomed in front of him; he fired his afterburners, turned right, and tried to pull his plane up. The T-38 was noticeably less responsive at low speeds, and the jet crashed into the roof of the building—the McDonnell plant, where their spacecraft was being assembled—then careered into the parking lot and exploded. Both men died on impact, just five hundred feet from their Gemini spacecraft.

Their deaths shook up the astronaut community—and the flight schedule. The backup crew for Gemini 9 was Tom Stafford and Eugene Cernan. Per Slayton's system, they were slated to be the prime crew on Gemini 12, the program's last mission. Now they were promoted to prime crew for Gemini 9. Their backups on that flight were Jim Lovell and Buzz Aldrin. With no more Gemini missions, it was a dead-end job. Lovell already had a Gemini flight under his belt, but Aldrin, a member of the third class of astronauts and not one of Slayton's favorites, would still be competing with dozens of others for a crew assignment of any kind, and at best, he'd get one of the later Apollo flights. The crash in St. Louis was tragic, but in practical terms, it didn't change much, and many of the men were just glad it hadn't happened during an actual mission; that might have had dire consequences for the program. But it was a lucky break for Aldrin: he and Lovell became Gemini 9's backup crew and, most likely, Gemini 12's prime crew, if all went well and no one pulled a Scott Carpenter.

Two days after their accident, See and Bassett's Gemini capsule was sent to the Cape for final preparations. They hadn't been the first astronauts to die. Ted Freeman, one of the third class of astronauts, had crashed his T-38 when a goose flew into his port-side air intake during a landing on October 31, 1964. At the funeral attended by their fellow astronauts, See and Bassett were buried in Arlington National Cemetery, close to each other and near Freeman's grave. And though Armstrong had been shaken up by See's death, which happened just sixteen days before Gemini 8 launched, he'd seen many friends and colleagues die during his almost two decades of war and flight research. He knew he had a job to do, and he would be ready to fly by launch day.

Four years earlier, in 1962, thirty-one-year-old Neil Armstrong had been in an enviable position for a test pilot. There was no more exciting or exacting cutting-edge aircraft than the rocket-powered X-15, based on a concept study by Walter Dornberger, von Braun's old boss at Peenemünde. Its sleek black body, needle nose, stubby, almost vestigial wings, and thick wedge of a tail fin looked like every boy's dream of a rocket ship. And it was, to some extent. Designed to explore the limits of an aircraft, and a pilot, at hypersonic speeds and extreme altitudes, it could reach speeds of—well, there was no telling how fast it could go, or how high. When the X-15 rose so high that the air became too thin for the terrestrial laws of aerodynamics to apply and its rudders, elevators, and ailerons became ineffective, small attitude-control thruster rockets on its nose and wingtips supplemented its aerodynamic controls. One pilot had already flown it more than four thousand miles an hour; another, forty-seven miles straight up into the atmosphere. That it could reach the edge of outer space—about sixty-two miles up—seemed entirely possible. An orbital X-15 space-plane had even been considered before Mercury was announced.

Armstrong was one of the few men chosen to fly the X-15, and he had reason to believe he would eventually become the program's chief test pilot. He might even fly the air force's X-20 Dyna-Soar, an even more ambitious space-plane, if it ever became operational. The X-20, which was designed to reach Earth orbit and glide down to a landing, was meant for aerial reconnaissance, satellite maintenance, and enemy-satellite destruction, if need be. An aerospace-engineering super-challenge, the X-20 program was just the kind of project Armstrong loved and had spent most of his life working toward.

Born August 5, 1930, in a farmhouse near the small town of Wapakoneta, Ohio, to Stephen Armstrong, and his wife, Viola, Neil Alden Armstrong was of Scots-Irish ancestry on his father's side and German on his mother's. From an early age, he was someone who thought before he spoke, and he avoided argument and confrontation. Others thought him shy, though he made friends easily—he had to, since his father's job, auditing the county books throughout the state, took about a year in each place, which meant a lot of moving around; Neil lived at sixteen different addresses in his first fourteen years. His family finally relocated back to

Wapakoneta, 115 miles north of Cincinnati, for good before his sophomore year in high school.

As a boy, Neil read almost constantly—he was "consumed by learning," remembered his younger brother. From his mother, he inherited a love of music, and he learned to play the piano and baritone horn. But science became his true passion, one he would never outgrow. When Neil was five, he and his father took a ride in a barnstorming pilot's Ford Tri-Motor, a durable early airliner that could carry a dozen or so passengers. That sparked a lifelong love of flight and of airplane models, made of anything he could find and usually powered by rubber bands; he also built a small wind tunnel in his basement. He was still in elementary school when he decided to become an aircraft designer. When Armstrong was a teenager, to learn all he could about aeronautics, he began hitchhiking the three miles to Wapakoneta's small, grassy airfield for flying lessons. Each one was nine dollars, and he worked several jobs after school, sometimes for as much as forty cents an hour, to finance his dream. He earned his student pilot's license on August 5, 1946, his sixteenth birthday, and soloed a few weeks later, all before he learned to drive a car.

In October 1947, when he was seventeen, Armstrong entered Purdue University, a few hours' drive from Wapakoneta, on a four-year navy scholarship. He wasn't particularly interested in a military career, but his parents didn't have the money for college. Aeronautical engineering was his field of study. He made time for model-airplane contests, both entering and attending. In the spring of 1949, at the age of eighteen, he reported for three years of military duty, after which he would finish his degree. He spent eighteen months in intensive flight training and then much of the next two years flying a Grumman F9F Panther, one of the first jet fighters, in a ground-attack squadron. The Korean War began in June 1950, and in mid-1951, his unit was sent to the center of the action. He still found time to fashion model airplanes from wood when he wasn't flying combat missions from carriers off the coast of Korea. He flew seventy-eight, but his seventh, on September 3, 1951, was almost his last.

He and his squadron primarily worked risky air-to-ground jobs that resulted in heavy casualties from thick antiaircraft fire; "bridge breaking, train stopping, tank shooting and that sort of thing," recalled Armstrong. While making a low-level bomb run over a hilly area in North Korea, flying at three hundred and fifty miles an hour, Ensign Armstrong's Panther hit a cable that ripped off six feet of its right wing. He lost aileron control, and his ordnance-heavy craft dropped to within twenty feet of the ground; he would have to bail out soon. Somehow, Armstrong pulled the plane up to fourteen thousand feet and nursed it more than two hundred miles along the Korean peninsula to friendly territory. He ejected just off the coast, but strong winds blew him inland and he came down in a rice paddy; he cracked his tailbone but was otherwise unhurt. A jeep arrived within minutes to whisk him to an American airfield nearby. The driver told him that the explosions audible off the coast were North Korean mines in the bay—right where Armstrong had been aiming to land.

That would not be the last time that his skill, quick thinking, and coolness under pressure combined with good luck to keep him alive.

Armstrong left the navy in 1952 and returned to Purdue the next year. He earned his aeronautical engineering degree in January 1955 and signed on with the NACA as a research pilot a few months later. Over the next seven years, mostly at the Flight Research Center at Edwards AFB in California, he flew the newest—and most dangerous—experimental aircraft at supersonic speeds. In 1956, he married Janet Shearon, a smart and lively Purdue student whom he had courted—in the carefully deliberate way he would approach many things in his life—for a few years. She was attracted to the soft-spoken navy veteran for several reasons: "He was a very steadfast person. He was good-looking. He had a good sense of humor. He was fun to be with. He was older." And although he was quiet, he was not meek; once he'd made up his mind about something, his course was set.

In his time at Edwards, Armstrong made more than nine hundred flights in dozens of the most advanced aircraft in the world, including the X-15, and he contributed much to their analysis and improvement. He was a skillful pilot and an excellent engineer. Those two attributes would make him very attractive to NASA. But Armstrong was happy where he was, on the cutting edge of aeronautical research.

On March 22, 1956, Armstrong was copiloting a B-29 mother ship flying over the Mojave Desert, and just seconds after it had released an experimental rocket plane, he felt a jolt and saw a propeller hub whiz by the cockpit. He looked over to see that the number-four propeller had disintegrated, and parts of it had damaged two other engines. The pilot's controls were gone, but Armstrong's still worked, barely, and with one engine, he managed to guide the big plane down from thirty thousand feet to a landing on a dry lake bed. Many of his flights were dangerous, but some were more dangerous than others.

Armstrong and his wife lived in an isolated cabin forty miles south of Edwards in the foothills of the San Gabriel Mountains. There they began raising a family, a boy and a girl. On their sixth wedding anniversary, January 28, 1962, he and his wife suffered the worst loss any parent can. Their two-year-old daughter, Karen—whom Neil had nicknamed Muffie—died after an eight-month struggle with an inoperable brain tumor that was diagnosed soon after a playground fall. Edwards grounded all of its aircraft on the day of her funeral.

The Armstrongs' four-year-old son, Rick, was a comfort to them, as were friends and family, and Neil tried to lose himself in his work. But the next few months saw some lapses in judgment. Once, during an X-15 flight, he bounced his plane too high in the thin air twenty-odd miles above the Earth. He couldn't turn properly, which resulted in his overshooting Edwards by forty-five miles and just barely making it back.

Early on, Armstrong had been unimpressed with the Mercury program. But soon after John Glenn's February 20, 1962, orbital flight—just a month after Karen's death—he realized that participation in NASA's manned space program would put him "way out at the margins of knowledge." When NASA announced openings for another group of astronauts, he applied. His application arrived a few days past the June 1 deadline. An NACA employee who had worked closely with Armstrong at Edwards and knew of his strong qualifications had recently transferred to Houston's Manned Spacecraft Center to oversee the astronaut training programs. All the applications came to him. He slipped Armstrong's late application into the pile with the rest before the selection panel's first meeting.

Three months later, after Armstrong had had many interviews and undergone two weeks of measuring, poking, and prodding, Deke Slayton called him with the news that he was now an astronaut.

The mission started off well. Despite a series of equipment problems in the two weeks before the launch, at 10:41 a.m. on March 16, 1966, Gemini 8 and its crew of Neil Armstrong and Dave Scott lifted off smoothly. After reaching orbit, command pilot Armstrong initiated the first of nine thruster maneuvers—burns—to catch the target, an Agena upper stage launched ninety-five minutes earlier and now in a higher orbit. Both Armstrong and Scott had spent extensive time in the much-improved, full-size Gemini simulator practicing rendezvous and docking with a full-size Agena. They were aided by Gemini's guidance computer, primitive but effective in determining the locations of the two spacecraft and calculating the best transfer arc. Less than six hours after liftoff, Armstrong braked his ship about a hundred and fifty feet from the silver-and-white, twenty-six-foot-long Agena, shining in the bright sunlight. Rendezvous was accomplished.

After a half an hour of station-keeping and inspecting the Agena for problems, Armstrong slowly approached to within three feet using the Gemini's small thruster jets. He received permission to dock, a job that required exquisite timing and a feather-light touch. A few moments later, like a giant shuttlecock nuzzling a huge thermos, his craft's nose eased into a docking collar in the front of the Agena and latched on. "Flight, we are docked. It's...really a smoothie," said Armstrong. In Mission Control, there were cheers, backslapping, and handshakes; even the reporters in the newsroom cheered. Armstrong and Scott had just achieved the first docking in space.

Flight controllers—and virtually everyone else in NASA—were wary of the Agena and had been even before its explosion five months earlier during the original Gemini 6 mission. They suspected the Agena's rocket thrusters might be faulty and had instructed Jim Lovell, CapCom at a tracking station on Madagascar, to warn the Gemini 8 crew. Just before they passed out of communications range, Lovell told them, "If you run into trouble and the attitude-control system in the Agena goes wild, just...turn it off and take control with the spacecraft."

Armstrong and Scott would soon be incommunicado. They turned up the cockpit lights and pulled out their flight books, then began doing docking chores and checking command links between the two spacecraft. In a little while, they could begin to relax and maybe even get some sleep.

Gemini 8 had moved into night, and since the lights were on in the cockpit, the crew couldn't see much through their two small windows. After a couple of hours of taking care of Agena operations and general housekeeping, they'd try to sleep. Scott, especially, needed a good rest. He was scheduled to do a two-hour-plus EVA the next day. While on a twenty-five-foot umbilical and a seventy-five-foot extension tether, he would float over to the Agena using a handheld maneuvering unit. Armstrong would undock and back away, pulling Scott, then move forward and dock again. Other delicate procedures would follow.

Twenty-seven minutes after uniting with the Agena, Scott looked up at the control panel and noticed that they were in a slow thirty-degree left roll. He told Armstrong, who used thrusters to correct it. After a minute or so, the roll started again. Remembering what Lovell had advised, Armstrong told Scott, who had all the Agena controls on his side, to turn off its attitude-control system. Scott did. The roll stabilized, but a few minutes later it began again, this time at a faster rate—and then even faster. Armstrong ordered Scott to switch the Agena on and then off again in case it was an electrical problem while he fought the motion with his attitude hand controller on the console between them, with little success.

They were spinning in space while connected to a rocket full of fuel, and they could not call anyone for guidance. This was an emergency situation that they had not practiced for and that no one had imagined. Something had to be done and quickly, before the gyrations broke them apart, caused the Agena to rupture or explode, or ripped the Gemini from one or both parts of its adapter section, which carried their power and life-sustaining essentials. Oxygen loss and quick death from asphyxiation would almost certainly follow. To make matters worse, Scott noticed that the fuel in one of their control systems was down to 13 percent.

Neither of them heard the loud cracking sound that would have meant their own thrusters firing. It had to be the Agena. "We'd better get off," Scott said to Armstrong.

"Okay, let me see if we can get the rates of rotation down so we don't re-contact. You ready?"

"Stand by."

Once they undocked, the Agena would be dead to ground control. Scott set the rocket's recording devices so a ground tracking station could pick up its data as it passed overhead and learn why it had malfunctioned.

"Okay, any time," he said. "We're ready."

"Go," said Armstrong, and as Scott hit the undocking switch, he quickly pulled them away from the Agena before the two spacecraft whirligigged into each other.

The Gemini rolled even more rapidly and began to tumble end over end, resembling more than anything a brutal MASTIF training session 160 miles above the Earth. Armstrong and Scott hadn't battled that machine—it had been discontinued after the Mercury Seven had undergone its tortures—but they had logged plenty of time on the human centrifuge, a study in sadism itself. That experience proved invaluable to them now. Brilliant sunlight glinted off the spaceship's black nose, and then darkness, and then sunlight—soon it was spinning at a rate estimated to be close to two full revolutions per second. As Armstrong put it later, "Physiological limits were being approached."

Test pilots had a phrase for flights that went bad: *go to worms.* The mission had swiftly gone to worms.

"Buddy, we've got troubles," Scott said.

"I gotta cage my eyeballs," Armstrong said drily. The two went to work trying to stabilize their craft.

About then, they came in range of another tracking station, *Coastal Sentry Quebec*, a ship in the western Pacific south of Japan with limited ability to communicate with Mission Control. The station crew could tell something was amiss. Their telemetry told them the Gemini had undocked, but they had no idea why. They would have only a few minutes to communicate before the spacecraft sped over them and out of range again.

"Gemini 8, *CSQ* CapCom. How do you read?"

"We've got serious problems here," Scott said. "We're tumbling end over end up here. We're disengaged from the Agena."

The *CSQ* CapCom could hear Scott, though the violent spinning distorted his speech, and scrambling antenna patterns fragmented the transmission. Voices faded in and out. The station could do nothing but acknowledge and ask what the problem was.

"We're rolling up and we can't turn anything off," Armstrong said, "continuously increasing in a left roll."

They were still spinning in roll, pitch, and yaw at more than a revolution per second. Everything that had been loose in the cabin—charts, checklists, flight plan—was bouncing against the walls. Both men were being thrown around, and they were becoming dizzy. They had trouble seeing the overhead dials and switches. Nausea was soon to come, from the contents of their stomachs sloshing around, as was vestibular nystagmus, a sickening, dizzying sensation that caused an uncontrollable movement of the eyeballs and blurred vision. Both were seconds away from passing out, and if they did, the chances of recovery would be remote. They could hear Flight Control cutting in from Houston and asking *CSQ* what was going on, then *CSQ* trying to explain, and then they were out of range again for another fifteen minutes.

They both knew there was only one option: the reentry control system and its two separate rings of thrusters in the nose of the spacecraft.

"All we have left is the reentry system," Armstrong said, his voice strained.

"Do it," said Scott.

There were half a dozen control panels around the interior of the spacecraft. The reentry control switch was in an awkward spot, right above Armstrong's head. After countless hours in the simulator, each man knew the position of every control by feel; as fighter pilots, they'd always gone through blindfolded cockpit checks, and they carried that over into their Gemini training. There were a dozen switches on the plate with the reentry control switch. Somehow Armstrong reached up and found the right one. He flicked it on, then threw the switches to activate the engines that would control the Gemini's reentry into Earth's atmosphere.

But when Armstrong tried the hand controller, he got no response. He asked Scott to give it a try—Scott got no response either. Without a hand controller, they wouldn't make it home. Still whirling and tumbling—the craft's thrusters were turned off, but there was no air to slow the capsule's movements—they started throwing switches again in case one was in the wrong position. Just then, the hand controllers began working. With a delicate pulsing of the thrusters, Armstrong managed to slow down the violent spinning and then, finally, stop it. He turned off the reentry control system to save fuel—they'd need it, and they had used about 75 percent of it just to stop the spinning. He reactivated his maneuvering thrusters one by one until he found the culprit: number eight, a yaw thruster, was stuck in the on position, probably due to an electrical short. They hadn't heard the thruster popping because it had been on the entire time. The Gemini, not the Agena, had been at fault.

A Gemini mission rule dictated that using the reentry system meant that the mission must be aborted; if these thrusters developed a leak, the crew would not be able to get the craft into position for the critical retrofire that would stabilize it and return them to Earth at the proper angle. Attitude control was essential to reenter the atmosphere safely. Flight director John Hodge knew he had to call an end to the mission. But where? And when? As soon as it was possible, of course, but could they find a prime or secondary recovery site?

After their twenty-six-minute ordeal, Armstrong said, "Sorry, partner"—he had planned to let Scott take the Gemini's controls later, and the EVA Scott had trained so long and hard for wouldn't happen. But Scott knew they had no choice.

Over the next twenty minutes, several groups of flight controllers—the calamity had occurred during a shift change—ran through the options. If they didn't bring Gemini 8 down very soon, they wouldn't have another opportunity for a full day—fifteen more revolutions. Too long, and too risky. Reentry in the seventh orbit, less than three hours away, was recommended. Hodge gave the go-ahead. If retrofire was nominal, the recovery point would be about 620 miles southeast of Japan. A navy destroyer, the USS *Leonard F. Mason*, began moving at flank speed toward the position.

As Gemini 8 passed over Africa, Armstrong was concerned that they'd land in a remote area far from civilization—and possibly on hard ground. The spacecraft was designed to handle that, but the impact would be excessive, even with their shock-absorbing contour couches, and since they had no control over the landing, it would be impossible to avoid ground obstacles or a steep hill or even a mountain. Scott worried when he saw the Himalayas getting larger below them as they reentered the atmosphere. But retrofire was nominal, and as the craft plummeted to Earth, the two astronauts were relieved to see the blue of water below them. Twenty minutes after they made a hard splashdown in rough seas, three frogmen dropped from an air force transport plane and secured the spacecraft. Three hours later, the destroyer winched Gemini 8 on deck. The crew was healthy but worn out after their ten-hour-and-forty-one-minute flight.

During the brief mission, camera crews had been camped out as usual in the front yards of the Armstrong and Scott houses near the Manned Spacecraft Center outside Houston, and more personnel were rushed there when news of the ordeal broke. The major TV networks interrupted their regular programs with emergency news bulletins, to the annoyance of some irate *Batman* viewers—more than a thousand called ABC to complain. The next day, the *New York Daily News* ran the headline "A Nightmare in Space!," and *Life* magazine ran stories about the mission in its next two issues, one of them under the title "Wild Spin in a Sky Gone Berserk." Neither Armstrong nor Scott appreciated the melodramatic approach, regardless of its accuracy. Armstrong downplayed the danger, as was his habit; a few years later, he would use the math/physics/engineering term *trivial,* meaning "easy to work out," to describe the crisis: "It was a non-trivial situation," he said.

Both Armstrong and Scott were commended for their calmness and professional performance under extreme conditions. There was whispering among some of the newer astronauts that the two had panicked, but no experienced astronaut thought that; they had followed the book and done what they'd had to do to survive—and done it well. Years later, Kraft would have the last word: "If we had heard about the problem when they were still docked, we would have told them to do exactly what they did, 'Get off that thing!'" Far from blaming the two astronauts, the NASA brass were impressed, especially with the commander. The flight only confirmed what they already knew: Armstrong was one cool customer in a crisis.

The failure rattled NASA and caused some newspapers and at least one congressman to demand a space-rescue system. Max Faget initiated plans for a study—one idea called for an extra seat to somehow be squeezed into the already cramped cabin of a Gemini spacecraft, which could be launched with a single astronaut to rescue two comrades—but due to red tape and a lack of funding, nothing came of it. If a spacecraft became stranded in orbit, there was still no way to save its occupants.

The same day as the Gemini 8 flight, halfway around the globe, two other space travelers made a safe landing in Central Asia. They had spent twenty-two days in space, much longer than anyone ever had, and had endured heavy doses of radiation from traversing the Van Allen radiation belts repeatedly. But they were fairly healthy, though weary, dehydrated, and suffering from bedsores. The two Soviets were retrieved from their capsule and whisked away to Moscow for a triumphant TV appearance. Later they both gave birth to healthy puppies.

The flight of the two female dogs, Veterok and Ugolyok, appeared, at least to observers in the West, to be a practice run for a Soviet shot at the moon. There was no other reason to send mammals into space.

Despite Gemini 8's near disaster and the other successful Gemini missions, the American public was uninterested. They began calling in to the TV networks broadcasting news about the flights to complain about interrupted football games and missed shows—even when the interruptions concerned troubled missions like Gemini 8. To the average American, the space program didn't seem to have much of a point. The early days of Mercury had been unprecedented—blasting a man into space in a capsule atop a rocket was dangerous and exciting. Ed White's space walk had been a refreshing change, and so was the rendezvous between Gemini 6 and Gemini 7. But after that, the flights just didn't seem important, and worse, they became routine. "Americans no longer half-expected the whole thing to blow up," wrote *Life* space correspondent Loudon Wainwright in an article entitled "All Systems Are Ho-Hum." Besides, everyone knew that Apollo was the big one, the program that would put a man on the moon.

But the missions continued, launching with almost metronomic precision every two months. Launches became so "businesslike," remembered Paul Haney, the NASA public affairs officer at the time, that it was "almost like working an airline terminal." The last four involved perfecting rendezvous, docking, and EVA skills. The rendezvous and docking went well, but none of the first three EVAs did. Each spacewalker had a difficult time, especially when he ran into Newton's pesky third law of motion (for every action, there is an equal and opposite reaction) and its peculiar effect in the microgravity environment of low Earth orbit. Every astronaut found that working for an extended time in a twenty-one-layer spacesuit, inflexible when pressurized, was much more difficult in space than it had been in the ground simulation.

None of them had a harder time than Eugene Cernan on Gemini 9. He spent two hours and eight minutes on an EVA, during which he exhausted himself trying to maneuver in a spacesuit that had, in his words, "all the flexibility of a rusty suit of armor." His heart rate zoomed to 180 beats a minute, and his visor fogged up so badly he could hardly see. Down in Mission Control, the controllers thought he might lose consciousness. Cernan barely made it back to the hatch, where his crewmate, Tom Stafford, had to help drag him inside. By the end of the three-day flight, he had lost thirteen and a half pounds.

Two handrails were added to the outside of the Gemini, and the astronauts were given a maneuvering gun, so Michael Collins's Gemini 10 EVA in July 1966 was an improvement on Cernan's. He was able to perform a few assigned tasks, but when he tried to spacewalk over to an inert Agena target vehicle on an extra-long tether, he cartwheeled and spun out of control between the two spacecraft. At one point he started sliding around the Agena and had to reach into it and grab a bunch of wires to stop himself. Later, he got so entangled in his fifty-foot umbilical that he needed help from crewmate John Young to unwind and get into his seat. His ramble was cut short after only thirty-nine minutes. Two months later, Gemini 11 offered little advance in that area. After a perfect docking with the target vehicle just eighty-five minutes after launch, Dick Gordon sallied through his hatch for a planned two-hour EVA. Forty-four minutes later, blinded by sweat and utterly exhausted, he was ordered inside by command pilot Pete Conrad. Gordon had discussed the EVA issues at length with his spacewalking predecessors, and he'd prepared obsessively for it, but no one had managed to make a completely successful EVA.

Conrad and Gordon had used the Agena's added power to boost them into an altitude of 850 miles, the highest orbit yet reached by a manned spacecraft. They also docked and undocked with the Agena four times. Finally, theirs was the first completely automatic, computer-controlled reentry. Spacewalking needed work, but other skills and techniques essential to Apollo did not.

As Gemini neared the end of its scheduled run of missions, the Soviet manned spaceflight program became suspiciously quiet. When CIA intel reports indicated the Soviets were constructing a giant rocket, it was thought that they might attempt a voyage to the moon sometime in 1967 to celebrate the fiftieth anniversary of their revolution.

NASA's confidence in Gemini was at its peak. Plans were made for an ambitious Gemini lunar mission: a capsule would dock with a large, fully fueled Agena—or possibly with one or more of the powerful Centaur upper-stage rockets—which would boost the spacecraft to an escape velocity of twenty-five thousand miles per hour and propel it to the moon. After a lunar flyby, the Gemini would slingshot back to Earth. It was even suggested that a small, lightweight "bug," an open-cockpit lander attached to the Gemini, could carry a single astronaut and drop him down to the lunar surface. Astronaut Pete Conrad supported the idea—naturally, he hoped to be one of the two men selected for the mission.

But Apollo, despite some setbacks, was also proceeding apace, and there was no need for competing moon-landing programs. Besides, NASA wasn't quite prepared to send humans 240,000 miles into space and return them to terra firma, especially if the journey involved an excursion to the little-known luna firma.

An earlier idea for Gemini was again proposed: as a rescue vehicle. If an Apollo craft became stranded in lunar orbit, an enlarged Gemini reentry module, beefed up with rockets for the return to Earth and life-support systems for the extra passengers, would rendezvous with it, and the Apollo crew could EVA to the rescue craft. Another suggestion was a version of Gemini, perhaps manned, perhaps unmanned, that could rescue an Apollo crew at any point during its mission, even on the lunar surface. These ideas were feasible but expensive, and NASA's budget had already reached its height; cost cutting would begin after 1966. Only Apollo would go to the moon, and there would be no rescue available if the astronauts ran into trouble far from home.

Gemini's final flight provided one last opportunity to solve the problems of EVA. If something went wrong on an Apollo moon mission during docking and two astronauts were stuck in the LM, they'd have to make their way over to the command module and climb in through its exterior hatch, so EVA expertise was essential.

Buzz Aldrin—who had indeed scored one of Gemini 12's prime crew seats—was determined to approach the space walk scientifically. A host of handholds, rails, foot restraints, and tethers were added to both the Gemini and the Agena, and Aldrin trained underwater in a spacesuit, using weights to achieve neutral buoyancy and approximate the microgravity conditions of Earth orbit. On November 11, 1966, he and Lovell lifted off in Gemini 12 and headed toward their Agena target vehicle. Then another Gemini first, an onboard radar, failed. Aldrin, who had written his MIT doctoral dissertation on manned orbital rendezvous, used a sextant, slide rule, charts he had largely prepared himself, and their small onboard computer to get them to the Agena. (Some in NASA joked that the radar failure was no accident.) They docked three hours and forty-five minutes later, then separated and docked again several times, approaching from different angles. Rendezvous and docking, clearly, had been mastered.

The next day, Aldrin opened his hatch and floated out on an umbilical for a space walk of two hours and twenty minutes. With the help of waist tethers, the hand- and footholds, and several rest periods, he performed a variety of complicated chores without difficulty. Everyone in NASA breathed a sigh of relief. The problems of EVA had finally been solved. For good measure, the astronauts used their onboard computer to handle both the guidance and firing of the attitude-adjusting rockets, another first.

The ten manned Gemini missions—each far more complicated than any Mercury flight—provided an opportunity for NASA to increase its knowledge and experience in manned spaceflight and to introduce and perfect techniques and equipment necessary to reach the moon. The Gemini program had garnered a total of 1,993 hours in space, valuable experience not only for the men in the spacecraft but also for the personnel on the ground, from the launch operations at Cape Kennedy to flight control and tracking operations. Apollo would not be possible without it.

The Mission Control Center especially had come into its own. Despite Gemini's smashing success, each mission had had its share of problems large and small—fickle fuel cells, cranky electrical systems, temperamental thrusters, patchy rendezvous radars, corrupted computer programs, and unreliable Agenas. Indeed, it seemed one of the few unfailing systems throughout the program was the human one, both in the void of space and on the ground. The pilots had excelled, and Kraft's resourceful teams of flight controllers and their backroom support groups had tackled multiple complications, and solved or found work-arounds for almost every one that mattered. Except for Gemini 8, every mission had continued to its end. More important, every astronaut returned to Earth safely. Spaceflight, it seemed to many Americans, wasn't so dangerous after all.

III.

OUT

CHAPTER NINE

INFERNO

We just became anesthetized by success.

CONGRESSMAN OLIN TEAGUE

GUS GRISSOM WAS PLEASED at being selected to command the first Apollo mission; making the first flight of a brand-new craft was every test pilot's desire, what he lived for. But Grissom wasn't surprised. Of *course* it would be a Mercury astronaut, and there weren't many of them left in the program. John Glenn had retired, and after his canceled Senate run, he was now an executive with Royal Crown Cola. Scott Carpenter had elected not to get involved in Gemini; instead, he'd developed a fascination with underwater explorer Jacques Cousteau, then requested detached duty to become involved with a navy experiment called Sealab. He'd broken an arm in a motorcycle accident—a grounding injury that resulted in his losing some mobility after surgery—and after that, it was likely only a matter of time before he would officially resign from the astronaut corps. Gordon Cooper was still annoying the NASA brass despite his superb handling of two compromised missions, and he wouldn't be on anyone's list for a shakedown flight. Deke Slayton and Al Shepard were both still grounded for medical reasons.

That left two of the original Mercury Seven: Wally Schirra and Gus Grissom. And since Gus had flown before Wally in both

Mercury and Gemini and since Wally preferred a later flight in a program, the choice seemed obvious. But Gus hadn't taken any chances. He was careful not to irritate any of his NASA bosses, and that extended to Chris Kraft. "I just close my eyes and kiss his ass," Grissom told Rene Carpenter one day while relaxing poolside at an astronaut party at Schirra's house. Kraft had gone after him, he told her, when the hatch problem occurred on his Mercury mission.

The Apollo spacecraft Grissom would command was the Block I, a prototype version of the craft that would ultimately journey to the moon and land men there: the more sophisticated Block II. The Block I was missing several features necessary for that mission—for instance, the docking tunnel that astronauts would use to move into the lunar module, an advanced hatch, and much more. The Block I booster rocket would also be different: the shorter, less powerful Saturn IB, which provided only enough thrust to boost an Apollo spacecraft into Earth orbit. Grissom's flight would be a shakedown cruise to check out every one of Apollo's advanced systems, from the booster and the spacecraft to launch operations, Mission Control, and others. Grissom, an engineering perfectionist, was looking forward to an open-ended mission that could last up to fourteen days.

But there were complications.

The launch was originally set for November 1966, the same month as the last Gemini flight. The date slipped to January 1967, then February 21, 1967. Developmental delays had plagued both spacecraft and rocket—according to Joseph Shea, the manager of the Apollo program in Houston, by the time the command-service module was accepted by NASA, in December of 1966, it had had approximately twenty thousand test failures. Test failures were a part of the process, but the endless fixes meant long delays, missed delivery dates, and, in some cases, shipments to Cape Kennedy of mediocre work that would have been unacceptable in another time. The idea was just to get it there and get NASA to accept it; problems would be handled later, during the craft's checkout process. Any modifications—and there were many—meant extensive redesigning, and that meant more difficulty meeting the lofty but essential safety and reliability standards. The list of things that could result in death for a crew was a long one.

The LM, especially, was behind schedule. No one at Grumman, one of America's finest and most successful aircraft manufacturers, had fully realized the difficulty of the job the firm had taken on. In 1959, the company's Mercury design had been chosen as the best of eleven entries, but the contract had gone to runner-up McDonnell; Grumman had been judged too busy with other military aircraft commitments. But the firm's engineers had been studying lunar-orbit rendezvous long before NASA had selected that mode in July 1962—in May 1960, Grumman representatives had had a meeting with Bob Gilruth, Max Faget, and James Chamberlin about a potential lunar lander—and their diligence helped win them the coveted LM contract. Grumman's design, like the other contractors' in the running, had been based on general specifications provided by Faget's office. But the company hadn't realized that NASA expected it to redefine and redesign the LM "item by item."

That process took much longer than anyone there had expected. The engineers had to think not only outside the box but beyond the immutable laws of aerodynamics underpinning every aircraft produced at their Long Island plant, since this vehicle would fly only in the vacuum of space and never feel the lightest breeze. That and other requirements meant constant and costly redesigns. And in addition to building a spacecraft that could do all the things asked of it in a vacuum and in low- or no-gravity, the engineers had to watch another important issue: weight. NASA offered a twenty-five-thousand-dollar bonus for every pound eliminated, but the ounces came off grudgingly, through hollowed-out joints, single window panels, and aluminum cabin walls milled to a hundredth of an inch thick. That barely met the minimum structural strength needed, and the fragile result could easily be damaged. Assembly and testing required complicated schemes that were costly and time-consuming and could only approximate, never duplicate, true operational conditions. Seats in the vehicle were deemed unnecessary, since the flight would be short and in one-sixth gravity; the two astronauts would stand like trolley motormen, tethers holding them in place. And combustion problems continued to plague the ascent engine, which would launch the astronauts off the lunar surface. It would have no backup, so it had to be as close to 100 percent reliable as possible.

Just as important was the LM's descent engine, which required something not even invented yet: a throttleable rocket motor. Rocket engines were ignited and remained at constant thrust until turned off. The astronauts flying the LM would need to throttle the engine over its full range of thrust while dropping to the lunar surface and moving over it to find a smooth landing spot among the moon's boulders and craters. That component had run into serious problems during testing, problems that hadn't been solved and wouldn't be for months. The lunar module was nowhere near ready, and it would not be involved in this test.

The Apollo hardware had been in development since 1961, even before NASA knew how they would get to the moon. That these components would help them get there had been a leap of faith on its own; whatever the mode, it would involve the Saturn booster at von Braun's Marshall Space Flight Center, the command and service modules at North American Aviation, and the lunar module at Grumman. By 1966, twenty thousand contractors were working frantically to keep pace with NASA's unforgiving schedule, and often failing.

The funding, fortunately, was not lacking. Years earlier, NASA administrator Jim Webb had made sure to overestimate the Apollo budget when he presented it to Congress, and since then he'd done a hell of a job keeping the program well funded and unbothered. That hadn't prevented cost overruns and delays in every aspect of the program due to the sheer size of the undertaking, and it was under more pressure and criticism than ever before. Justification had also become a problem. The United States had clearly surpassed the USSR in manned space accomplishments, and besides, the Soviets hadn't launched a man or woman into space in almost two years, not since Alexei Leonov had performed the first EVA. Nikita Khrushchev and the occasional Russian scientist had claimed a lunar landing would be too expensive and too dangerous, and that position was seconded by a prominent British astronomer, Sir Bernard Lovell, who, after an extended July 1963 visit to the USSR and its observatories, had announced that a platform in Earth orbit, not a moon landing, was now the focus of the Soviet space program. Still, maybe they were planning to put their own man on the moon, or around it, this year, the fiftieth anniversary of the 1917 Russian revolution. After all, the Soviets liked to link their space ostentations to anniversaries for extra punch in the prestige area.

Few at NASA believed the Soviets' claims that they weren't trying to reach the moon. On a wall at MSC was a large sheet of paper displaying several dates that represented the next launch window for a manned Zond; the Soviets had been sending an unmanned version of the small probe into space, with mixed results. And there were clues pointing to a Soviet moon-landing attempt. In 1965, after Leonov's EVA, the Soviet press reported that his spacesuit was a prototype for one to be used on the lunar surface. And CIA analysts studying spy-satellite photographs of construction at the Baikonur cosmodrome concluded that a huge new booster was being prepared, one that could only be aiming for the moon. Jim Webb and other NASA officials had conceded that there was a fifty-fifty chance that the USSR would reach the moon first—"a worldwide propaganda disgrace," *Missiles and Rockets* called that possibility. The magazine proposed a solution: Change the national goal to an even more impressive target in space before a manned lunar landing. "Then," its editors suggested, "if the Russians were the first to arrive on the Moon, we would be in a position to gracefully acknowledge their achievement while pointing out that to us the Moon was but a way-station en route to a more distant objective." There was no official response from NASA.

No one in the West knew it, but the Russians had, in fact, run out of space spectaculars for the moment. Sergei Korolev had been pulling rabbits out of his *ushanka* fur hat for years to satisfy the demands of his country's political leaders. The USSR might have committed to a lunar landing, but the corresponding budgets were, as always, threadbare and consisted of more than two dozen different programs within a program, or "design bureaus." Each one was dedicated to a specific task—communications satellites, earth-observation satellites, military missiles, interplanetary probes, and more—and they were all fighting for those meager rubles. Operating on the theory that competition would bring out the best ideas, administrators sometimes assigned two or more design bureaus to the same task. One of Korolev's rivals, engineer/designer Vladimir Chelomey, was also working on a moon rocket.

Korolev had suffered from heart and kidney problems for a long time, the result of his Gulag stay. After he had a heart attack in December 1960, doctors had warned him that if he didn't reduce his workload, he would face an early death. Hospital visits for gallbladder issues, intestinal bleeding, and more heart problems followed in the next few years, but he continued to work eighteen hours a day, six days a week. Early in January 1966, he was admitted to a Moscow hospital for what was considered routine surgery to remove benign polyps in his large intestine.

In his absence, Vasily Mishin, a competent engineer and Korolev's deputy, assumed control of the program. A few days later, Mishin and his design bureau were severely criticized in a board meeting. He decided to resign. Only a phone call from Korolev, who had heard about the matter, persuaded him not to. "Just wait till I get back from the hospital," the Chief Designer told him, "and we'll decide who is right and who is wrong."

Korolev would never get the chance to discuss it. On January 14, 1966, while doctors were removing the polyps, they found a large, cancerous tumor in his abdomen. Complications ensued, including severe hemorrhaging, and he died that night, two days after his fifty-ninth birthday. His cremation ashes were interred in a niche in the Kremlin Wall, the highest honor for a Soviet hero and a deserving one for the guiding force behind the Soviet space program.

Korolev's shoes proved too big for Mishin to fill. He did not have Korolev's genius for finding compromise solutions to large-scale problems, or his gift for leadership, or his ability to deal with the Soviet bureaucracy, and the deeply flawed moon rocket program he inherited was far too much of a challenge, as it might have been even for Korolev. Mishin would also find Leonid Brezhnev's regime more repressive and difficult to work with and less interested in grandiose space plans that had no direct impact on Soviet defense than Khrushchev's had been. (In October 1965, the cosmonaut corps had sent a letter to Brezhnev complaining of the lack of support for Russia's manned space program. They received no answer.)

In his obituary, published in *Pravda* with a photograph two days after his death, Korolev was identified by the title Chief Designer for the first time; his death was blamed on a "long and fatal illness." It would be decades before his importance to Soviet manned spaceflight was fully appreciated inside and outside Russia.

* * *

In addition to the ongoing (but decidedly lessened) Soviet threat, Americans had serious matters closer to home to worry about. The country had changed dramatically since those halcyon years of the late fifties when the nation was flush from the postwar industrial boom and free of major social disturbances. Those seemingly happy days had hidden underlying stresses that blew up in the sixties, and by 1966, deadly race riots rooted in poverty and inequality were frequent, and protests against America's involvement in the Vietnam War were escalating.

Despite the increased criticism and apathy toward it, the Apollo program continued, though NASA's budget, adjusted for inflation, was about to start shrinking. It was still massive—for 1966, it was 4.4 percent of the federal budget—but the message from Congress appeared mixed; they were essentially saying, *Go ahead and land on the moon, but don't count on another program of equal size after that.* More than anything, what kept NASA going was its commitment to a beloved leader cut down in his prime less than three years into his administration, a president who had been indifferent to the space effort at first but who had come to embrace it, at first for political purposes, then with genuine enthusiasm. His death had only strengthened that promise, and the people of NASA and others in government would keep it, despite the naysayers in Congress and the scientific community who said that the dangers of manned spaceflight outweighed its benefits and that machines and robots could do the same things as their human counterparts and more—and for a lot less money.

By late 1966, Go Fever had taken hold at both NASA and the factories of its thousands of contractors. The rush was on to get everything done as quickly as possible, and that meant overlooking potential small issues, since dealing with them might lead to missed delivery dates. The end of the race was in sight. And though Grissom was well aware of the command module's problems, he was caught up in Go Fever too. Because if this flight went well and the next few did also, he was convinced he'd be the first choice to land on the moon. After all, NASA's upper management, which included Slayton and Kraft, believed that it should be a Mercury astronaut if at all possible, and he was the only one left that they trusted to get the job done right. Following his successful Gemini flight, he felt he'd redeemed himself after the bad ending of his Mercury mission, and his relationship with the press had much improved; from then on, one newsman said, he was "a reporter's delight." If he and his crew could just get through these tests and NASA could get those fixes done on his spacecraft, they'd be all right.

Grissom's crewmates were thirty-six-year-old Ed White, the first American spacewalker and already a national hero, and thirty-one-year-old Roger Chaffee, a member of the 1963 group of astronauts. White had recently told his father, who had been a barnstorming pilot in the thirties, that his goal was to make the first flight to the moon. Chaffee, though he hadn't flown in Gemini, was highly regarded; he was a former navy pilot of fighter jets and spy planes—he had flown reconnaissance missions over Cuba during the missile crisis in October 1962—and a perfectionist as an engineer. Chaffee and Grissom were both Purdue graduates, and the two had become close—Chaffee had even picked up some of Grissom's habits, like salting his speech with an occasional profanity. Gus, who would soon turn forty-one, was fond of the young pilot and referred to him as "a really great boy."

Grissom hadn't been able to ride herd on the Apollo spacecraft from its earliest manufacturing stages as he'd done with Gemini. Because the two programs were developed concurrently, other astronauts had been involved in the early assembly and testing of the command-service module at North American Aviation's plant in Downey, California, and they hadn't been allowed the input Grissom had with Gemini. To make matters worse, the contractor had been unwilling to share data and drawings with NASA flight controllers and astronauts. But Grissom was doing his best to catch up, and he wasn't happy about how things were going. None of the Apollo components was progressing smoothly or on schedule.

In fact, if the module had been a horse, "they would have shot it sometime in 1966, perhaps as early as 1965," said Walt Williams, the former Mercury operations director. What would be Grissom's craft, AS-204—labeled as such because it was Apollo-Saturn, launched into space by the fourth booster produced in the second Saturn series, the Saturn IB—was particularly rife with problems, from its communications and propulsion to its environmental systems and beyond. This resulted in an unruly accumulation of electrical wiring—there was some twenty miles of it in the spacecraft—that could barely be squeezed in. Apollo was several orders of magnitude more complex than Gemini, and everyone was beginning to find out what that meant for schedules.

Slayton had assigned Grissom the first Apollo flight soon after Gus's March 1965 Gemini 3 mission. Gus and his crewmates began spending weeks away from home, either at North American Aviation's factory in California or at Grumman's on Long Island, though most of their time was spent at the former—Jim McDivitt, who was off Gemini 4 in early June, had been assigned to the LM, and Grumman was in charge of that. They spent long days attending countless meetings, monitoring design and manufacturing reviews, making inspections, and testing the spacecraft, which mostly meant sitting in it for hours on end while reporting design and operational flaws to one or more engineers or technicians. Some North American Aviation engineers had dubbed Grissom "the Nitpicker" for his thoroughness. Grissom's home life and that of his crewmates and their backups consisted largely of spending a single weekend night with their families to remind their kids that they had fathers and their wives that they had husbands.

At the Cape, the main mission simulator was so far behind in incorporating the latest developments that Grissom hung a large lemon from it a few days before the test. And in a December press conference, Grissom had stated that a successful flight would be one in which he and his crew made it back alive. The reporters laughed, thinking he was joking. It wasn't entirely clear that he was, particularly in view of what he told Al Shepard in private: "This is the worst spacecraft I've ever seen." He told his wife that his crewmates were not spending enough time on the command module—"He thought they should be working instead of playing," she remembered. But he was careful not to gripe too loudly. "They'll fire me," he told his old Gemini 3 crewmate, John Young.

The pressure to get these components finished and shipped to Cape Kennedy was intense, and despite some shoddy workmanship and incomplete inspections, they did. There was just too much involved for the astronauts to stay on top of it all. Grissom gave Slayton and Shepard a long list of problems, and they assured him they'd be fixed before the actual launch. But Go Fever had taken over, and there wasn't enough time to do things right or fix what needed to be fixed *now*. NASA had three manned Apollo missions scheduled for 1967 and a total of fifteen Saturn V rockets on order, though it was hoped that a lunar landing would be accomplished by the ninth or tenth launch—and before the end of the decade. Keeping such a tight schedule depended on a good, solid shakedown flight to find all the problems in the command-service module.

Though the actual mission was set for February 21, 1967, there were several important tests scheduled before then. One was a plugs-out test: a simulated full countdown, at the end of which the spacecraft would be switched to internal power, almost identical to actual launch conditions, to test the compatibility of all systems and make sure that the spacecraft could function on internal power alone. It would involve only the command and service modules, no booster, so it would be safe, a routine dress rehearsal that was scheduled to run for about five hours. Inside the cabin, the environment would be 100 percent oxygen, not Earth's sea-level atmosphere of 80 percent nitrogen and 20 percent oxygen, in order to avoid the bends that nitrogen in the blood could induce. A single-gas atmosphere also eliminated the need for complex plumbing, which was required to maintain the proper mix, and that plumbing's extra weight. Pure oxygen, though highly flammable, had been used in Mercury and Gemini with no complications.

Wally Schirra and his backup crew had been in the cone-shaped command module two days before performing a similar countdown test, this one plugs-in, using external power with the hatch left open. They had done it at sea-level atmosphere, breathing ambient air, and without spacesuits. That test had become a twenty-three-hour marathon that had ended at three a.m. the previous day. Afterward, Schirra told Grissom that he had a bad feeling about the spacecraft. "You're going to be in there with full oxygen tomorrow," he said, "and if you have the same feeling I do, I suggest you get out."

At about noon on a chilly Friday, January 27, at Cape Kennedy's launchpad 34, Grissom, White, and Chaffee, in their white flight suits, took the elevator two hundred and twenty feet up to level eight and went across the twenty-foot catwalk to the White Room, a protective enclosure surrounding the command module during installation and checkout. Deke Slayton was with them—he had considered lying down at their feet in the cabin during the test to try to figure out some of the communications problems dogging the command module, but Grissom vetoed the idea. By one p.m. they were strapped into their couches, familiar from hours spent in vacuum-chamber tests in Houston, and Slayton left for the blockhouse, where he would monitor the test. The command-service module sat atop the unfueled Saturn IB booster.

Technicians sealed the three-part entry hatch—first the inner hatch, then the outer hatch, and finally the booster cover cap. The original design had called for a one-piece hatch that would be released by explosive bolts, but when Grissom nearly drowned after splashdown in the *Liberty Bell* 7, the design had been changed to one that could never be accidentally opened. None of the astronauts liked it, since it eliminated the possibility of an EVA from the command module. A simpler, hinged hatch was in the works, though it wouldn't be available on Grissom's Block I version. You needed a wrench to loosen the six bolts on the inner hatch (in simulations, no one had been able to do that in under ninety seconds), and the hatch couldn't open unless the pressure inside and out was equal. The cabin was pressurized to 16.7 pounds per square inch, slightly higher than sea-level atmospheric pressure of 14.7 pounds per square inch.

The crew sat three abreast, their shoulders almost touching: Grissom on the left in the commander's seat, senior pilot White in the middle, and pilot Chaffee on the right. Above them and in front of them were multiple gauges, switches, dials, lights, and toggles.

The crewmates had been at the Cape all week, but they'd spent the previous Sunday night with their families. Grissom and his wife had discussed the big party scheduled for all the astronauts and their wives for the day after the test, Saturday, back in Houston. One of the last things he'd done was pluck a lemon from the tree in his backyard for the simulator. The crew hoped to finish this plugs-out test—and a practice emergency egress that Grissom had insisted on—at a reasonable time so they could fly their T-38s back to Houston, get a good night's sleep in their own homes, and try to let some steam off at the party.

But the command module wasn't cooperating. The astronauts slowly worked their way through the preflight checklist and waited through several holds while the ground crew labored to fix a radio glitch; constant static marred communication between Mission Control and the spacecraft. After Grissom had to repeat himself several times to be understood, his frustration boiled over: "I said, Jesus Christ, if we can't communicate across three miles, how the hell are we going to communicate when we're on the moon?"

The day wore on. At 4:00 p.m., one shift of technicians left and another came on. At 5:40 p.m., near sunset, another hold was called at T minus ten minutes to deal with one more communications problem before the simulated liftoff, when the plugs would be pulled. This, everyone hoped, would be the final delay. After it, they could proceed with the last ten minutes, finish it up, get through the emergency-egress practice—the three astronauts would take the gantry's high-speed elevator down to a fireproof truck waiting at the base of the pad—and get out of there. Someone suggested that the test be postponed, but that was overruled. Redoing the test would cost more time, and time was something they didn't have.

A few seconds before 6:31 p.m., as the crew members once more ran through their checklist, there was a slight surge in voltage.

Nine seconds later, one of the crew yelled, "Hey!"

A moment passed, then a voice—maybe White's—rang out: "We've got a fire in the cockpit!"

Seven seconds of silence followed. Then a garbled transmission, possibly from Chaffee: "We've got a bad fire—let's get out...we're burning up."

There was a final howl of pain, and nothing more.

The twenty-seven men of the pad rescue team rushed across the catwalk. Fourteen seconds after the first shout of alarm, the command module's hull ruptured, spewing flames and gases. The shock wave knocked them down, and some of them ran across the catwalk to the elevator, believing that the command module had exploded or was about to. Several grabbed fire extinguishers, ran to the White Room, and struggled to open the module's hatch, but the heat and smoke drove many of them back. They returned moments later, some with gas masks. While the pad leader called for firefighters and ambulances, five men took turns with a hatch-removal tool, working by touch in the dense, dark smoke and making several trips in and out of the White Room to breathe. About five minutes after the first report of fire, they finally got all three hatches open, but by then it was too late. The fire had lasted just twenty-five seconds, but the three astronauts were gone, asphyxiated by the toxic gases in the cabin. There was no fire extinguisher inside.

A quarter of a mile away, in the concrete blockhouse, Deke Slayton, Grissom's best friend, was sitting next to rookie astronaut Stu Roosa, the CapCom, and talking to Rocco Petrone, the no-nonsense director of launch operations at Cape Kennedy. Slayton jumped up from his seat when he heard the first shout. He and everyone else there turned to the video monitors and watched helplessly as flames in the spacecraft built to a white glare and then subsided. Slayton thought he saw a movement in the cabin. A few seconds later, they heard someone at the launchpad yell for a doctor. Slayton and two physicians rushed to the pad, rode the elevator up to level eight, and hurried into the White Room, where the hatch was already open. When Slayton peered in, he saw a blanket of black ash covering everything. "It looks like the inside of a furnace," he said; the *Washington Post* used those words as a headline a few days later.

It would be determined later that a spark below and to the left of Grissom's couch—probably a short in a bundle of wires somewhere in the many miles of wiring in the command-service module—had reached something flammable and ignited a fire that had raged through the cabin, burning anything and everything in its path: belts and straps, nylon netting, spacesuits, helmet covers, oxygen hoses, aluminum coolant tubes, and the many Velcro fasteners and patches scattered everywhere. The pure oxygen was almost instantly replaced by carbon monoxide and toxic black smoke that invaded the crew's oxygen lines. The official cause of death was suffocation, although the men had also suffered serious but not life-threatening burns. It was later estimated that the interior temperature reached at least twenty-four hundred degrees Fahrenheit—the melting point of stainless steel, found melted inside.

In the last seconds before he died, Grissom had moved out of his seat, presumably to try to help White open the hatch bolts. The heat and the melted material had welded the astronauts to various parts of the cabin and, for Grissom and White, to each other. When Slayton looked inside the blackened shell, he couldn't tell which head belonged to which body. After all the doctors, firemen, and other emergency personnel had arrived, the scene was extensively photographed, inside and out, to aid in the forthcoming investigation. At about 12:30 a.m., they began removing the bodies; it would take ninety minutes to complete the job. Twenty-seven pad technicians were taken to the hospital and treated for smoke inhalation.

Sometime later, after the escape rocket was disarmed, Slayton left for his office. As word of the tragedy spread through NASA's ranks, Deke and Chuck Friedlander, the gregarious chief of the astronaut support office at Cape Kennedy, spent hours calling everyone who needed to know. Deke alerted astronauts in the Houston area and gave them a tough assignment: They or their wives were to get over to the Grissom, White, and Chaffee homes as quickly as possible to tell the families what had happened before they heard it on the news or got calls from inquiring reporters. Michael Collins had the task of driving over to Nassau Bay to tell Martha Chaffee. A few astronaut wives had arrived at her house earlier but hadn't told her; when she saw Collins arrive, she knew.

She and Collins retreated to a back bedroom to discuss it out of earshot of the children. When Martha Chaffee told her eight-year-old daughter, Sheryl, that she wouldn't be seeing her father again, Sheryl thought her parents were getting divorced. Her mother explained that there had been a fire and that her father was dead. Then her mother gave her a necklace with two hearts that he had planned to take up to space with him.

Jo Schirra walked over to the Grissom house through a hole in the backyard fence the families shared (they had deliberately made the hole so they could visit without alerting newsmen or sightseers in front). Another neighbor, whose husband was a NASA engineer, was already there—her husband had told her to go over and have a drink with Betty, but he didn't tell her why. As soon as she saw Jo's face, Betty Grissom knew it was bad. Jo told her there had been an accident and that Gus had been injured, though she said she didn't know how badly. A few minutes later, when a black NASA car pulled up outside, Betty knew. Her two sons—Mark, thirteen, and Scott, sixteen—were in their rooms. She went to tell them. Neither one cried. She didn't either.

Jan Armstrong, who lived next door to the Whites, was standing in their driveway when thirteen-year-old Ed White III rode up on his bike. Ed and his ten-year-old sister, Bonnie, were sent to another neighbor's house. Astronaut Bill Anders, who lived three blocks away, soon arrived to tell Pat White of her husband's death. She became distraught. Other families—the Bormans, the Staffords—showed up to offer support.

About two a.m. at Cape Kennedy, Friedlander picked up a ringing phone. It was Lowell Grissom, Gus's brother. He was calling from his parents' house. No one from NASA had phoned to tell them—Betty had. Friedlander handed the phone to Deke, who talked to Lowell for a few minutes. Slayton also called Tom Stafford and his crew, who were preparing for the second manned Apollo flight. They and their backups were running tests on their Block I spacecraft at the North American Aviation plant in California. Slayton told them all to get back to Houston. There would be no further testing for a while.

Earlier in the day, five astronauts—Gordon Cooper, Scott Carpenter, Neil Armstrong, Jim Lovell, and Dick Gordon—had spent the afternoon at the White House with Bob Gilruth, President Johnson, and other dignitaries to witness the signing of a space treaty outlawing the militarization of space and forbidding the staking of a land claim on the moon or any other heavenly body. The treaty was signed simultaneously in London and Moscow. Afterward, Carpenter took a taxi straight to the airport, but the others went to their hotel, the Georgetown Inn, where they were told of the accident. Gilruth joined them a short time later. The men gathered in a private suite and tried to eat dinner, then stayed up late into the night with a bottle of Scotch, discussing what they thought had happened at the Cape and pondering the future of Apollo. They wondered whether they would make Kennedy's end-of-the-decade deadline or if the manned space program would continue at all. There was also anger at NASA for accepting such a flawed spacecraft from the contractor.

Nine hundred miles away from the Cape, at Mission Control in Houston, John Hodge and a roomful of flight controllers had been monitoring the countdown through voice and radio only. When several of them suddenly lost telemetry, heard something about a fire, then lost communications with the Cape, it was unclear what was happening, but they knew it wasn't good. Hodge called Kraft from his office, and they soon found out.

Gene Kranz had been at home, getting ready to go out with his wife, when his next-door neighbor, a NASA branch chief, knocked on his door and told him of the fire. He raced back to the MSC, but there was nothing he or anyone there could do. One flight controller broke down in the parking lot outside the control center, sobbing and saying, "It's horrible! It's horrible!" A few others walked across the highway to the Holiday Inn bar. Kranz and most of the others gathered at the Singing Wheel, the two-story, barnlike building a mile or so west of the MSC that had become the favorite watering hole for Mission Control. The proprietor, Nelson Bland, sent out all the other customers. The evening wore on and they ordered pitcher after pitcher of Lone Star beer and mourned the astronauts; eventually, wives came in looking for husbands and congregated in a back room.

Near midnight at Cape Kennedy, a weary and numb Slayton made the last of many phone calls. Friedlander pulled out a fifth of Scotch from a locked cabinet. He and Slayton sipped the whiskey until four in the morning. "It was," Slayton remembered, "a bad day. Worst I ever had."

After sixteen manned missions in which twenty-six astronauts had returned to Earth safely, three had died in a spacecraft on the launchpad during a routine test, and in a fire—the greatest fear of every pilot. The death toll could have been worse; if the flames had reached the solid fuel of the escape tower rocket atop the command module, many more people might have perished. They might well have lost everyone on that level of the gantry, something the technicians were quite aware of while fighting to get the hatch open. And earlier in the day, not only Slayton but Joe Shea had considered joining the crew—lying on the floor beneath them—to see the problems inside the spacecraft firsthand.

After the success and safety of the Mercury program and the Gemini program, the American public had forgotten how dangerous spaceflight was. The fire was a deadly reminder that men could die, and without even leaving Earth. Astronauts had perished previously in air crashes, but that was different.

The fire would change everything in NASA. "After this, nothing would be the same again," Kranz remembered later.

The day after the fire, NASA announced the formation of a nine-person accident investigation board to be chaired by the head of Langley Research Center, Floyd Thompson. The board also included Max Faget, three other NASA managers, and one astronaut, Frank Borman, chosen by Gilruth and Slayton for his unwavering integrity and thoroughness. Borman flew to Cape Kennedy the day after the accident. After viewing the charred insides of the Apollo command module, even a straight arrow like Borman was shaken to the point of imbibing. That night he, Slayton, and Faget went out to a nightclub, where they drank several toasts to Grissom, White, and Chaffee, then had several drinks more. At some point, Faget began doing handstands while a go-go dancer gyrated a few feet from him. They ended the evening by throwing their glasses against a wall—a fighter pilot's tribute that a submariner like Faget could understand.

A few months before the Apollo 204 fire, Grissom's parents had come from Indiana to visit. Gus asked Chuck Friedlander to give them a tour of Cape Kennedy. Since Gus's mother was so worried about this mission, Gus had asked Friedlander ahead of time to make sure he pointed out all the safety features along the way. Friedlander took Gus's parents up the elevator to level eight and walked them over to the command module. Mrs. Grissom, who had glasses and short, wavy gray hair, turned to him and asked, "Can I touch it for luck?" Friedlander held her around the waist while she reached out and put her hand on the command module. For the rest of his life, Friedlander would wonder how often she thought of that moment.

CHAPTER TEN

RECOVERY

We were given the gift of time. We didn't want that gift.

NEIL ARMSTRONG

ON A COLD, GUSTY Tuesday four days after the Apollo 204 fire, Gus Grissom, Ed White, and Roger Chaffee were buried with full military honors.

Grissom and Chaffee were buried at Arlington National Cemetery. Against a background of bare trees and a hazy blue sky, Grissom was the first to be laid to rest. On hand to pay their respects were the remaining members of the Mercury Seven, along with Bob Gilruth, Chris Kraft, and other NASA people, most of them dressed in dark wool winter coats. President Johnson, clearly grief-stricken, was there as well, doing his best to comfort Grissom's widow, children, and parents. After a rifle volley and Taps played by a lone bugler, four air force fighters soared low overhead, and then one pulled up and away, leaving an empty slot—the missing-man formation. A few hours later, at one p.m., a similar ceremony for Chaffee commenced. The two crewmates were buried next to each other. Joining the president were members of the Fourteen, the 1963 group of astronauts.

The third astronaut, Ed White, was laid to rest at West Point, his alma mater, on a wooded bluff overlooking the Hudson River.

The remaining members of the 1962 astronaut class, the New Nine, were present for White's service, as were Lady Bird Johnson and Vice President Hubert Humphrey. In the chapel before the service, Ed's widow, Pat, had given Chuck Friedlander an envelope and pointed to the closed door behind which her husband's casket lay. Friedlander went in, opened the casket, and placed the letter on White's bandaged remains, over his heart. He left the room and nodded to her. That day would have been the Whites' fourteenth wedding anniversary.

After Grissom's burial, Alan Shepard and a few others had a drink at the bar of their hotel, the Georgetown Inn. With tears in his eyes, Shepard said, "I hate those empty-slot flyovers."

In President Johnson's official statement following the fire, he had reaffirmed the nation's commitment to his predecessor's lunar challenge. But NASA officials were knocked for a loop. The end-of-the-decade target was now "a reasonable possibility" and "no longer a sure thing," said one. On February 3, 1967, the agency announced it had suspended all manned spaceflights until the cause of the fire was known. That included the three manned Apollo missions scheduled that year.

Regardless, the bloom was off the rose of the U.S. space program. Criticism of spaceflight—particularly the manned kind—had increased over the past few years, and support in Congress and among the public had cooled. Almost a decade after Sputnik and the panic it engendered, open hostilities between the two superpowers appeared far less likely, making the political-prestige argument less vital. In one poll, nine out of ten Americans said that it wouldn't matter who got to the moon first. Billions were being spent on an unnecessary space stunt and its artificial deadline. Was it worth the exorbitant cost and, now, the loss of life? There was the perception that the billions granted NASA were being shot straight into space. In reality, at the program's peak, they were funding regular paychecks for 420,000 Americans, and the much-ballyhooed benefits from it were rarely tangible things that one could point to as being valuable. To many, manned spaceflight seemed a gargantuan waste of money that could be put to far better use domestically; there were pressing problems on Earth to deal with, such as poverty, crime, disease, pollution, unemployment, and more. Technology, which NASA embodied, didn't inspire the same gee-whiz admiration it had a decade ago. Prominent members of the scientific community were also calling for the cancellation of manned spaceflight, since they believed unmanned efforts would be cheaper, safer, and more productive. A burgeoning distrust of the technology that had produced nuclear bombs and deadly chemicals such as DDT and thalidomide further drained the enthusiasm for manned space exploration.

The command-module fire increased the criticism, and Apollo's cancellation became a distinct possibility. NASA's response to the tragedy and its handling of the investigation would be crucial to its survival.

Many in NASA—particularly the higher-ups—engaged in self-examination: Could they have said or done anything to prevent the tragedy? Gene Kranz, deputy director of Flight Control under John Hodge, went beyond introspection. After Friday night's unofficial wake at the Singing Wheel and a weekend spent looking for answers and soul-searching, Kranz decided something more needed to be done.

He called a meeting on Monday of everyone in flight operations as well as all the spacecraft contractors they worked with in Houston. A few outsiders, like Max Faget and Frank Borman, attended. After Hodge opened the session with a summary of the fire and the response to it, Kranz walked onto the auditorium stage and took the microphone.

It was up to the people in that room, he said, to make sure that Grissom and his crew had not died in vain. He placed the blame for the accident squarely on himself—and everyone else there. They had all known there were problems, he pointed out. "Not one of us stood up and said, 'Dammit, stop!'" he said.

"We did not do our job," he continued.

No one in the audience said a word.

"From this day forward, Flight Control will be known by two words: 'Tough and Competent.' *Tough* means we are forever accountable for what we do or what we fail to do. We will never again compromise our responsibilities. Every time we walk into Mission Control we will know what we stand for.

"*Competent* means we will never take anything for granted. We will never be found short in our knowledge and in our skills. Mission Control will be perfect."

He ordered them to go to their offices and write *Tough and Competent* on their blackboards—and never erase it. "These words will remind you of the price paid by Grissom, White, and Chaffee. These words are the price of admission to the ranks of Mission Control."

Not all flight controllers obeyed that order, even if they agreed with Kranz's evaluation. Some felt it unnecessary to adopt a public relations campaign to show they could do better. "If you are really tough and competent, you don't need to advertise it," Cliff Charlesworth, another flight director, said on the way out of the auditorium. But Kranz's speech, both critical and inspiring, helped transform the culture of Mission Control, and its tenets were embraced by NASA as a whole. Psychologically, it provided a way forward, channeling the anger, grief, and guilt into a higher level of accountability and excellence.

* * *

The day after the fire, Jim Webb met with the president at the White House and persuaded him to allow NASA to handle the investigation. He didn't tell him the agency had already begun that process the night before, forming a review board and picking members.

The Apollo 204 Review Board's investigation was thorough. An uncannily perfect reenactment of the fire was conducted with the remaining Block I command module, virtually identical to Grissom's. Twenty-one panels were set up to cover various systems and subsystems, and fifteen hundred technicians and experts in those areas and others tore apart both modules piece by piece, wire by wire. For the first few weeks, the investigators worked literally around the clock.

Some in the press doubted that an in-depth, impartial investigation and analysis could be made by the committee, since most members were NASA employees. But they underestimated the investigators' scientific desire, compounded by their grief and guilt, to find out what was wrong and figure out how to fix it. The board's members felt strongly that they were the only ones who could effectively investigate the disaster.

A Soviet newspaper—*Trud*, the official publication of Soviet trade unions—blamed the Apollo 204 deaths on the "careless haste" of U.S. space officials in the race to beat the USSR to the moon. There was no mention in *Trud* or any other Soviet news source of a similar misfortune their own space program had experienced.

What no one in the West would know until it was revealed in the Russian newspaper *Izvestia* fifteen years after it happened was that a deadly fire had occurred just a few weeks before Yuri Gagarin became the first man in space. On March 23, 1961, cosmonaut Valentin Bondarenko, at twenty-four the youngest cosmonaut trainee, was on the tenth day of his fifteen-day isolation-chamber test. He tossed a cotton pad soaked with rubbing alcohol toward a garbage pail but missed, and it landed on a hot plate that had been left on. The pad ignited in the oxygen-rich atmosphere, and when Bondarenko tried to put out the flames with the sleeve of his woolen coveralls, his clothing caught fire and burned quickly, spreading to his skin and hair. By the time technicians got the chamber open, he was burned over 90 percent of his body and barely alive. Eight hours later, after whispering a few words of apology, he died.

Had the news of that tragedy been released at the time, would it have prodded NASA to reexamine its use of a 100 percent oxygen system? Maybe—but it's unlikely. There had been four oxygen fires involving military American personnel in the five years before the Apollo tragedy, none of which had resulted in changes at NASA.

The grief and shock felt throughout NASA—particularly in Houston, where so many knew the three men personally—were palpable. For their part, the astronauts knew they had to put the tragedy behind them and get on with it; the test pilots, especially, were accustomed to death. They had lost friends before and knew there was nothing to do but continue the job.

But the nature of the catastrophe and its primary cause—an oversight that was not the fault of one person but of many—meant that getting past it was very difficult for some people, especially the engineers and managers associated with the command-module design, testing, and checkout. During a briefing in Bob Gilruth's office, one blank-faced engineer walked over to a blackboard, drew a large box with lines leading to smaller boxes below it, and announced it was an organizational chart of heaven. "At the top is God, whom we'll call Big Daddy," he began, then lapsed into incoherency. He was flown home in a straitjacket, although he eventually recovered with the help of psychotherapy and electroshock treatment. A McDonnell adviser to the review board descended into severe depression and spent three weeks in a mental institution. Another engineer, one of NASA's, also went off the deep end, exhibiting outlandish behavior, but he never received treatment and left soon after.

Joe Shea, the brilliant, youthful, high-energy director of the Apollo program, counted several of the astronauts as good friends—he often played handball with them and would get in punning duels with Wally Schirra. Shea had been warned about the dangers of 100 percent oxygen, but he eventually concluded that the safer two-gas setup would be too complicated and only worsen the ongoing weight problem. He blamed himself for the fire and began drinking heavily—a habit quite a few at NASA took up. He became so despondent, he was almost unable to function, especially when it became apparent to everyone that he had been, in Frank Borman's words, "a poor administrator who had simply let North American's design mistakes pile up like unnoticed garbage."

Hearings in both houses of Congress began a week after the fire and continued through May. Webb, some of NASA's top administrators and managers, and several astronauts were grilled at length in the government's efforts to ascertain the exact reason for the accident. Webb told a committee that the agency had taken technical risks because of an "austere budget," which didn't go over well. More than anything, the congressmen appeared determined to fix blame somewhere—and Webb was just as determined to protect his agency. He told his subordinates that he planned to take the brunt of the criticism, and he did so, especially after a year-old internal NASA report criticizing the work of North American and some of its subcontractors came to light; few of the recommendations in it had been carried out. Webb knew nothing about the report and had to admit that fact, which didn't help; if he truly hadn't known about it, he should have, went the reasoning. He clammed up after that and was criticized for being evasive and trying to control the information flow from his agency. The hearings became so harsh and the potential fallout so devastating that one respected astronaut, Tom Stafford, sent a private message to Webb through a mutual friend. "Tell him," said Stafford, "that if something is not done to straighten out the problems down here, several of us will pull out of the program. I want you to get Webb to do something." The message emphasized that it wasn't just external forces Webb had to satisfy but his own people. In addition to defending NASA before Congress, he had to make serious changes in the Apollo program, and fast.

The review board's thirty-three-hundred-page report, delivered to Congress on April 5 and made public a month later, was an impressively impartial analysis of the fire and a scathing criticism of NASA and North American Rockwell methods. It revealed a litany of mistakes, rampant carelessness, and the administrators' appalling inability to recognize the dangers inherent in an overpressurized, full-oxygen spacecraft loaded with flammable materials. (There were five thousand square inches of Velcro in the command module's cabin, ten times what should have been there.) Under the conditions of the plugs-out test, the spacecraft had been a death trap. Many observers had predicted that the report would be a whitewash. The harsh self-criticism made clear that it wasn't.

Yes, some of North American Aviation's work was sloppy and hurried—the report cited the "ignorance, sloth, and carelessness" of the contractor. But that had been a response to pressure from NASA and its rigorous, unforgiving timetable. Besides, the agency was supposed to be overseeing and approving every step, and it had relaxed its standards on quality control, particularly in final inspections, in its zeal to keep to its schedule.

The review board's extensive recommendations included the near-complete replacement of combustibles, a quick-opening hatch, improved pad-emergency procedures, and a safer gas atmosphere. The exact cause of the fire was never ascertained, though the likely culprit was a bare wire under Grissom's couch that had rubbed against something and sparked.

Borman and four other astronauts testified before the House committee in April and expressed confidence in NASA management, which swayed some of the congressmen who had been ready to call for a delay in the Apollo program, as much as five years. After another month, the congressional hearings finally petered out; they hadn't gotten any further in fixing blame than the review board had. The Senate space committee wouldn't release its report until January 31, 1968, more than a year after the fire. It contained plenty of criticism and called for further discussion of NASA's ability to make the end-of-the-decade deadline.

Webb had gradually lost tight control of the program as he had entrusted his administrators with more responsibilities. They had not done their jobs well enough. He regained control with a sweeping series of high-level personnel changes announced in early April. Shea was shunted to a different job at the agency's Washington, DC, headquarters and replaced by George Low, a longtime NACA and NASA project manager who was smart, dependable, and soft-spoken. Until then, Low had been deputy director of MSC—the number-two man at Houston—and this would be a demotion, running a program office, but he took the job for the sake of the agency. After a couple of months in his new job, Shea would resign. North American Aviation also underwent a management shakeup at the highest level and incurred heavy financial penalties; the company spent millions of dollars to fix the issues that had resulted in the accident. They promised to bend over backward to fix and improve the Apollo command-service module and make it the best spacecraft ever built.

The fire provided NASA with the time and the determination to build a safer, more reliable ship and implement improved safety practices and higher levels of quality control. NASA investigators compiled a list of 8,000 potential problems that needed to be resolved, and 1,697 changes were recommended. The NASA configuration control board approved 1,341 alterations. The command module eventually underwent 1,300 changes. Among other improvements, the hatch was redesigned so that it opened outward, easily and in three seconds, flammable materials were replaced with fireproof Beta cloth, and the twenty miles of wiring were consolidated and insulated with Teflon. Further, while the spacecraft was on the ground, its cockpit atmosphere would be a much safer mix of 60 percent oxygen and 40 percent nitrogen; at liftoff, that would gradually change to 100 percent oxygen. Even the astronaut spacesuits would be made from fire-resistant materials.

Webb was quoted as saying he didn't think NASA would fully recover "until we make a couple of these birds fly." Everyone in the organization and every contractor associated with the Apollo spacecraft was more committed than ever to that goal. The fire had made it personal, and now they had an emotional investment in this endeavor.

On April 8, 1967, three days after the Apollo 204 Review Board's report was released, Slayton gathered eighteen astronauts in a small conference room at MSC. Only one Mercury Seven astronaut, Wally Schirra, was there; the others were the remaining seven of the 1962 group and ten of the 1963 class. NASA had added two more groups since then—six scientists in 1965, for whom the pilot requirement was waived, and nineteen more engineering pilots in 1966—for a total of fifty men in the astronaut corps.

After a few brief comments—"Gentlemen, we won't make the same mistake twice"—Deke got straight to the point. "The guys who are going to make the first lunar landing are here in this room." As heart rates accelerated and men looked around at their comrades, Slayton told them who would fly the next few missions and who would back those pilots up. Apollo 7, the first manned flight, would be crewed by Schirra, Walt Cunningham, and Donn Eisele (backed up by Tom Stafford, John Young, and Gene Cernan). Apollo 8, the first test of the LM, would be crewed by Jim McDivitt, Dave Scott, and Rusty Schweickart (backed up by Pete Conrad, Dick Gordon, and C. C. Williams). And the crew for the Apollo 9, the first test of the entire package in a high Earth orbit, would be Frank Borman, Mike Collins, and Bill Anders (backed up by Neil Armstrong, Jim Lovell, and Buzz Aldrin).

Assignments for the flights to the moon—there would be at least three—would come later. Slayton finished with, "Be flexible. This stuff will change."

After the fire, many had predicted that the USSR would overtake the United States in the space race. In the midst of the congressional hearings, on April 23, 1967, the Soviets launched their first manned mission, Soyuz 1, in twenty-five months. Vladimir Komarov, the handsome commander of the three-man sardine can called Voskhod 1, was the cosmonaut chosen to test a new, larger spacecraft model, the Soyuz, in Earth orbit—Korolev had been working on it before his death. Like Gemini, it had a rocket engine that enabled the pilot to maneuver it in space.

For several months leading up to the flight, rumors had been leaking from the USSR of a major new mission in the works. Soviet officials couldn't resist bragging about the impending spectacle, though one general issued a word of caution: "We do not intend to speed up our program," he said. "Excessive haste leads to fatal accidents, as in the case of the three American astronauts last January." The day before the launch, the Associated Press reported that the mission would "include the most spectacular Soviet space venture in history—an attempted in-flight hookup between the two ships and a transfer of crews."

The flight was part of a plan to vault the Soviet space program past the Americans' in one ambitious mission. Soyuz 2, carrying three cosmonauts, would launch from the Baikonur cosmodrome, then dock with its sister ship the next day. Two of the crew members would EVA to Soyuz 1 and return to Earth with Komarov—an entire Gemini program plus the first crew transfer between spacecraft all in one.

The Soyuz 1 launch went well, as did the first day in space. On the second, complications started; solar panels would not unfold, and stability problems led to the pilot using too much fuel to get his craft under control. During the craft's seventeenth revolution, ground control attempted to land it, but retrofire did not achieve the proper orientation. On the next orbit, Komarov managed to make a precisely timed manual retrofire and survived reentry through the atmosphere. But his parachute lines became entangled, and Soyuz 1 hit the ground near the edge of the Ural Mountains on the Kazakhstan border. Komarov died on impact, and the retro-rockets blew up, leaving his body a small, black, molten mass. Soyuz 2 was canceled, and the tragedy caused the Soviet program to pause and undergo its own reassessment. Jim Webb suggested that "full cooperation" between the United States and the Soviet Union might have helped save some or all of the two countries' lost spacefarers. Komarov's remains were displayed in an open-casket funeral—he had requested it before the flight and had told others the spacecraft wasn't ready—then interred in the Kremlin Wall, close to Korolev. NASA sent condolences, but privately, many in the agency were relieved. The war might have been a cold one, but it was still a war.

Despite this setback, it was known that Soviet spacemen had been training on helicopters for at least a year; a Russian cosmonaut, Pavel Belyayev, had let that out in May 1967 during a vodka-filled private meeting with astronauts Mike Collins and Dave Scott at the Paris Air Show. The only reason for helicopter work was to prepare for a lunar landing in a craft similar to the Americans' lunar module. Belyayev had also revealed that he expected to make a circumlunar flight in the near future.

There were other clues to their moon plans; statements over the previous few years, inadvertent or deliberate, truth or propaganda, made by cosmonauts and academicians had mentioned either orbiting the moon or landing on it. Details were scarce, as usual. But there was no question that there was a Soviet lunar program and that they hoped to beat the Americans to some of the firsts involved.

And there were those rumors of a massive rocket. In July, during the Senate Appropriations Committee hearings, Jim Webb mentioned a giant booster that the Soviets were working on, one that was larger than the Saturn V. News sources began calling it "Webb's Giant" and referring to it as if it were a mythical creature.

Although its name was unknown to the West, the huge rocket was Korolev's pet project—the N1. The development of the three-stage behemoth had started in 1964, with thirty clustered engines powering the first stage alone, and it had finally begun to take shape. Recent American spy-satellite photos had revealed evidence—massive launchpads, a larger erector/transporter, and construction facilities—indicating that its thrust might be eight to sixteen million pounds, possibly twice as powerful as the Saturn V. A moon rocket, unquestionably.

Several months after the fire, Marge Slayton, the de facto leader of what had been dubbed the Astronaut Wives Club, concluded that a party was needed to cheer up the troops. She and the rest of the wives organized a dinner fete for astronauts and their spouses only. Marge and Joan Aldrin and Pat Collins, the finance committee, decided there would be an open bar after dinner. The party was held at an upscale restaurant called the King's Tavern, and everyone had a fine time—until the end of the night, when the bar bill came. It was more than two hundred dollars, much higher than the women had estimated and more than they had collected; the wives had grossly underestimated the number of drinks a large group of astronauts could down. Deke Slayton, who had consumed his share, wrote a check for the difference.

For NASA, there were more tragedies to come. In June 1967, just when the agency was beginning to regain its confidence, it suffered another blow. Shortly after midnight on June 6, near the town of Pearland, south of Houston, astronaut Ed Givens, of the nineteen-man astronaut class of 1966, missed a turn and drove into an embankment. He was killed; two other men in his Volkswagen, also USAF officers, survived. They had been at a party given by an exclusive fliers' fraternal organization. Four months later, on October 5, Astronaut Clifton "C.C." Williams, a member of the 1963 Final Fourteen group and on the backup crew for Apollo 9, was flying a brand-new T-38 near Tallahassee, Florida, when one of his aileron controls became stuck. The plane slammed into a dirt road and disintegrated. Williams ejected but at too low an altitude to survive. As the lunar-module pilot for Apollo 12, he would probably have walked on the moon alongside command pilot Pete Conrad. That brought the total number of spacemen killed in one year to six, five Americans and a Russian.

Grissom's mission was retroactively named Apollo 1; the first manned Apollo mission was designated Apollo 7 (the intervening numbers represented unmanned test flights), and on May 9, 1967, Jim Webb announced its crew: Schirra, Cunningham, and Eisele. They would man a Block II command module that would be placed atop a Saturn IB booster—it had enough thrust to reach orbit but not enough to attain the escape velocity of about 25,000 miles per hour necessary to leave Earth's gravity and journey to the moon. The LM wasn't ready yet, so it wouldn't be aboard, but the crew would remain in orbit at least ten days on a straight shakedown cruise. Jim McDivitt's crew had been next in line, but NASA brass decided that they were better suited for what became the more complex Apollo 9 mission than Schirra, whose training habits had become lax.

But first, there was plenty of work to do. The command module needed to be made shipshape, and Schirra took it upon himself to ensure it was done right. Gus, his good friend and next-door neighbor on Timber Cove's Pine Shadows Drive, had been killed in the Block I version, and he was taking this challenge personally.

Schirra had originally been named commander of the second Apollo manned mission, which was slated to go after Gus Grissom's and planned to be a carbon copy of that one but with lots of science experiments. Schirra hated experiments. Like Grissom, he preferred a straight engineering test flight, unencumbered by what he called "junk." He had agreed to the mission only because Deke Slayton, still grounded due to his atrial fibrillation, hoped to be cleared for a flight and take it on himself—Wally was just a "caretaker" commander.

But when Slayton's request to fly was turned down in October 1966, Schirra was left holding the bag—and the mission. He began a steady barrage of complaints about its redundancy and criticisms of the many experiments planned for it, even going so far as to send a two-page list of ultimatums to the Apollo program office. Once again, Wally was determined to show that he was a naval commander and captain of his ship. His crewmates on the mission, Donn Eisele and Walt Cunningham, went along with him; they were both still military, after all. And if they had done otherwise,

Schirra might have accused them of mutiny, like a spacefaring Captain Bligh, and possibly booted them from his crew.

His bluster backfired. The powers that be at NASA, including Slayton, were unamused. The crew learned indirectly, from a press release, that the mission was canceled. The next manned flight, scheduled for early 1967, would involve the first Apollo rendezvous between the LM and the command-service module—the kind of engineering shakedown flight Schirra lived for.

Only it wouldn't be Wally's. The reliable Jim McDivitt would command that one. Schirra and his crew were made backups to Grissom and company. It was humiliating, especially for Schirra—during both Mercury and Gemini, he had initially been a backup, and now it was the same on Apollo. By this time, it was mid-November. Schirra at first refused. Only after Slayton and Grissom practically got on their knees and begged him to take the job did he agree to accept this lesser position.

But following the death of his comrade and the subsequent decision to replace Grissom's crew with his own, Schirra was no longer the lighthearted Jolly Wally of the endless puns and elaborate gotchas but the Carry Nation of the command module, stalking the halls of North American Aviation's Downey, California, factory, ordering changes left and right, large and small, some of them necessary, some of them definitely not. This mission, scheduled for late 1968, would be his last, he decided—he would make that announcement a few weeks prior to the launch—but before he left, he would by God make sure that this one would be done right. Apollo 7 became "Wally's mission." And Wally's mission, as he saw it, was to save the space program. Everyone knew he was deadly serious about it because he quit smoking in January 1968.

"We labored day and night getting the first spacecraft ready," Schirra remembered. He and his crew, and the backup crew, and the support crew (a recent addition—three astronaut trainees who did anything the other two crews couldn't get to) practically lived at North American Aviation (which merged with Rockwell in September 1967, becoming North American Rockwell) until their Block II command module was ready.

But an astronaut was already in charge of the command-module redesign. After his exemplary work on the Thompson Committee investigating the Apollo 1 fire, the no-nonsense Borman had been assigned by Bob Gilruth to be the official NASA overseer of North American Aviation's changes. After a few clashes with Schirra, Borman got Gilruth to rein him in and forbade any astronaut to visit North American Aviation without his approval.

North American Aviation had never been as open as Grumman to the astronauts' requested changes. Even after the fire, there was still a lack of cooperation. One day, Borman climbed into the command-module simulator to try out the controls. When he found that the stick worked exactly the opposite of how it did in an airplane—when he pulled back on it, the nose went down instead of up; when he pushed it forward, the nose went up instead of down—he asked to see the engineer in charge.

"You've got the polarity reversed," Borman said.

"It's not reversed, Colonel. It's the way it should be."

Borman said, "Reverse it so I can fly the damned thing."

"But this is the way Apollo is going to fly," the engineer said.

"Not with me or any other astronaut in it," Borman said. "Fix the goddamned thing or nobody'll fly it."

The controls were reversed to conform with the training every astronaut had received since flight school.

Schirra was adamant about one other matter: he wanted the return of pad leader Guenter Wendt. A former Luftwaffe flight engineer during World War II, Wendt had been a McDonnell engineer in charge of supervising final launchpad preparations and closeout procedures during the Mercury and Gemini programs, both of which involved McDonnell spacecraft. He had insisted on complete control before taking the job, and his passion for safety—and his willingness to stand up to anyone who threatened it—was legendary. One day, for instance, Jim McDonnell, aka "Mr. Mac," the owner-founder of McDonnell, decided to drop by the White Room. When his extended visit began disrupting work, Wendt strongly suggested he get on the gantry elevator. He did.

Wendt kept a two-inch-thick metal pipe at hand, just in case there was an emergency and he had to clobber someone blocking the emergency exit. The slight, bespectacled man in the white cap and coat and black bow tie holding a clipboard had become a good-luck charm to the astronauts, and his presence overseeing the White Room team was reassuring. Wendt was obsessed with the safety of the mission's crew and his own; even nights when he relaxed on his fishing boat on the Banana River, Wendt would play the what-if game, running through every problem scenario he could think of. John Glenn had nicknamed him "der Führer of der Launchpad" for his strict style and German accent. It was a moniker of affection and appreciation.

But North American Aviation, not McDonnell, was the Apollo command-service module contractor, so Wendt hadn't been present for the Apollo 204 tests. Now Schirra insisted that Wendt come back, and he lobbied Deke Slayton and North American Aviation's vice president for his return. When North American Aviation agreed to hire him, Slayton called Wendt personally to ask if he'd do it. Wendt agreed. The Pad Führer was back, to everyone's relief.

The Atlas rocket that launched John Glenn into orbit had already undergone ninety-one unmanned flights; Gemini's Titan II, thirty-four. The Saturn V, its first stage alone twenty-one times more powerful than the Atlas, would have just two. But everybody trusted von Braun's team. After all, he and most of his top supervisors had been working together, and very successfully, for thirty years or more, since Peenemünde and the V-2, the world's first long-range guided ballistic missile.

On June 20, 1944, under the auspices of the German army, von Braun and his rocket team at Peenemünde had launched a forty-six-foot-high V-2 rocket that became the first to soar into outer space, reaching an altitude of 109 miles. Twenty-three years and four months later, early in the morning of November 9, 1967, the first Saturn V test flight would be attempted at Kennedy Space Center. The Apollo "stack" stood 363 feet tall—sixty-two feet higher than the Statue of Liberty—and was two hundred and fifty times more powerful than the V-2. It weighed 6.2 million pounds, and its first stage alone generated 7.6 million pounds from its five huge Rocketdyne F-1 engines—the power necessary to lift the spacecraft from the surface of the Earth—and would burn 212,000 gallons of kerosene and 346,000 gallons of liquid oxygen, a total of 2,200 tons of propellants, if it cleared the launchpad. The next two stages would burn almost half a million more gallons of fuel.

Just transporting the Saturn's massive components to Kennedy and assembling them had been a lengthy and complicated process. The first and second stages had been shipped from Huntsville by river barge to the Gulf of Mexico, then down and around Florida and up through the Banana River to Merritt Island. The smaller third stage was flown to the Cape in a specially converted freight aircraft called the Super Guppy; a similar model delivered the command and service modules and the LM. They were joined in the Cape's Vehicle Assembly Building, a 525-foot-high structure with 1.5 million square feet of floor space that could fit four Saturn V launch vehicles, each in its own bay. When finally constructed, the spacecraft and its launch tower were moved to pad 39A, three and a half miles distant, on a three-thousand-ton tractor called the Crawler that traveled at a speed of one mile an hour.

The test was a year behind schedule. It would have been even further behind if not for a radically new approach.

The Germans at the Marshall Space Flight Center were deliberate and cautious in their work and in their testing. They had planned the initial live test of the Saturn to include the booster's first stage with dummy upper stages. If that was successful, the next flight test would consist of live first and second stages and a dummy third stage—and so it would continue, with just one major change between flight tests. At least ten tests were planned.

But NASA budget cuts would not permit that extravagance—and the Saturn V fabrication delays meant that such a plodding schedule would dash hopes of achieving a lunar landing before the end of the decade. Enter the forward-thinking George Mueller, NASA's associate administrator for manned spaceflight, hired by Jim Webb in September 1963 specifically to reorganize the agency's unwieldy management structure and improve efficiency. Early in 1964, after studying the Apollo schedule, he decided to implement a testing concept he had successfully used in his previous job at a civilian space technology company: instead of step-by-step trials, all the functional components of the Saturn V would be tested at the same time. After Mueller explained his thinking, most of NASA's top managers agreed—but not von Braun. When he first heard the idea and realized Mueller also wanted to include a live Apollo command-service module as a payload, he was resistant. If the rocket blew up before the first stage was jettisoned two minutes into the flight, how would they be able to ascertain which component was faulty? The risk of failure seemed much too high. But von Braun's Teutonic deliberateness eventually wilted, and he bowed to the logic, and the necessity, of the "all-up" approach. "It sounded reckless, but George Mueller's reasoning was impeccable," he wrote years later. All-up testing would become integral to the program.

By November 1967, it was clear that Congress had lost all of its early enthusiasm for spaceflight, manned or unmanned. The country was waging a costly, bloody, unpopular war in Vietnam. Young U.S. soldiers were dying every day, and few Americans understood what exactly they were dying for. And at home, things were chaotic—there were campus protests over the war that climaxed in a one-hundred-thousand-strong march on Washington, race riots in cities across America that included twenty-six dead in Newark and forty-three dead in Detroit, and multiple problems with the president's well-meaning but expensive and inefficiently administered Great Society programs. The Apollo 1 fire had been the last straw for Congress. The president refused to increase taxes, so he looked to the space program to provide relief. NASA's 1968 budget had been slashed by $420 million. Virtually every program except Apollo was put on hold, and that included planetary probes such as the ambitious Project Voyager, originally scheduled to launch in 1973 and land a life-detection capsule on Mars. (The successful Viking program would grow out of Project Voyager, and the Voyager name would be given to a new deep-space probe launched a decade later.)

A lot was riding on the successful flight of this new vehicle and its millions of functional parts, more than just fulfilling Kennedy's end-of-the-decade directive. If NASA failed to meet its self-imposed deadline, it would be a black eye for American political commitment, technological competence, and prestige. Thousands of NASA employees would be furloughed or fired, and the agency's centers would be put on standby status. Another disaster would not only ruin any chance of meeting Kennedy's deadline but probably finish Apollo for good.

At exactly 7:00 a.m., on November 9, 1967, ignition occurred, and seconds later the monster rocket began to slowly lift off pad 39A. The noise was louder and the sound pressure it generated was greater than anyone there had ever experienced. As it continued higher and then arced to the southeast, an eight-hundred-foot-long flame spewed from the first stage. Three orbits and almost nine hours later, the command module floated down into the Pacific near Midway Island. The mission was deemed a complete success. It had achieved every one of its goals, including a simulated lunar trajectory that had taken the spacecraft eleven thousand miles into space and resulted in a plunge into the atmosphere at 24,900 miles per hour. The big service propulsion system (SPS) engine in the rear of the service module did its job and propelled the spacecraft back down to Earth, and the guidance system navigated it through precise reentry maneuvers. The Saturn V worked—and so had Mueller's daring all-up approach. NASA might still be able to make Kennedy's deadline.

CHAPTER ELEVEN

PHOENIX AND EARTHRISE

To see the Earth as it truly is, small and blue and beautiful in that eternal silence where it floats, is to see ourselves as riders on the Earth together.

ARCHIBALD MACLEISH

THE FIRST UNMANNED TEST flight of the Saturn V had been a resounding success. The second, made on April 4, 1968, and intended to reaffirm the booster's reliability, was close to a disaster. The first stage had pogo problems similar to the Titan's—it vibrated badly. The second stage, complicated by engine malfunctions, pogoed even worse, and had there been any astronauts aboard, they might have been shaken into unconsciousness, or even to death. The third stage ignited, then shut down and failed to reignite. Only skillful work by Gene Kranz's Mission Control team prevented all threewas at t stages from busting into pieces; they managed to get the command-service module into orbit and through a successful reentry and splashdown.

Wernher von Braun and his team leaders observed the launch from the firing room a safe distance away. They returned to Huntsville with a massive diagnostic challenge ahead of them. Teams immediately began working around the clock, seven days a week, to pinpoint the problems and correct them. The diagnosis took twenty days. Von Braun reassured his NASA bosses that the issues were minor—for instance, one rocket engine quit because of a simple human mistake in wiring—and would be fixed.

NASA accepted the Apollo 7 spacecraft from North American Aviation in May 1968 and began strict checkout tests. When Schirra and his crew flew it into space, it would be a different, and more reliable, animal.

By the scheduled launch date in October, Schirra's crew had trained for more than a year, including spending almost six hundred hours together in the command-module simulator with its 725 manual controls arrayed before, beside, and above them. They were ready for the real thing.

A few weeks before the flight, Schirra announced that he would retire after it. By 1968, thirty-six thousand people worked directly for NASA, and it was fast becoming a bureaucracy; Schirra thought it was overly influenced by politics and special interests, and he wanted out. Besides, it seemed highly unlikely that he, or anyone, would get to command two missions in the Apollo program. There were too many good, experienced astronauts now, and too few seats, and Schirra felt that at the age of forty-five, he had about outlived his usefulness to the agency.

"Three Astronauts Ready to Face Challenge Three Others Died For" ran the *Miami Herald* headline before the launch. Von Braun was at the Cape, and in interviews he brought up the Russian threat, pointing out the "spectacular performance" of Zond 5 and rating Russia's chances of a circumlunar flight before Christmas as better than America's. Landing a man on the moon, he said, would be "a photo finish."

The USSR had launched Zond 5 less than three weeks earlier, on September 22. Its Proton booster, with roughly the same thrust as the Saturn IB, was capable of sending a cosmonaut or two around the moon on a free-return trajectory but not actually orbiting it, and its Soyuz spacecraft—much improved since Vladimir Komarov's death in April 1967—was large enough to transport them there. Soviet scientists had described the Zond program's mission as "testing new systems in distant regions of extraterrestrial space," and four previous unmanned Zonds had been sent out as probes to Mars, Venus, and the moon, but few in the West believed that was Zond 5's only purpose—especially when U.S. tracking ships had picked up a Russian voice issuing from the spacecraft reeling off instrument readings. It soon became clear that it was only a recording. The spacecraft splashed down safely in the Indian Ocean—the Soviets' first water landing—after seven days in space and a loop around the moon. A month later, the Soviets announced that its biological payload of flies, worms, plants, seeds, bacteria, and two tortoises had been recovered and was safe and sound, though the turtles had lost 10 percent of their body weight. Many observers suspected a manned Zond was next, perhaps by the end of the year. And even if the craft just made a loop around the moon, the USSR could claim that it had won the space race—after all, the Russians had never committed publicly to a lunar landing the way the Americans had. In any case, it would take at least some of the starch out of a U.S. landing.

And there was also the looming specter of Webb's Giant, the large Soviet booster Jim Webb continued to use as a threat against budget cuts in congressional hearings. He didn't provide specifics, since his information was top secret. The previous December, CIA reconnaissance satellites had produced the first photos of the rocket, and it was indeed the monster that analysts had expected. The CIA's National Intelligence Estimates predicted the earliest Soviet lunar landing in mid- to late 1969—about the same time as NASA's planned landing.

On September 16, 1968, a few weeks prior to the Apollo 7 launch, Webb made a visit to the White House to discuss the Russian space threat and NASA's future. President Johnson, dismayed by the widespread animosity toward him that the Vietnam War had engendered, had already decided he wouldn't run again in November. Webb had always planned to leave the agency when Johnson left office—he felt NASA needed to be depoliticized, and his departure would be a good way to start. Webb suggested his new deputy, Thomas Paine, as his successor and hoped he would be acceptable to the next president, whether he was Republican or Democrat. Johnson agreed on Paine, and to Webb's astonishment, he insisted on making the announcement about Webb's retirement, which would go into effect in three weeks, that afternoon. The president liked and admired Webb, who had served him loyally, and the reason for his decision to move up Webb's resignation date, four months ahead of the new administration, would remain a mystery.

Webb's early resignation took everyone by surprise, and many in NASA had their own opinions of why he was leaving. Some thought it was the fire and the extensive grilling he had received over it in the congressional hearings. Besides having to defend the agency ad nauseam, he'd been confronted with that damaging internal report he'd known nothing about. Some thought the public embarrassment had knocked the wind out of him...or maybe he just didn't want the responsibility of another tragedy. "He was never the same after the fire," remembered Bob Gilruth.

Thomas Paine would prove himself to be a more than capable administrator. But Jim Webb, with his Washington insider's knowledge and experience and his willingness to fight tooth and nail on the Hill for NASA, would be missed.

On the morning of October 11, 1968, four days after his official resignation, Webb breakfasted with Deke Slayton, the Apollo 7 crew, and a few other NASA officials. As the astronauts ate the traditional steak and eggs, technicians were pumping super-cold liquid oxygen and hydrogen into the rocket's tanks. At 11:03 a.m., Webb

“Missile Row” in 1964 on the Cape’s Merritt Island, looking north. NASA shared Cape Canaveral with the air force, and the first few (lower) sites visible were air force nuclear launchpads, followed by Mercury and Gemini sites, with midconstruction Apollo complexes in the far distance.

John Young (left) and Gus Grissom in the Gemini simulator.

Grissom and Young crewed the first manned Gemini flight on March 23, 1965. During three revolutions around the Earth, they demonstrated that, with its more powerful thruster jets, the new spacecraft could change orbit.

Voskhod 1, launched October 13, 1964, was the first spacecraft to carry more than one crewman. The cosmonauts are shown here after their twenty-four-hour mission (left to right): Vladimir Komarov, Boris Yegorov, and Konstantin Feoktistov. The craft was a Vostok cabin with two extra seats added and the ejection seat thrown out; there was so little room that the cosmonauts did not wear spacesuits.

Alexei Leonov, during the flight of Voskhod 2, became the first human to conduct an EVA (extravehicular activity) on March 18, 1965. After the twelve-minute space walk, his suit became overinflated and he couldn't fit through the tight air lock; he had to bleed off some air to get in. *(Courtesy Fédération Aéronautique International)*

On June 3, 1965, just seven weeks after Leonov's EVA, Ed White spent twenty minutes outside his Gemini 5 spacecraft, maneuvering with a gas-jet gun. He enjoyed it so much that he had to be coaxed back in.

On December 15, 1965, Gemini 7 and Gemini 6 (shown here in a photograph taken from the window of Gemini 7) performed the first space rendezvous of two spacecraft, another important step needed before NASA could land a man on the moon.

With the objective of studying the effects of long-duration spaceflight on humans, Gemini 7 orbited the Earth for fourteen days. When they splashed down safely on December 18, 1965, crewmen Jim Lovell (left) and Frank Borman were no worse for the wear save for some initial weakness after two weeks of inactivity.

Neil Armstrong, commander of the Gemini 8, had also flown the experimental X-15 (shown here) seven times.

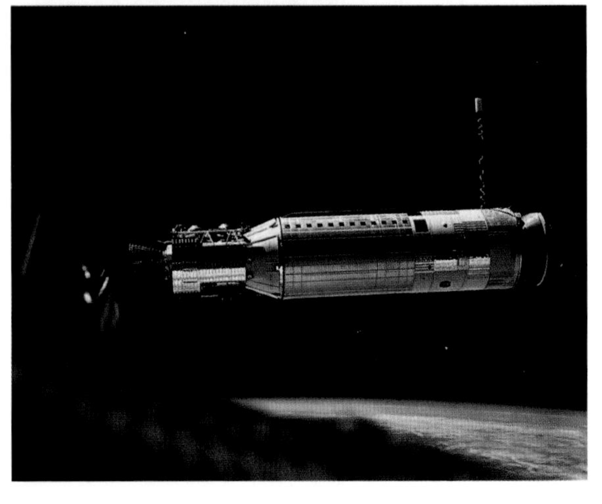

Gemini 8, crewed by Armstrong and Dave Scott, performed the first hard docking in space with this Agena target vehicle (shown here forty-five feet away) on March 16, 1966. Thirty minutes later, they went into a wild corkscrew spin from which they barely escaped.

Armstrong (inside capsule, on right) and Scott had to make an emergency landing in the Pacific. With the hatches open, the crew and three air force pararescuemen await the recovery ship.

Michael Collins found his July 20, 1966, EVA during Gemini 10 more difficult than he'd expected.

John Houbolt, an engineer at NASA's Langley Research Center, was the leading proponent of the lunar-orbit rendezvous mode of landing on the moon when almost everyone else doubted it could work. His tireless evangelizing for LOR eventually won over even his harshest critics.

Gemini 12, launched November 15, 1966, was the final mission of the program. Buzz Aldrin used the experience of previous EVAs, his underwater training in neutral buoyancy (shown here), and many hand- and footholds and rails to perform three successful EVAs for a total of five and a half hours.

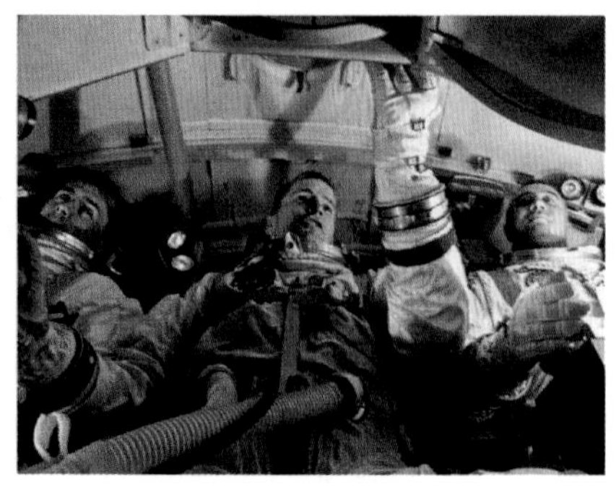

Gus Grissom (right) would once again command the first voyage of a new spacecraft. He and his crewmates, Ed White (center) and Roger Chaffee—shown here in the Apollo command module—knew their ship had problems but soldiered on.

All three astronauts died in a cabin fire during a plugs-out rehearsal on the launchpad on January 27, 1967, just three weeks before the mission's official launch date (spacecraft interior shown at left). The tragedy almost scuttled the Apollo program.

At the Senate committee hearings on the fire, several top NASA officials testified (left to right): Deputy administrator Dr. Robert Seamans; administrator Jim Webb; associate administrator for Manned Space Flight Dr. George Mueller; and Apollo program director Major General Samuel Phillips.

Three months after the Apollo 1 fire, the Soviet program endured its own tragedy. Vladimir Komarov was selected to crew the first manned test flight of the new Soyuz spacecraft, launched on April 23, 1967. He was killed during reentry when his parachutes and retro-rockets failed and the capsule crashed. Fellow Soviet officers viewed his remains at an open-casket funeral. *(Courtesy RIA Novosti)*

Von Braun with the five massive F-1 engines of a Saturn V first-stage test vehicle, which would generate 7.6 million pounds of thrust. (The fifth F-1 is out of sight at upper right.)

The Apollo 7 crew named to replace Grissom, White, and Chaffee on the first manned flight of the redesigned Apollo spacecraft: Walt Cunningham, Donn Eisele, and Commander Wally Schirra. Their successful October 1968 mission proved the overhauled Apollo command-service module was spaceworthy.

This U.S. Air Force September 19, 1968, reconnaissance satellite photo of the Baikonur cosmodrome confirmed the existence of the massive N-1 booster designed to take a cosmonaut to the moon.

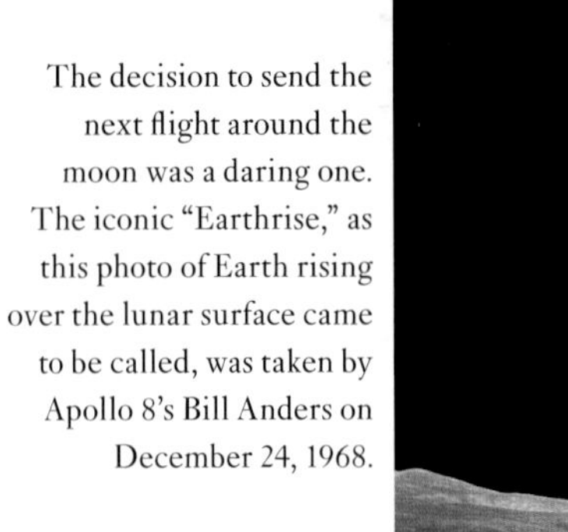

The decision to send the next flight around the moon was a daring one. The iconic "Earthrise," as this photo of Earth rising over the lunar surface came to be called, was taken by Apollo 8's Bill Anders on December 24, 1968.

Apollo 9, launched March 3, 1969, was an Earth-orbit mission whose primary objective was to test the lunar module in space. This photo taken from the LM shows the command-service module just prior to docking.

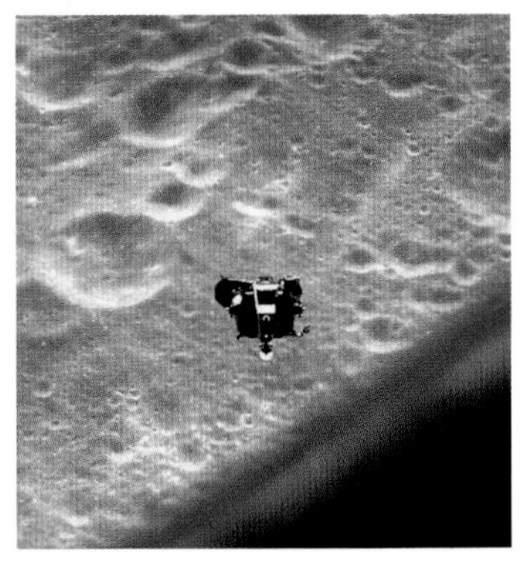

Apollo 10, launched May 18, 1969, was essentially a dress rehearsal for Apollo 11—everything except for the lunar landing. This photo taken from the command module shows the LM approaching for redocking.

would watch his first manned launch as Apollo 7 blasted into space twenty months after the scheduled liftoff of Grissom's Apollo 204. Pad leader Guenter Wendt was the last person the crew saw before they were sealed into their new, improved command module.

The flight went smoothly from a technological standpoint. The mission experienced only minor hardware problems, mostly glitches in the electrical system, and Apollo's every system was thoroughly tested during the eleven days the spacecraft was in Earth orbit. On this manned flight, unlike the previous sixteen, it was not the rocket thrusters or rendezvous radars that were balky but the crew members.

It started before the launch, when Schirra had objected to the number of biomedical sensors the doctors were attaching to him; Slayton overruled him. On the second day in orbit, Schirra developed a severe head cold. The nature of weightlessness made it difficult for him to clear his sinuses—drainage, after all, depends on gravity—and decongestant tablets didn't seem to help much. The cold, his unhappiness over NASA's decision to placate the American taxpayers with regular live transmissions from the crew, and the plethora of experiments planned, some of them dreamed up during the flight—all these contributed to his irascibility. A naval officer, Schirra, began exhibiting more of his uncooperative "I'm captain of this ship" attitude. Whether it was rejecting suggested experiments and insulting their originators ("I wish you would find out the idiot's name who thought up *this* test. I want to talk to him personally when I get back down") or just complaining ("I've had it up to here today"), Schirra set a surly tone that never completely disappeared during the flight. He seemed oblivious to the fact that the information on his ship's screens and gauges was just a fraction of the amount of data Mission Control had, and important decisions depended on it.

On the second day out, Schirra refused to go through with a

scheduled TV transmission, meant to be the first live broadcast from space, "without any further discussion"—even after Slayton, acting as CapCom, pleaded with him just to turn the TV switch on. When he and his crew finally went on air in glorious black-and-white the next day and then six more times during the eleven-day flight, Schirra turned the charm back on, at least temporarily. *From the lovely Apollo room, high atop everything,* read a card opening the show, and all three crew members became ebullient tour guides during the seven- to nine-minute broadcasts. "The Wally, Walt, and Donn Show" was a big hit with viewers, and after the astronauts returned, each of them was presented with a special Emmy Award.

It played well on TV, but Chris Kraft gave the mission a different name: the Wally Schirra Bitch Circus. Some of the flight controllers half joked about purposely landing the spacecraft in the middle of a hurricane swirling near Hawaii. Years later, flight director Gene Kranz—who had claimed Schirra was one of his favorite astronauts to work with—reported that there wasn't anyone in Mission Control who didn't want to "pull the plug on Wally Schirra and leave him to circle the Earth on his own there without communications for a while."

The normally easygoing Eisele acquired not only Schirra's cold, though a less severe one, but also his attitude. On an open microphone, he criticized a test, saying that he wanted "to talk to the man, or whoever he was, that thought up that little gem." That man turned out to be flight director Glynn Lunney, who was not amused. Eisele also criticized at length a Mission Control error with the onboard computer that forced a system restart—an understandable response but one that could have been handled without the ad hominem remark "Somebody down there screwed up royally."

Cunningham, a pugnacious ex-Marine fighter pilot turned scientist, never caught their colds and wasn't as uncooperative. Schirra claimed they all had colds, either to bolster the strength of his complaints to Mission Control or to refute the idea that two rookies were doing better in space than the old pro. He also accused an argumentative Eisele of threatening mutiny. Despite using ten boxes of tissues and downing plenty of Actifed tablets, Schirra never completely shook his cold symptoms during the flight.

The clashes with Houston continued right up to reentry, when Schirra flouted an important mission rule. He insisted that his crew would not wear their helmets during the dangerous plunge through the Earth's atmosphere, claiming that the pressure buildup might cause their eardrums to rupture in the pressurized suits. The flight surgeon disagreed, and Mission Control, worried about a life-threatening pressure leak, insisted they wear the helmets. Slayton again got on the mike to talk to Schirra, but Wally was adamant. To avoid an unpleasant public confrontation, Flight Control allowed them to leave their helmets off. Reentry occurred with neither burst eardrums nor pressure leaks.

After the mission, the three astronauts were all tarred with the same brush and considered hard to get along with. Schirra was given a good talking-to by Slayton, but he was retiring, so he didn't care, and he made no attempt to keep his crewmates from going down with him when Chris Kraft swore that neither Cunningham nor Eisele would fly again. Despite Deke's protestations, several apologies by the two crew members to Mission Control, and a personal appeal to Kraft by Cunningham, they never did. (Kraft heard Cunningham out and acknowledged that he should have another chance, but the astronaut left the program before that could happen.) Slayton was in charge of the crew selections, but he was hamstrung. "I wasn't going to put anybody on a crew that Kraft's people wouldn't work with. Not when I had other guys," he said. The blackballing sent a message to the rest of the astronauts, and it was one they got: Individualism or, put another way, disrespect to Mission Control would not be tolerated. This was a team effort.

Despite the human problems and a few minor electrical issues, NASA called the mission "101 percent successful." Every system had been thoroughly tested, and any faults found were minor and could be easily fixed. The service module's SPS engine, which would play an important part in a trip to the moon, had been test-fired eight times. Schirra's suggestion of the *Phoenix* as the name of the spacecraft might have been overly dramatic—and it was quickly shot down by NASA—but the textbook flight had allayed many fears both inside and outside NASA. Eleven days was more than enough time to complete a lunar-landing mission, and after 163 orbits and more than four million miles, this had been the shakedown cruise to end them all. An unsuccessful flight—or, worse, one with a tragic ending—might have prompted more calls to cancel the program. It almost certainly would have dashed any hopes of landing on the moon by the end of the decade. But faith had been restored—in Apollo, in NASA, and in the ability of man and machine to fly to the moon.

The Apollo timetable called for four more progressively complex flights before an attempted landing by Apollo 12 late in 1969—if, that is, the problematic LM was ever finished. But the Apollo 7 mission had gone so well that one of NASA's administrators decided on a big change in the schedule. The seed of the idea had been planted two months earlier.

One Sunday early in August 1968, two months before the Apollo 7 launch, Frank Borman was at the North American Rockwell plant in Downey involved in some testing when he got a call from Deke Slayton summoning him back to Houston. After a quick T-38 flight to Ellington AFB and a ten-minute drive to MSC, he arrived at Slayton's office.

Jim McDivitt's mission was next in the schedule, so Deke had told him first: The next manned Apollo flight was going to the moon. McDivitt wasn't interested—he and his crew would rather stick with the flight they'd been training for, a test of the command-service module and LM in low Earth orbit, with an EVA and docking thrown in. Like most test pilots, they wanted to be the first to fly a craft, and they preferred the challenges offered by their own mission. The moon journey provided fewer of those—"You were just a passenger," McDivitt's LM pilot, Rusty Schweickart, said later of the Apollo 8 crew. "You weren't really *doing* anything." That was fine with Slayton; while McDivitt's crew had been working extensively with the LM, Borman's crew had barely started their LM training.

Now, Slayton told Borman—a "tightly wound little sumbitch," in the words of another astronaut—the same thing: The next flight was going to the moon. Instead of flying the mission after McDivitt's—blasting into high Earth orbit, around four thousand miles up, to test the command-service module and LM in deep space—did he want to go around the moon instead? Not only would this enable the agency to leapfrog a testing step, but it would scoop a possible similar Soviet flight. NASA higher-ups had just received word from the CIA that the Soviets were planning a lunar flyby before the year ended. That seemed to confirm what was already suspected—the previous fall, two unmanned Russian satellites had linked up, and two more had docked just a few months ago, in April. The rumors of a monster rocket more powerful than the Saturn V with a total thrust estimated in the fourteen-million-pound class—Webb's Giant—appeared to be true. Though the Soviets denied it publicly, it was clear that they had the moon in their sights. It looked doubtful that Webb's Giant would be ready to go to the moon very soon, but the Proton rocket, which had recently launched an unmanned Zond around the moon, could probably do the same thing while carrying a cosmonaut or two.

There were two other important reasons for sending Apollo 8 around the moon. First, the flight would enable the crew to photograph potential landing sites for the lunar-landing mission. That information would prove invaluable. Second, the flight would add important data about *mascons*, for "mass concentrations"—areas on the moon, always in the *mares* (the large, dark plains mistaken by early astronomers for actual oceans), where regions of denser material under the surface cause an increase in gravitational attraction. The resulting orbit perturbations had first been noticed in the Lunar Orbiter missions sent around the moon in 1966 and 1967. The mascons' origins, and almost everything else about them, were unknown—some scientists postulated meteorites that slammed into a thick primordial mush billions of years ago and remained intact; others believed it was a long sedimentary process. More information was needed before men on precisely calculated trajectories tried to land, since a slight change could mean a huge difference to a LM with a limited fuel supply. The moon still held many mysteries, and some of them had to be solved before a lunar landing could be deemed safe.

In 1961, when Kennedy had issued his challenge, no one knew exactly what the surface of the moon was like.

Galileo Galilei, the early-seventeenth-century Italian polymath, was not the inventor of the telescope—at least one man, a Dutchman named Hans Lippershey, had applied for a patent for one in 1608. But the more powerful telescope Galileo constructed the next year was the first to be aimed skyward. One of his first objects of study was the moon, and his detailed, accurate ink renderings of its surface—its mountains, craters, and seas—would not be significantly improved until two centuries later; even the most powerful terrestrial telescopes improved only topography, not composition. At the time of Kennedy's moon speech, no one knew for sure whether its surface was hard or soft, rock or dust, thick or thin. This information, of course, was essential to have before engineers,

Russian or American, could plan a lunar landing. Toward that end, both countries soon initiated unmanned programs specifically designed to analyze the moon's surface.

The Russians began sending unmanned probes toward the moon in 1959; this was their Luna program. In January of that year, Luna 1 was intended to hit the moon, but it missed and became the first man-made object to orbit the sun. In September 1959, Luna 2 crashed onto the lunar surface, the first man-made object to hit the moon. Luna 3 circled the moon in October 1959 and sent back the first photos of its far side, which can never be seen from Earth. The next five Luna missions attempted to achieve a soft landing using retro-rockets, but each one either missed the moon or unintentionally crashed into it. Finally, in February 1966, Luna 9 achieved a soft landing, the first on another heavenly body, and returned several panoramic photos—the first close-up images of the lunar surface. Also that year, three more Lunas went into orbit around the moon; two of them gathered scientific data, and one returned high-resolution photos. In December 1966, Luna 13 soft-landed and transmitted more panoramas.

The first American probes launched toward the moon were part of the Ranger program, which began in 1959; they were designed to transmit photographs of the moon's surface until they hit it and were destroyed. The first six Ranger flights failed, but Ranger 7, in July 1964, and two follow-ups in 1965 were all successful. They sent a total of 17,225 images of the lunar surface, sharp, up-close photos with details of its pockmarked seas, craters, and plains, magnified about a thousand times more than ever before. The solar-powered Ranger spacecraft were the first of a new generation of sophisticated probes that would explore not only the moon but the planets beyond it. Most experts, after analyzing the Ranger photographs, were of the opinion that the moon's surface could support a spacecraft landing, though no one was absolutely sure.

The next U.S. program provided a more definite answer. Surveyor was designed to make a soft landing, take close-up photos, and analyze the hardness and composition of the lunar soil. On June 2, 1966, the first Surveyor—more than twice as large as the two-foot-tall Luna lander—cut its three small rockets off at thirteen feet above the moon, fell the remaining distance, bounced once on its tripod landing gear, and settled safely onto a demonstrably hard surface with only a thin covering of dust. That eased a lot of fears at NASA; for starters, it gave the engineers a better idea of what kind of landing gear they'd need. Four more Surveyor landings over the next twenty months confirmed a solid basaltic crust, one that would support a lunar lander…at least in the areas examined by Surveyor. And the moon's color, to almost no one's surprise, was various shades of gray.

One more item was necessary—a photo map of the surface so a suitable landing site could be selected. Getting that map was the job of the Lunar Orbiter program, which from August 1966 through November 1967 sent five unmanned probes into orbit around the moon. All five worked perfectly, and they also provided information about radiation levels near and on the moon. Their photos, with a resolution of two hundred feet or better and sometimes as close as three feet, enabled NASA to map 99 percent of the moon, both the far and near sides, and Apollo 8 photographs would, it was hoped, improve on even these. By the time NASA was ready to send a man to the moon, sufficient mapping would be available.

The change in plans for the Apollo 8 mission was due largely to George Low, the newly hired Apollo program manager. While looking for ways to keep the Apollo momentum going and maybe counter yet another Soviet space spectacular, he had come up with a plan.

The original idea was to test the combined operations of all modules—command, service, and lunar—in Earth orbit. But since

Grumman's LM still wasn't ready, the mission would essentially be a repeat of Apollo 7's journey. Instead of an Earth-orbital flight, how about making the next flight, Apollo 8, circumlunar? Sending a crew to the moon without the LM would provide Mission Control with some much-needed experience in working a mission's navigation, tracking, and communications in deep space. It would also allow NASA to skip a step in the Apollo flight sequence, which would make an end-of-the-decade landing more likely.

One morning, Low talked to Kraft, Gilruth, and Slayton about it. "It would ace the Russians," he told them, "and take a lot of pressure off Apollo." They responded enthusiastically. In fact, they decided to fly up to Huntsville that afternoon to run it by von Braun. He too was excited at the thought of one of his Saturn Vs going to the moon and promised his full support. His rocket would be ready, despite the troubles that had cropped up on its second test flight. Jim Webb, however, was harder to convince and insisted on a successful Apollo 7 flight first—that was still to come. But preparations had to begin immediately.

The flight dynamics branch had seen Low's bet and raised it. When Kraft asked his key people what they thought of the idea, they said they wanted a day to mull it over. When Kraft arrived at work the next day, they were in his office waiting. They had decided it was a great idea—but they didn't want to circle the moon just once; they wanted to go into orbit around it. "An order-of-magnitude difference of risk right there," remembered Kraft. If they could do that and use the same orbit parameters they were planning on using for the lunar landing, well—they'd have them ready. Kraft considered that wrinkle and finally presented it to Low and the other administrators. At first, they were shocked—Webb especially, since he was still raw from the Apollo 204 fire and its aftermath—but they came around. Eventually it was agreed: Apollo 8 would orbit the moon.

Now Slayton wanted to know: Was Borman up for a round-the-moon mission, and could he and his crew be ready two or three months sooner? If Borman decided to stay with the mission he and his crew had been training for—an Earth-orbital flight—his backup crew of Neil Armstrong, Buzz Aldrin, and Fred Haise would be given the assignment. It took Borman less than a second to say yes. He and his crew would be ready.

But the plan was full of risks. So many things had to go right, things that had never been attempted before outside of simulations on the friendly surface of Earth. Borman and his crewmates—Jim Lovell, who had replaced Mike Collins after he underwent spinal surgery that summer, and Bill Anders, who would sorely miss the LM he had become quite attached to during training—estimated that the chances of a fully successful flight were only one in three: "A one-third chance of success, a one-third chance of a survivable accident and a one-third chance of not coming back," Anders remembered thinking at the time.

Four days after the Apollo 7 splashdown, forty-seven-year-old cosmonaut Georgy Beregovoy, a former test pilot and World War II ace, flew the first successful mission of the revised Soyuz spacecraft, designed to carry up to three men. His Soyuz 3 orbited the Earth for four days and achieved Russia's first space rendezvous, with Soyuz 2, and, just as important, he made a safe reentry and landing. (The information that he had failed at two docking attempts was withheld and would not be known in the West for years.) Like the Americans, the Soviets had a space program that was up and running again.

On December 9, 1968, almost six weeks after Richard Nixon was elected the nation's next president, Lyndon Johnson held a formal dinner and ceremony at the White House. It would be his final chance to offer tribute to his cherished space program and to the man who had led it since 1961. In the State Dining Room, one hundred and forty-three people gathered, among them twenty-three Apollo astronauts and their wives, Jim Webb, Wernher von Braun, Thomas Paine, Charles Lindbergh, and many other distinguished guests, including Vice President Humphrey and members of Congress. Johnson awarded Webb the Presidential Medal of Freedom—"The highest civilian honor the president can bestow on any individual whose work advances a great cause," he said—and praised him generously. He also saluted the Apollo 8 crew, wished them Godspeed, and said, "I hope none of you take cold." The president retired early, but Vice President Humphrey kept the Marine band playing long after midnight as the guests drank and danced. When the party finally wound down, a tipsy Gene Cernan capped off the night by sliding down a long brass banister on the way out.

The first time that the Apollo spacecraft and its massive and massively complex Saturn V booster left Earth's gravitational field and ventured into the black void of space, it would carry three humans. It would disappear behind the far side of the moon and remain out of communication for thirty-six minutes, and there, the service module's SPS engine—which had no backup—would have to perform a perfectly timed retro-brake burn to allow the spacecraft to fall into orbit just sixty-nine miles above the moon's surface. After it made ten orbits, another precisely timed burn would alter its trajectory and send Apollo 8 back to Earth. Then it would shed its service module and enter the atmosphere at Mach 32, nearly 25,000 miles per hour. It would have to do this at just the right reentry angle of attack and upside down, with the blunt end of the gumdrop-shaped command module slowing its speed and the heat shield's ablative material absorbing much of the five-thousand-degree heat and flaking off. If it entered at too steep an angle, g-forces would tear the spacecraft apart, and the friction-created heat would incinerate everything and everyone inside. Too shallow an angle, and it would bounce off the atmosphere like a rock skipping over a lake and resume orbit without the power necessary to return; its occupants would eventually run out of oxygen. Their corpses would continue to circle the Earth in a kind of space crypt until the craft's orbit deteriorated and it plunged down into the atmosphere, unintentionally cremating the astronauts' remains.

With all that in mind, the three crewmates—Borman, the genial Lovell, and Anders, who was, like Borman, an intensely focused, serious-minded type—practiced the procedures over and over in simulators, running through some eight hundred emergency situations until they could do them all in their sleep. Navigation and trajectory throughout the 579,606-mile trip (that included the trip to the moon, the orbits, and the trip back) would have to be pinpoint accurate, and, since it involved three rapidly moving objects—the Earth, the moon, and the spacecraft—MIT's computers would need to calculate various in-course velocity corrections. In short, there was little room for error.

At 7:51 a.m. on December 21, 1968, Apollo 8 lifted off from launch complex 39A perfectly. After two orbits around the Earth, Michael Collins, acting as CapCom, told the crew, "You are go for TLI," and the third stage's J-2 engine—which had failed to restart in its most recent unmanned test flight—ignited and accelerated the craft to 24,259 miles per hour, enough to achieve translunar injection. After jettisoning the third stage, the command-service module was on its way to the moon, and its three occupants had already traveled farther away from the planet than any other human. While pulling away from Earth's gravity, the spacecraft slowed to 2,200 miles per hour, then it accelerated as it left Earth's sphere of influence and the moon's gravity began to take hold. Two and a half days after launch, the astronauts turned the craft's hind end forward, and disappeared behind the moon and out of radio contact. "We'll see you on the other side," Lovell said just before they lost communication.

In Houston, Mission Control conversations became hushed as most of those present stared at their consoles, then at the large screens, then at the two clocks, smoking and fidgeting as the seconds ticked away. Bob Gilruth and Chris Kraft sat next to each other in the back row, waiting with everyone else for confirmation from Apollo 8 that the burn had occurred at exactly the right moment and lasted exactly four minutes and seven seconds to decelerate the craft to 3,700 miles per hour, the right speed to drop them into lunar orbit without crashing them into the surface or hurling them into deep space. After they'd waited for thirty-six excruciating minutes, Lovell's garbled voice was picked up: "Go ahead, Houston. Apollo 8." One flight controller jumped up without thinking and yelled, "The Russians suck!" The three astronauts were in orbit around the moon, at their lowest point skimming 69.5 miles above the surface. All the controllers stood and cheered, then got back to work. Gilruth wiped tears from his eyes. Kraft reached over and squeezed his arm, and they shook hands firmly.

It was Christmas Eve. When they weren't making frequent navigation checks and firing their attitude-control thrusters to keep the windows facing the right way, the crew spent much of the first eight orbits taking hundreds of photographs, charting potential landing sites, and describing what they were seeing like a boatful of tourists circling Manhattan Island. "Oh my God, look at that picture over there. There's the Earth coming up," Bill Anders said a few seconds before he took a color photo of the scene. "Wow, is that pretty!" On the ninth and next-to-last revolution, as millions of Americans watched the first televised photos of the moon, the three astronauts took turns reading from the Bible, the first ten verses from the book of Genesis: "'In the beginning, God created the heaven and the earth.'"

On the next revolution, when they were again behind the moon, the engine fired for three minutes and twenty-three seconds to boost the spacecraft out of lunar orbit and into a perfect trajectory toward home. The reentry went smoothly, and just before dawn on December 27, Apollo 8 splashed down safely in the Pacific, six hundred miles northwest of Christmas Island and just three miles from their recovery ship, the carrier USS *Yorktown.* This flight, like the previous one, had gone almost perfectly save for some minor anomalies that were more irritating than life-threatening. Mission Control had proven it could handle a spacecraft almost a quarter of a million miles away, even when the vehicle disappeared behind the moon and contact was lost. Just as important, the spacecraft had flown there almost completely autonomously and extremely accurately—a tribute to Apollo's computer-controlled navigation, guidance, and tracking systems.

The photo Anders snapped of a luminous blue-and-white Earth rising over the lunar horizon would become one of the most reproduced and best-known photographs of the twentieth century. "Earthrise" and other photos taken by the crew—the first color images to show humans their home as a complete globe—would change the way people felt about their planet. "We flew all the way to the Moon," said Anders, "and the most important thing we discovered was the Earth."

By any standard, 1968 had been a rough year, one of upheaval and pain in the United States and elsewhere, and there'd been plenty to distract Americans from the space program. Civil rights leader Martin Luther King Jr. was assassinated in April; Robert F. Kennedy, in the middle of what might have been a successful presidential campaign, was murdered in June. Rioting, looting, arson, and murder in the nation's streets continued. In Vietnam, the Tet Offensive in the early part of the year resulted in thousands of American deaths and the public's realization that the war it had been told was being won wasn't. Support for the war eroded even more, as evidenced by large protests on college campuses across the nation. In Biafra, a new African republic, a million people died of starvation. The Soviet Union and its Warsaw Pact allies invaded Czechoslovakia to halt liberal reforms in Prague. Red China exploded a thermonuclear device, and in the Middle East, Arabs and Israelis were about to engage in another war. Global peace had never seemed so unlikely.

Yet, for a brief time at the year's end, Apollo 8 gave people a reason to look up to the heavens and hope. The mission was hailed around the world. Ten Soviet cosmonauts sent a congratulatory cable, and *Pravda* called it "an outstanding achievement." Even Radio Havana, which had never broadcast anything from the Voice of America, the U.S. external radio and TV service, replayed a VOA description of the flight. Much of the American public's apathy toward spaceflight disappeared, in large part because of the live TV transmissions, which allowed the nation to share in the adventure. President Johnson, nearing the end of his presidency and more than anyone the driving force behind the American space program, felt vindicated. *Time* magazine named Borman, Lovell, and Anders its Men of the Year. One anonymous telegram received by Borman summed it up: "To the crew of Apollo 8. Thank you. You saved 1968."

CHAPTER TWELVE

"AMIABLE STRANGERS"

Both Neil and Buzz had more than the Right Stuff—they were magicians of confidence.

ALBERT JACKSON, LM SIMULATION INSTRUCTOR

YOU DIDN'T HAVE TO like the men you crewed with; that had been the case for eons, whether the craft was a boat or a wagon or a tank or a spaceship. But you'd better be able to trust them with your life in a dangerous situation, and that all-important element of trust usually increased if there was friendship involved.

For the better part of a year—and in addition to their duties as backup to another mission—members of an Apollo crew spent countless hours together preparing and training for their flight, their days often stretching into fourteen-hour marathons that barely gave them enough time to eat and sleep. Their families were the poorer for it, since an astronaut was rarely home for more than one weekend night a week. And since the men were on the road so much—at the various contractors that were spread across the country designing and fabricating their spacecraft and spacesuits and, over the last few months of training, at the Cape—they usually spent their few off-hours together also. If they weren't close before they were crewmates, they became close after.

Some crews took it to the extreme. The Apollo 12 crew, close friends and navy pilots, were Pete Conrad, spacewalker Dick

Gordon, and rookie Alan Bean. As a show of solidarity, all drove matching Corvettes, gold '69 coupes with the initials *CDR, CMP,* or *LMP* (commander, command module pilot, and lunar module pilot) painted on the front fender in a red, white, and blue panel. (Even the Apollo 12 wives occasionally dressed alike, once parading out of the Conrads' house in identical off-white pantsuits with red, white, and blue sashes.) The Apollo 15 crew also owned matching Corvettes—one red, one white, one blue. Each car had red, white, and blue stripes down its center. Their choices were made easier by the fact that the sports cars were leased from an admiring local dealer for one dollar annually—and each man got a new model every year. Apollo 10 was another Three Musketeers crew: "We were old friends and had total confidence in each other," observed Gene Cernan.

Not the crew of Apollo 11. When the men didn't have to spend time together, they went their separate ways. They were, as Mike Collins described them later, "amiable strangers." As much as they had in common—they were all superb test or fighter pilots, with engineering skills to match—each was his own man. And none of them was what you'd call "one of the boys." Except for the occasional Astronaut Wives Club get-togethers, their spouses weren't close either.

Six years after he'd become an astronaut, thirty-eight-year-old Neil Armstrong still looked younger than his age. In spite of his soft-faced, youthful appearance, Armstrong had been highly respected from the beginning by both NASA officials and other astronauts. In meetings, he was rarely one of the first to talk; when he did, he usually began with "In my view." When people heard that, they all paid extra attention. And he was dependable: "If you had to ask somebody and then count on it," said Alan Bean, "he would be the guy." He wasn't a gung-ho leader like Frank Borman, and he never raised his voice. He led by example and inspired his crewmates in other ways. No one in the astronaut corps was surprised when Deke selected him to command an Apollo mission.

This mission commander, curiously, had little actual command experience. When he'd reported for flight training in 1949, he'd chosen single-engine fighters, not multi-engine bombers and their crews, because, as he'd told his mother, "I didn't want to be responsible for anyone else." As a young fighter pilot during the Korean War and then a test pilot over the next decade, he'd had virtually no opportunity to command anyone. NASA was a civilian agency, and no one was going to be court-martialed for disobedience, but crews were still expected to observe the command structure, and because all the astronauts were military or ex-military, it was second nature for them. Starting with his selection as backup commander on Gemini 5, Armstrong had begun to develop a low-key, unforced leadership style that inspired respect and cooperation from his crews and approval from his superiors.

In mid-1968, the moon landing was just in the planning phases. The LM hadn't been tested in space yet, and there were those at NASA who thought Apollo 10 should give it a try if Apollo 9 went well. Nobody at the agency—not Bob Gilruth, not Chris Kraft, not Deke Slayton—knew for sure which mission would be the one to land on the moon. If Gus Grissom had still been alive, all agreed that he would have been the one to command it. But he wasn't, so for potential commanders, Deke favored McDivitt, Borman, Stafford, Armstrong, and Conrad, and those five, in that order, would helm Apollo 8 through Apollo 12.

"You're it."

That's what Slayton said to Armstrong, Aldrin, and Collins on January 6, 1969, after he called them into his office at MSC and closed the door. He told them they would constitute the crew of Apollo 11, scheduled to make the first lunar landing in July. Aldrin and Collins were both surprised, especially Collins. Aldrin had been on the Apollo 8 backup crew; he knew Slayton's usual selection process, and half-expected it. Armstrong had known for weeks.

Two weeks before, on the afternoon of December 23, Neil had been in Mission Control while Apollo 8 was halfway through its translunar coast to the moon. As backup commander for the flight, he'd been at the Cape for the launch and had flown back to Houston later that day. He'd spent most of his time since then in Mission Control, monitoring the progress of the flight, fielding questions from the flight controllers about the crew. Then Slayton had led him to a back room to discuss his next assignment.

As he usually did, Slayton got right to the point. He told Neil he was going to command Apollo 11. Was Neil okay with Mike Collins and Buzz Aldrin as his crew? Deke was a fan of the hardworking Collins, and he wanted to get him back in the mix on the next possible crew. Collins already had extensive training as a command-module pilot and had served in that role on Apollo 8 until a spine injury had taken him off active status and threatened to end his astronaut career.

Collins had first noticed the problems in the summer of 1968, during handball games. Many of the astronauts played the sport for an intense hour-long workout. Before the January 1967 fire, Grissom had been the best handball player of the Mercury Seven—though there was a rumor that he'd let Shepard win once, just to keep him happy. But since his selection in October 1963 with the Fourteen group, Collins, easygoing, nonaggressive Mike Collins, had been the acknowledged champ of the astronaut corps. Ed White, the natural athlete, had been his toughest competitor before the fire, and there were other good players—Al Worden, Walt Cunningham, Rusty Schweickart—but no one had managed to dethrone him. (Some suggested that his being left-handed gave Collins an advantage.)

During a game one day that summer, Collins noticed weakness in his legs. Then other physical difficulties arose, and soon he felt the weakness spreading up his left side. Finally, in July, he told the NASA flight surgeon. A visit with a specialist revealed a herniated disc between his fifth and sixth cervical vertebrae that was pressing on his spinal cord. Surgery was the only answer. He underwent the operation nine days later, but it meant months of recuperation even though the operation was successful. The tight Apollo schedule waited for no one, and Jim Lovell took his place on Apollo 8.

By November, Collins was back on full flight status, and Deke wanted him on Apollo 11. Deke had a hard-and-fast rule that the command-module pilot had to be a spaceflight veteran—he would, after all, be flying an extremely complicated spacecraft all alone around the moon for more than a day—and Collins would be a perfect fit. Fred Haise was currently on Armstrong's crew as the backup LM pilot, but he hadn't flown yet, and Deke had some reservations, since this might be the first landing. Anyway, Collins had seniority, so Deke planned to bump Haise off Apollo 11 to a later flight if Armstrong agreed to the switch. He did.

Aldrin had been the other member of Armstrong's Apollo 8 backup crew, though as command-module pilot. He'd move to LM pilot if Armstrong accepted him. Then Deke told Neil that he knew Aldrin wasn't that easy to work with, and he said that if Armstrong wanted, he could have the reliable and affable Jim Lovell, at that moment on his way to the moon, instead of Buzz.

Armstrong had been working with Aldrin for several months, and he hadn't had a problem with him. Armstrong's quiet, non-confrontational manner didn't require much social interaction and buddy-buddy camaraderie—not Aldrin's strong suit—and that probably helped. But Neil understood what Slayton meant. He

asked for some time to think about it. The n
he told Deke that he'd rather keep Buzz. B
a command of his own. Collins had been tra
mand module and knew it well. The luna
number-three position in an Apollo crew,
ally do any piloting; the commander woul
instead act more as a systems engineer, monitoring
Buzz was perfectly suited for that job.

It was settled; the Apollo 11 crew would be Armstrong, Collins, and Aldrin, with Lovell, Anders, and Haise backing them up. Borman had decided to retire from astronaut duty rather than endure the exhausting training for, first, a backup crew, then prime. Some claimed that Susan Borman had put her foot down. She knew how dangerous a trip to the moon was—before her husband's Apollo 8 flight, Chris Kraft had terrified her by describing it as "the riskiest one to date." That ordeal had been enough, and she didn't want him risking his life once more and forcing her and their children to go through that emotional wringer again.

After the aborted flight of Gemini 8 in March 1966, Armstrong had been named to the backup crew for Gemini 11—his third Gemini assignment. When that mission was over, he switched to Apollo. Years earlier, as a test pilot, he had been involved in the early developmental work on the lunar-landing research vehicle (LLRV), an odd-looking apparatus built by Bell Aerosystems that was designed to mimic, as closely as possible, a landing on the moon and its one-sixth gravity. Once the LM began taking on its final shape, NASA decided to use the LLRV and an improved version, the lunar-landing training vehicle (LLTV), as trainers for the planned lunar landing. Armstrong was heavily involved in the transformation, making the LLTV as close as possible to the LM itself, at least in its basic handling and control features. By early 1967, both

-two LLRVs and three LLTVs—had been shipped to ...on and installed at Ellington Field, near the Manned Space-...ft Center, and were in use by potential moon-landing astronauts. Though a helicopter didn't provide a very good simulation, there were some similarities between helicopters and the trainers, mostly in trajectory and visual fields, and each astronaut attended a crash course on helicopters and spent time in a ground simulator before taking a flight in one of the trainers.

Armstrong had been flying the LLRV since March 27, 1967, soon after it was shipped to Houston. Over the next year and a half, he would make many flights in it. Nicknamed the "flying bedstead" for its ungainly appearance—basically a conglomeration of aluminum-alloy struts with a pilot's seat, a jet engine underneath, and a fuel tank on each side of the vehicle—the machine was extremely dangerous, since the pilot had to take it up to five hundred feet for its six-minute flight. The engine would provide enough steady lift to simulate one-sixth gravity, and small side-mounted thrusters powered by hydrogen peroxide approximated the LM's descent-control characteristics. If the main engine or the thrusters failed, the LLRV had no wings to provide lift, so it could not glide to a landing. In an emergency, the rocket-powered seat would eject the pilot and allow him to parachute to the ground. But there was little room for error, and the vehicle could turn deadly in many ways.

On the windy afternoon of May 6, 1968, Armstrong was at Ellington piloting an LLRV for the twenty-first time. In the final one hundred and fifty feet of a descent, his thruster system malfunctioned, leaving him unable to control the machine's attitude. It began to sway from side to side—"Like a kid's balloon at a party, where they blow it up and let it go," remembered Anders, who had just finished his turn on the machine—then turned and plunged toward the pavement below. Over the loudspeaker Armstrong said,

"Got to leave the vehicle," and ejected at about fifty feet, a second before the LLRV crashed to the ground and exploded in a fireball. The ejection seat blasted him a few hundred feet into the air at about two hundred miles an hour and with a force of fifteen g's. After automatic separation from the seat and deployment of the parachute, Armstrong floated safely down into the tall grass bordering the area, the brisk wind wafting him away from the flames. He landed, stood up, and walked away, his only injury a bloody tongue where he'd bit it and a large bruise on the base of his buttocks. After a debriefing, Armstrong got in his car and drove over to his office. He had just escaped death, and he was disappointed that he'd lost an expensive machine, but he had work to do.

An investigation concluded that a badly designed thruster system had allowed propellant to leak out and cause the thrusters to shut down. The high winds had also required Armstrong to use more fuel than usual. Despite protests from Bob Gilruth and Chris Kraft, the LLRV and LLTV would remain a vital part of training for the LM. Armstrong, who would fly the contraptions more than two dozen times, would later call them "absolutely essential" to the success of the mission.

No other astronaut of the Apollo era seems to have experienced as much good luck in his flight assignments as Buzz Aldrin. He had gained a Gemini mission because of the T-38 deaths of Charlie Bassett and Elliot See; their Gemini 9 spots were immediately filled by Tom Stafford and Gene Cernan, and that had allowed Aldrin and Lovell to move up from the backup crew of Gemini 10 to Gemini 9 and, eventually, to fly the last mission in the series, Gemini 12.

But Aldrin came close to losing his seat on Gemini 12 not once but twice. During Gemini 9, he had stunned Bob Gilruth in Mission Control by loudly and insistently suggesting a dangerous solution to an in-flight problem—it involved wire cutters wielded during a space walk, and this was back before EVA had been mastered. Four days after that, Gilruth had told Slayton to replace Aldrin. But Deke stood up for Buzz and insisted to Gilruth that he had confidence in him, so Buzz remained on the crew. (Aldrin later insisted that he had only wondered if the option had been considered.) Deke's confidence had its limits, however. When serious thought was given to using the astronaut-maneuvering unit (AMU) backpack, which would allow for untethered space walks, during an ambitious EVA planned for Gemini 12, Slayton was ready to replace Aldrin with Cernan; Aldrin had no experience with the AMU, and Cernan did. But the AMU was dropped from the mission, and Buzz kept his seat.

Perhaps it was the gods of aviation who interfered on his behalf, for no one in the astronaut corps could claim a closer kinship to the pioneers of flight and rocketry. Buzz's father, Edwin E. Aldrin Sr., was an aviation pioneer who had studied under Robert Goddard, taught flying during World War I, and associated with such flying legends as Orville Wright, Charles Lindbergh, Howard Hughes, Jimmy Doolittle, and Billy Mitchell. He'd even traveled across the Atlantic on the *Hindenburg* a year before it caught fire in 1937. Like his son after him, he got the military to send him to MIT for his doctorate.

Aldrin Sr. was one of the early proponents of air travel, writing an occasional newspaper column called "Sagas of the Skies." While an aide to Mitchell in the Philippines, he met his future wife, Marion Moon, the daughter of a Methodist army chaplain. During the mid-1920s, he and his new bride toured Europe in a single-engine Lockheed Vega, a trip that included skimming over the Alps at fourteen thousand feet. In 1929, after the arrival of two daughters, the Aldrins bought a three-story, seven-bedroom white stucco house in Montclair, New Jersey. On January 30, 1930, Edwin Eugene Aldrin Jr. was born. The family called him Brother; younger daughter Fay, only two years old, pronounced it "Buzzer." That soon became Buzz, which stuck.

Aldrin Sr.'s job as an executive with Standard Oil often took him away on long trips, and Buzz grew up in a household of women. Besides his mother and sisters, there was Anna, the cook, and Alice, the housekeeper, who shared the third floor with Buzz. For a good chunk of Buzz's childhood, an aunt, uncle, and grandmother on his mother's side also lived with them. Much of Buzz's young life and character were shaped by his desire to please a remote but strict father who had great expectations for his only son: "He planted his goals and aspirations in me," Buzz would write later.

When Buzz was two, his father flew most of the family down to Florida for a vacation; the future astronaut got sick during the flight. Soon after, Edwin Aldrin Sr. drove his family the fifty miles from their Montclair home to Hopewell, New Jersey, to visit his friend Charles Lindbergh, whose child, just five months younger than little Buzz, had recently been kidnapped. When they arrived at the house, it was decided that Buzz would remain in the car with the rest of the family, lest the Lindberghs be reminded of their own toddler.

Young Buzz read science fiction books and magazines and the *Buck Rogers* and *Flash Gordon* comic strips, and he constructed model airplanes and hung them in his room during World War II. He had plenty of pets, including a squirrel, several white mice, and, for a short time, an alligator named Agamemnon. In high school, at a hundred and sixty pounds, he played center on his school's state champion football team. He had a few close friends, but he was a quiet teenager, and at some point in those years he discovered a love of science and began to apply himself to his studies. The boy did everything he could to gain his father's approval, but when Aldrin Sr. wanted him to attend the Naval Academy at Annapolis, Buzz defied him and went to West Point, where he graduated third in his class of four hundred and seventy-five with a BS in mechanical engineering.

With graduation came another disagreement with his father. Graduates had the choice of joining the newly formed air force, which had no academy at the time, or the army. Both father and son agreed he would join the air force and become a pilot, but Aldrin Sr. wanted Buzz to choose multi-engine school and eventually command a bomber crew—a tried-and-true method of scaling the officer ranks. Buzz chose fighter-pilot school, which had fewer command opportunities. The Korean War was still raging, and after eighteen months of flight training, Aldrin arrived in Korea in December of 1952. By the time the armistice was signed, in July 1953, he had flown sixty-six F-86 combat missions and shot down two Soviet MiG-15s.

In December 1954, he married Joan Archer, a young blond actress from New Jersey who had appeared in theater productions and landed a few small TV parts. In late 1955, in the middle of the Cold War, Aldrin was assigned to a fighter squadron in West Germany, where he became good friends with another young air force pilot, a former West Point track-team buddy (Buzz had been a good pole vaulter) named Ed White. Several assignments later, Buzz earned a ScD in astronautics from MIT in 1963 with a dissertation entitled "Line-of-Sight Guidance Techniques for Manned Orbital Rendezvous." His dedication read: "In the hopes that this work may in some way contribute to their exploration of space, this is dedicated to the crew members of this country's present and future manned space programs. If only I could join them in their exciting endeavors!" When he applied to be an astronaut that year, his dissertation topic couldn't have hurt. Aldrin was assigned to work on mission planning, specifically orbital rendezvous, and some of his ideas were incorporated into NASA procedures. (Ironically, he found that some parts of his theories were wrong.)

But behind the brilliant mind and high-achieving exterior was a man battling with insecurity and a family history of depression that had escaped NASA's battery of psychological tests. Shortly after his triumphant Gemini 12 flight, Aldrin took to his bed and didn't leave it for five days. Later he would recognize this as a sign of depression, but he didn't understand it at the time. Eighteen months later, on May 24, 1968, Aldrin's mother deliberately took a fatal dose of sleeping pills; the reason most often cited by Aldrin for this was her reluctance to face the attention that would result from her son's probable upcoming trip to the moon. Her father, a minister, had put a gun in his mouth and killed himself; an uncle of hers had also committed suicide. Aldrin self-medicated his demons with excessive alcohol and extramarital flings, but since drinking and fooling around were common astronaut behaviors, they didn't draw attention like they might have in other professions, especially since at the time, newspeople just looked the other way. The regimented discipline of his job also prevented the bad habits from getting too out of hand. He was a workaholic and too goal-oriented to allow that to happen.

Aldrin had been socially awkward in high school, and he never really overcame that. When he arrived at his astronaut interview in a suit and tie wearing his flight wings and Phi Beta Kappa key, Gus Grissom said, "We've already seen your résumé. Why are you wearing it?" Small talk was a foreign language to Buzz, and one he never mastered. Even fellow astronauts dreaded sitting next to "Dr. Rendezvous" at dinner, since the conversation usually became a one-sided lecture on Aldrin's favorite topic, orbital mechanics. He once spent hours lecturing an astronaut's wife on the subject. "Aldrin," said one friend, "is a professor who is always on." One newspaper referred to him by a nickname that some at NASA, and even Aldrin, had used: the Mechanical Man. "I sometimes think he could correct a computer," one flight planner commented. If a computer could talk, he might have sounded like Aldrin. After the Gemini 9 misunderstanding, Slayton had suggested that next time, Buzz should let him translate.

Sometimes he seemed to communicate via telepathy. Once during training, when he was walking with Jim Lovell, his Gemini 12 crewmate, he said, "Isn't that right, Jim?"

Lovell was dumbfounded—he had heard nothing before that. He said, "What are you talking about?"

"Oh, about the rendezvous," said Buzz, and he began to elucidate.

Lovell said, "Buzz...please speak up when you're doing that."

Aldrin was a loner who was participating in a team sport, and even he admitted that he didn't work well as part of a team. "I just wasn't an organization man," he wrote later. Most of the other astronauts didn't care for him much, but they respected his keen scientific mind.

Mike Collins had required no such luck to gain a seat on the crew. Slayton was impressed with his work ethic, his attitude, his intelligence. And though every astronaut was as smart as a whip, it's doubtful that any of them were as literate and cultured as Collins. When asked for his five desert-island books, he named *Don Quixote*, *The Rubáiyát of Omar Khayyám*, an anthology of English verse, the Bible, and a contemporary novel by Jan de Hartog, *The Spiral Road.* He also loved poetry, and not just John Magee Jr.'s "High Flight," the one poem every pilot knew, at least the first line: "Oh! I have slipped the surly bonds of Earth." Despite these questionable traits—questionable in the tough-guy test-pilot world, anyway—and others, such as a passion for rose gardening, Collins was well liked. He also possessed a self-deprecating wit. If he had a weakness, it was his insistence on finding the humor in any situation, even serious in-flight ones. Not every astronaut appreciated *that* much humor.

Like his Apollo 11 crewmate Aldrin, Mike Collins had been raised in privileged circumstances, although his parents were not wealthy. His father was a two-star general, but no one, not even a general, became rich in the army. Collins's father had spent time in Mexico chasing after Pancho Villa with General John J. Pershing's horse cavalry, and had been awarded a Silver Star in France during World War I. Military service ran in Mike Collins's family; one uncle was army chief of staff during the Korean War, another uncle was a brigadier general, his brother was a colonel, and one of his cousins was a major—all army.

General Collins's assignments took him to several locations around the world, including Rome, where on October 31, 1930, his son Michael was born in an apartment just off the Borghese Gardens. Like most military families, the Collinses never spent more than a few years in one place; they lived in Oklahoma, Texas, Puerto Rico, New York, and elsewhere. It was in Puerto Rico that young Mike, about ten, took his first airplane ride, in a Grumman Widgeon, a twin-engine amphibious aircraft. His father had made his own first flight in 1911 on an early Wright flying machine, sitting on the wing next to the pilot.

In 1945, at the end of World War II, General Collins retired, and the family settled in Alexandria, Virginia. Mike had inherited a love of art and literature from his mother, and like almost every other American boy, he read *Buck Rogers* comics and watched Flash Gordon serials at the movies. He dreamed about going to Mars, his favorite planet.

A life in the military seemed preordained but actually wasn't; his parents wanted their children to do whatever they wanted, so young Collins wasn't pushed into a service career, or pushed at all. Consequently, in school he was more interested in making mischief than in getting straight A's. In high school, he decided to try to get into West Point, not because it would lead to an army commission but because it was a free, and excellent, education. He earned an appointment and graduated in 1952 with a bachelor of science degree. To fulfill his military obligation, Collins chose the air force. That wasn't due to a lifelong passion for flying, for his interest had been sporadic, but to avoid nepotism, or its appearance. As he later put it, "I felt I had a better chance to make my own way [in the air force]."

It didn't take long before he fell in love with flying. After instrument and formation flying, then jet indoctrination, he graduated to day-fighter training in F-86 Sabres in preparation for battling MiG-15s in Korea. But the armistice removed that option. Over the next few years he was assigned to fighter units in California and then France, where he became a flight commander and trained to fly against the enemy behind the Iron Curtain. He developed a love of fine food and wine while stationed in France, where he also learned how to drop nuclear bombs. Soon he "could tell a good wine from a bad wine," he later claimed, and, with tongue in cheek, "roughly what district it was from and maybe what chateau or vineyard." One night on the base in the officers' mess, he met a young woman named Patricia Finnegan, a smart, attractive brunette from suburban Boston. Eager to see some of the world, after college she'd taken a civilian job with the air force that brought her to Europe. The two hit it off. He loved her warmth and vivacity; she loved his voracious appetite for life and his sunny approach to it. They married in April 1957.

That was about the time Collins developed a yen to be a test pilot and attend the air force's school at Edwards AFB. In August 1960, along with another air force pilot named Frank Borman and several other top aviators, he began the grueling eight-month course. After graduating in the spring of 1961, he was assigned to fighter operations, which was what he had yearned for. At Edwards, he developed a passing acquaintance with a hotshot test pilot named Neil Armstrong.

But these less desirable jobs he was given left him unfulfilled. In

1962, two months after John Glenn's three-orbit triumph, NASA called for an additional group of astronauts, and Collins applied, despite the sneers of some of his Edwards comrades about the lack of actual flying done by the capsule passengers. He underwent the five days of physical exams and the weeklong battery of psychological and stress tests. He didn't make the final cut. He had gotten the highest score on the Miller Analogies Test, which measured verbal abilities, but he had scored lower in mathematical reasoning and engineering tests. Over the next year, he worked to improve his knowledge of the new cutting-edge aircraft, and in June 1963, when the call went out for another group of astronauts, he applied again, and his scores were better. Deke Slayton phoned him in mid-October and casually asked if he might still be interested in joining the group. Collins somehow managed to tell him that yes, he was.

On January 9, 1969, Armstrong, Aldrin, and Collins were announced publicly as the crew of Apollo 11.

Five days later, Soyuz 4 launched into space with one cosmonaut, Vladimir Shatalov, aboard. Rumors about its intent immediately began to circulate in the West; it was said that another Soyuz craft would rendezvous and even dock with it, perhaps to effect a crew transfer or to begin the process of assembling a space station from which a moon mission might launch. The first guess was accurate. Soyuz 5 lifted off the following morning carrying three cosmonauts. The two spacecraft docked the next day. Just two orbits later, two tethered crewmen from Soyuz 5 EVAed over to Shatalov's craft and entered. After closing the hatch and repressurizing the cabin, Shatalov and his new crew undocked, and they returned to Earth the next morning in a smooth landing. Perhaps desperate for some positive propaganda, the Soviets claimed it was "the world's first experimental space station."

The return of Soyuz 5 and its commander, Boris Volynov, the

following day did not go as well, though the West would not learn of its difficulties for decades. Because of a module that did not attach as it should have, the spacecraft reentered nose-first, with its heat shield at the rear. That left nothing but the ship's inch-thick insulation to protect Volynov from inferno-like temperatures, and as outside fuel tanks exploded, Volynov watched in dread as the hatch buckled outward and then inward from the extreme pressure. Just before the heat reached its peak, the struts connecting the modules ripped off, and the craft flipped around to its proper attitude. The main parachute deployed only partly, and as the memory of his comrade Komarov's fate flashed through Volynov's mind, the chute lines disentangled and the retro-rockets fired and slowed the spacecraft. It hit the frozen ground hard. The impact bruised Volynov badly and broke every tooth in his upper jaw—but he was alive, and he recovered after several months in a hospital.

Another Soyuz disaster would surely have scuttled the program, at least temporarily. But it appeared that the Soviets had leapfrogged Gemini and achieved another first with its crew transfer. To observers in the West, they seemed primed for a journey to the moon.

That assessment was reinforced in December 1968 when CIA satellite photos showed a massive Soviet rocket on a new launch complex at the Baikonur cosmodrome, a rocket as large as the Saturn V—almost certainly a lunar-mission booster.

NASA had scheduled an ambitious five Apollo flights for 1969 and two more if necessary after Apollo 11's July launch. Most of the astronauts thought Apollo 12 would be the first to actually land on the moon—too many things could still go wrong with the earlier missions, and there were plenty of unknowns, even with the extensive planning and training.

No one had ever trained for any kind of voyage as much as an

Apollo astronaut did for a lunar-landing mission. The crews practiced endlessly for every move they would make in every phase of the flight in the command module, in the LM, and on the surface of the moon. They became familiar with the hundreds of switches they would throw and every control they would push as well as some they never would; they went through hundreds of urgent situations that might come up, from a fuel leak to a dead engine to an emergency rendezvous at a dangerously low altitude, and they reviewed all the experiments they would conduct in space and on the ground and every step they would take there. They even practiced eating, drinking, sleeping, defecating, and urinating in the weightlessness of space and in their cramped quarters. In the six months between January, when they were officially assigned the mission, and the July 16 launch, the three crewmen of Apollo 11 trained fourteen hours a day six days a week and often seven, and they spent much of any time they had left reading reports, procedures, and mission rules—and there were a lot of mission rules, a thick book of them.

Sometimes the three of them got together for an integrated simulation, often connected to Mission Control in the closest approximation of the real thing. But Collins usually trained alone in the command-module simulator.

Armstrong continued to make runs on the LLTV, whose predecessor, the LLRV, had almost killed him the previous May. After his session on June 16—the last of thirty-four hours he spent on the dangerous contraptions—his superiors gave a collective sigh of relief. After Armstrong's last-second escape, none of them had wanted the astronauts to continue training on it, but Armstrong and the other Apollo commanders had insisted. To train for docking, Armstrong and Aldrin regularly flew up to Langley, where full-scale replicas of both the LM and the command-service module hung from cables in a large hangar. He and Aldrin practiced fully suited for their EVA in a reduced-gravity simulator at Langley. All three astronauts took turns on the centrifuge at MSC, going over breathing techniques they would use during the expected ten g's of reentry.

Armstrong and Aldrin spent much of their time together, as they would be manning the LM, and in the simulators, they stood as they would in the LM itself, practicing the most important phases of the mission, and the most dangerous: the descent, landing, and ascent. In between simulations, they sat on a shelf behind them. Other Apollo crews practicing for a landing, like Apollo 12's Pete Conrad and Alan Bean, talked and joked during breaks; Collins often chatted with the sim instructors. Armstrong and Aldrin said little to each other or to anyone else, and their silence often tempted the sim instructors to walk over and look in to make sure they were awake. The simulations were grueling, and after a long day involving a dozen intense sessions, the participants were wrung out. At the end of one such day, Armstrong climbed out of the LM simulator, lit up a cigarette, and said, "Well, that's my one cigarette for the year." It was the only time they ever saw him smoke anything but an occasional cigar.

The flight simulators were remarkable pieces of equipment, although they didn't look like much from outside—John Young dubbed the command-module simulator the "Great Train Wreck." It appeared to be a large, tossed-together jumble of boxes with a carpeted staircase leading up to the entrance, an exact replica of the spacecraft's hatch. Inside, the resemblance was even more uncanny; every dial, gauge, button, and switch was there and in the right place, and they all worked. These simulators were far removed from the simple Link trainers that pilots had practiced on since the first one's invention in 1929.

* * *

Twenty-six-year-old Steve Bales worked almost constantly, and he had no social life to speak of outside of NASA, but he couldn't have been happier.

Every day, he got to MOCR at eight in the morning and worked till seven or eight at night, with just a short lunch break in the cafeteria. Then he'd make the fifteen-minute drive home to his small rental house on Galveston Bay, eat dinner, decompress for a while, go to sleep, and get up and do it again. He wouldn't have traded it for anything.

Beginning with the Apollo 7 mission in March 1968, Steve Bales had been part of the guidance (GUIDO) team, but he was also assigned to follow and study LM lunar ascent and descent. He had a lot of essential issues to work out: How were they going to align the LM platform when it was docked with the command-service module, since the LM radars would be blocked from the stars they navigated by? And how would they align the platform from the lunar surface? And would GUIDO have the responsibility of calling an abort during the descent? If so, what were the criteria? Not every member of the Trench—those flight controllers who sat at consoles on the first row in MOCR, the ones who monitored the mechanics of the flight, not the systems—could call an abort that would stop the mission.

Fortunately, he didn't have to figure this out alone. Other people were working the lunar-landing problem. He and the rest of the MOCR crew assigned to the Apollo 11 mission labored long hours on ascent and descent. Bales got to know the LM's guidance systems well. The primary guidance and navigation system (PGNS, pronounced "pings") handled the LM's descent, ascent, and rendezvous using the LM guidance computer. While the command module's guidance system was powered up and aligned by a ground crew of hundreds, the LM's system had to be started up and initialized by a crew of two who were two hundred and forty thousand miles away from Mission Control—and it had to be done twice, once during lunar orbit for descent, and again from the lunar surface for ascent. There was also the backup abort guidance system (AGS, pronounced "ags"), which was much smaller. It could perform only abort and rendezvous, though it could navigate during the landing if PGNS failed.

Bales was overwhelmed at first. But in late summer and fall, he began to get more of a handle on it, largely by attending Howard "Bill" Tindall's Lunar Landing Flight Techniques Panel, which met every month and sometimes more frequently.

Tindall, chief of Apollo Data Priority Coordination, a catchall term that gave no hint of his broad involvements, was one of the unsung—at least to the public—heroes of the Apollo program. As a mission planner, particularly in the area of orbital mechanics, he was brilliant; for Gemini, he had figured out rendezvous—not just theoretically, with equations on paper, but how to actually do it. He had also played an important part in developing the Apollo guidance system and the onboard computer. But it was his ability to bring together a wide-ranging group of individuals to determine mission techniques—in plain English, how a particular thing was going to get done—that was most valuable. Chris Kraft would later write that no one had contributed more to the success of Apollo.

Tindall was a family man, slender, in his early forties with brown hair and glasses, and he liked to drive his Ford Pantera fast on Houston's freeways. He led weekly "Black Friday" meetings that tackled any number of important Apollo matters and involved up to a hundred people—astronauts, contractors, flight controllers and planners, engineering specialists, and others. (Later sessions would involve subgroups of ten or twelve in his office conference room.) He generally kicked off a meeting with "Why are we here? What are we trying to do?" Anyone was allowed to speak his mind without being judged. Then he'd offer an opinion or summarize—a variation on the Socratic method—and allow others to respond. Ideas and approaches were introduced, discussed, attacked, defended, abandoned, and combined until what was left was pretty close to how something would be done; the process resembled a sculptor chipping away until a work of art revealed itself. The problem wasn't completely solved, but usually the meetings cleared a path to solving it. When he thought he'd gotten enough for a particular subject, and the requisite assignments for further study or action had been made, he'd move on to the next.

The first Lunar Landing Panel, held in August 1968, was a two-day marathon of fifty or sixty people. The meeting had only one subject: What needed to be done in the first hour after the crew entered the LM? The response was "a total cacophony of opinions," remembered Bales. "People talking at once, sometimes shouting. Once in a while someone would get so mad they would have to leave the room." By noon of the first day, there was some consensus on what the major tasks would be. After lunch, they tackled the crew's second and third hours in the LM; the next day, hour four and more. "By the end of day two," recalled Bales, "we had baselined a big picture of what was needed to be done." Each group had action assignments to perform—more than fifty—before next month's meeting.

Tindall's summaries and his opinions on and concerns about any number of important matters were often recapped in weekly reports that became known as Tindallgrams. Concise, well-written, straightforward, and occasionally playful and even folksy, they were a refreshing change from NASA's bureaucratese and its endless acronyms and abbreviations. In one Tindallgram, he wrote: "Maybe I'm an 'Aunt Emma'—certainly some smart people laugh at this concern, but I just feel that the crew should not be diddling with the DSKY [the onboard computer's interface] during powered descent, unless it is absolutely essential. They'll never hit the wrong button, of course, but if they do, the results can be rather lousy."

By the time simulations began for Apollo 11's launch, in February 1969, the procedures were in place.

Over the next several months, Bales spent hundreds of hours in flight techniques and mission rules meetings and many more hours in the room dubbed the "Guidance Officers Training School" running through every type of guidance failure conceivable, hundreds of them. There were also trips to the MIT Instrumentation Lab, where the Apollo Guidance Computer had been designed, to go through all the software what-ifs, and to the Griffith Observatory in Los Angeles with astronauts; because they would navigate by the stars, he had to know them also. In Houston, because so many flight teams were practicing a lunar landing, there were shifts throughout the day and evening. Sometimes, after a long night shift, Bales and some other controllers would relax with a round of golf in the morning, then return to the MSC, catch a few hours of sleep in the bunk room on an upper floor of the Mission Control Center, and get back to the work that evening.

A couple of months or so before the Apollo 11 launch in July, Bales began to notice something.

He had heard a story about the great boxer Joe Louis and how, near the end of his long career, he had told a reporter, "I have to think to throw my right hand"—before that, it had just come, without conscious thought. The young GUIDO realized that, for the first time, he didn't have to think to make all the calls: "I knew what to do," he recalled. "I knew which ones to worry about and which ones not to…I realized I didn't have to consciously think what's the next most important thing to worry about—it just came naturally."

That was the good thing. The bad thing, or at least what kept Bales up at night, was that after weeks of discussion, NASA had decided that GUIDO—in this case, Bales—had the power to abort during descent to the lunar surface. Bales hadn't wanted this responsibility, and neither had some of the other flight controllers. Now GUIDO could abort the landing even if it could be completed, under certain circumstances. And if the abort led to a botched rendezvous—entirely possible, since it would result in a changed trajectory—and if that led to the two crewmen stranded in the LM with no chance of rescue...Bales tried not to think of what could happen.

In addition to figuring out what to worry about, Bales had also developed the ability to follow several conversations at once. While in the MOCR, he wore an earpiece on his left ear, and he could pick the loops—the radio conversations between different parties involved in the mission—he wanted to take part in. He regularly listened to six: his staff support room (SSR), the flight director's loop, the air-to-ground loop, the MOCR trajectory-dynamics loop, the computer-dynamics loop, and the systems loop. Some controllers listened to more. He had learned to talk to one person on one loop and monitor the others at the same time. The talent carried over from work; he could walk into a party or a restaurant and pick up on several conversations at any volume and comprehend them all. It was a kind of superpower but one that, he would recall, was "not a terribly socially endearing thing."

Twenty-four-year-old Jack Garman was knee-deep in his work too. Over most of the previous year, he had been the key software support for the Apollo 8 and Apollo 9 missions, learning the details of the command module and LM software programs. Bales thought he was the most knowledgeable computer person he'd ever met, and he impressed others too, so much that he soon earned the nickname "Gar-Flash." Garman started flying up to the instrumentation lab at MIT every other week to learn all he could about the Apollo computer, and the recent college grad started asking the older guys there to show him what they were doing and why. He found that they loved to talk about the Apollo computer—and why shouldn't they? They'd invented it.

Then Garman found a new interest. During the Apollo 8 mission in December, he'd struck up a conversation with a smart, young math aide, one of several young women in the Mission Control Center who plotted mission analyses. They began seeing each other and soon fell in love. They became engaged, and planned to marry soon after Apollo 11.

But except for the occasional Sunday-afternoon barbecue at Glynn Lunney's house, where the MOCR people would drink beer and eat burgers and oysters—and talk shop—there just wasn't much recreational time. The Apollo program was like a fully loaded freight train hurtling down a mountain, and there was no chance to get off. The young controllers wouldn't have had it any other way.

But before Apollo 11 was given the go for a landing, two other fully crewed Saturn V missions had to be perfect—and both would involve dangers no human had faced before.

CHAPTER THIRTEEN

A PRACTICE RUN AND A DRESS REHEARSAL

The crew of Apollo 9 are prepared to make up with their personal courage any shortcomings of their spacecraft.

PRAVDA

THERE WERE NO GUARANTEES that Apollo 11 would attempt to land on the moon. First, Apollo 9 and Apollo 10 had to go off without a hitch. Each mission had its challenges and its dangers.

Apollo 9, scheduled to launch in March 1969, would be the first manned test of the LM in space. To Commander Jim McDivitt and lunar-module pilot Rusty Schweickart would fall the tricky job of firing the LM, flying it out of sight of the command-service module and its pilot, Dave Scott, then maneuvering back to it and docking successfully. There would also be an EVA involved, a dry run to prepare for an emergency. The LM had no heat shield and could not withstand the tremendous heat of reentry without incinerating, so if something went wrong during rendezvous, McDivitt and Schweickart would be stranded in space with no hope of rescue. If they couldn't dock, they could probably spacewalk over, but they'd have to be close, so the LM *had* to rendezvous with the command-service module. But because the flight would largely be an engineering one and the spacecraft would not leave low Earth orbit, most of the American public had little interest in the mission. After all, the previous one had gone around the moon.

If there were complications, the crew would be up to them. McDivitt had flown one hundred and forty-five combat missions in the Korean War, and he was as reliable as they came—he had impressed Slayton so much that he had been the first man in Gemini to command his initial spaceflight. Scott had acquitted himself well during the near calamity of Gemini 8. The red-headed Schweickart had logged more than thirty-five hundred hours in high-performance jet aircraft, and he knew the LM inside out. McDivitt and his crew were well trained and well prepared; they'd been together as a crew for three years and they'd worked with the LM that entire time. None of them were bothered by the fact that Apollo 8 had jumped them in the schedule and stolen their thunder—they regarded Borman's crew as mere passengers on an unchallenging and quasi-political mission designed to get America to the moon before the Soviets. They considered their own a "connoisseur's flight."

Mission complexity had increased significantly since the man-in-a-can Mercury flights, and that meant much more preparation. McDivitt's crew had been training up to eighteen hours a day, spending most of that time in simulators, but the simulators broke down frequently, so the crew had to train hard until the last minute, leaving them little time to rest and recuperate. The physical and psychological strain finally caught up with them. All three developed serious head colds and sore throats a few days before the scheduled launch, so it was postponed three days to allow them to recover. By March 3, they were healthy enough to fly.

Von Braun's Saturn V boosted Apollo 9 into space at exactly 11:00 a.m., and a few hours after liftoff, Scott separated from the Saturn's third stage, maneuvered the command-service module around, and linked it with the LM, packed chrysalis-like inside a launch adapter just below the service module, despite—shades of Gemini 8—a few stuck thrusters. After remotely opening the LM's four-piece housing shroud, Scott carefully lined up an extending device, the docking probe, with a circular hole on the LM that was in the center of a conical recess, or drogue, and moved forward. Three latches on the probe grabbed the inside of the drogue, a retract system pulled the two vehicles together, the probe retracted, and twelve latches on the sides of the LM's docking tunnel snapped into place to form an airtight silicone seal—a hard dock. Then the craft delicately pulled away from the Saturn. The first docking attempt had been successful.

Despite some serious space-sickness on the part of Schweickart, the astronauts pressurized the LM, and Schweickart and McDivitt collapsed the probe and drogue, entered the LM through a tunnel, powered it up, and pulled free. While Scott piloted the command-service module alone, the fragile LM ranged up to a hundred and eleven miles away. Over the course of six hours, it passed every maneuvering test with flying colors. Even the all-important ascent engine—which had no backup—worked smoothly. It powered the LM back to the command-service module, where the two vehicles rendezvoused and docked once more. The piloting skills of both McDivitt and Scott were put to the test, but especially McDivitt's, since he was the one who guided the LM first by radar and then by sight back to the command-service module. The space-sick Schweickart's first EVA—a risky crawl from the docked LM to the command-service module and back—was canceled. The next day, he felt better and performed a seventy-seven-minute space walk that fully tested the reinforced lunar spacesuit and its four-hour independent oxygen supply.

Ten days after leaving Earth, Apollo 9 returned safely. One essential test flight was done. The mission had been so successful that there was talk of allowing the next one to attempt a landing—but the talk didn't last long. The LM was still just a bit too heavy to try that within the safety margins. Apollo 10 would return Americans to a lunar orbit, nine miles above the surface, and just to make sure, the LM's fuel tanks were only partially filled. If the LM did happen to land on the surface, there might not be enough to get it back into lunar orbit. Besides, the crew hadn't been trained for a landing.

And there were still too many unknowns to think about a lunar landing. In addition to the issues of tracking and communications with two spacecraft orbiting the moon, the delicate dance of orbital mechanics and rendezvous and docking in the moon's weaker gravity, the untried landing radar, and many other problems, there were the mascons, those lunar regions that for unknown reasons exerted a stronger gravitational pull on orbiting objects than other areas did. They had changed the orbit of Apollo 8 dramatically. Twelve mascons had been charted, but there might be more of them—or there might be some other kind of anomaly that could wreak havoc on a lunar landing.

In 1957, several years before NASA revealed its plans for a lunar landing, a science fiction writer named Poul Anderson wrote a short story entitled "The Light." It involved three American astronauts—a commander-pilot, an engineer, and an instrument man—in a spaceship on mankind's first trip to the moon. After they land—presumably having employed the single-spacecraft direct-ascent mode—and prepare for a two-week stay, they draw lots to see which two will go for a walk. The engineer remains inside, and the commander and the instrument man go outside; the instrument man has the privilege of going out first. Soon he and the commander are standing in the shadow of their ship. Since the instrument man was the first one out, the commander tells him to make the speech, which was written earlier. The instrument man defers to the commander. Neither has any desire to give the speech, so neither does; later, the commander will write in the ship's log that the speech was delivered. They begin to perform the first job—collecting rock samples—while the commander takes photographs.

The story provided amusing parallels to the upcoming landing attempt, but there were two differences: the utter lack of importance attached to being the first man to set foot on the lunar surface, and the complete lack of interest in his official first words.

An early timeline of the LM landing and surface activity written by mission planners years before, when the LM existed only on paper, indicated that the lunar-module pilot would be the first crewman to egress through the hatch, climb down the ladder, and step onto the moon's surface. That followed a tradition in Gemini, where the command pilot, the left-seat astronaut who did almost all the flying, remained in the spacecraft and continued piloting while the junior crewman in the right seat performed the EVA.

Apparently that early plan was still in effect on February 26, 1969, six weeks after Apollo 11's crew was announced, when a "top NASA official" told the *Chicago Daily News* that the flight plan called for lunar-module pilot Buzz Aldrin to do the honors. This was a new development; when Deke Slayton had been asked about it at a press conference on January 16, he replied, "I don't think we've really decided yet," and Armstrong chimed in that the choice would not be "based on individual desire, but on how the job can be best accomplished on the lunar surface."

But newspapers throughout the country ran stories trumpeting the choice of Aldrin. Nine days later, after the successful launch of Apollo 9, NASA changed its tune again. At another presser, General Sam Phillips, the Apollo program manager, said, "The decision really hasn't been reviewed by all of us who expect to take a look at the details of that mission plan. From my standpoint, the decision hasn't been made."

Cloudy as it was, that was the public version of the story. Behind the scenes, it was more complicated.

When Aldrin heard a rumor that Slayton had decided that Armstrong would be the first to walk on the moon, he was not happy. He also heard that Neil's civilian status was a reason for the choice—NASA wanted to make a clear statement about the nonmilitary nature of the landing and of the American space program as a whole. Aldrin decided to confront Armstrong about it. According to Aldrin, Neil "equivocated a minute or so, then with a coolness I had not known he possessed he said that the decision was quite historical and he didn't want to rule out the possibility of going first." Aldrin approached a few other lunar-module pilots and used charts and graphs and statistics to show why he—and they—should step out onto the moon before other crewmen. When he tried to discuss it with Mike Collins, Mike cut him off. Aldrin also bugged Apollo 10 commander Tom Stafford, who was involved in mission planning. He pushed for the final mission plans to show the LM pilot exiting first.

Slayton heard about Aldrin's evangelizing and decided a talk with him was necessary. He explained that since Neil had seniority, it was only right that he be the first. Aldrin would later claim that this satisfied him—it had been the ambiguity, he said, that he found unsettling. Buzz may have been okay with the explanation, but his father wasn't. Soon after Buzz told him about it, the elder Aldrin contacted high-placed friends with connections to NASA and the military and tried to get the plan changed, with no luck.

Meanwhile, the Apollo 11 crew continued to train. Armstrong and Aldrin spent many hours in a full-size LM mock-up practicing the lunar landing—including their egress onto an elaborate fake moonscape—and the more times they ran through it, the more obvious it became that Armstrong exiting first made more sense.

At a NASA press conference on April 14, it was officially announced that "plans called for Mr. Armstrong to be the first man out after the moon landing." Slayton provided Aldrin with another explanation of why Armstrong should be the first out, and it seemed to placate him: In the LM, the hatch leading outside was on the left, where Armstrong, the commander, would be standing, and it opened inward, with the hinges on the right. It would be impractical for the lunar-module pilot, on the right and in a fully pressurized spacesuit and EVA backpack, to maneuver around his similarly suited commander in the tight confines of the small LM cabin, then get down on his knees and crawl backward through the small opening. There were too many switches and circuits that might be broken off, activated, or deactivated during such a dance. (Of course, the astronauts could have switched places before donning their bulky backpacks and helmets and pressurizing their suits.)

Buzz put up a good front in public, but privately, he was "devastated," according to his wife, Joan. After the press conference, Collins noticed a significant downturn in Aldrin's mood. It might have sunk even lower had he known of another reason—the main reason—for the decision.

Shortly before the announcement, there had been another meeting attended by four NASA officials. Early on, Chris Kraft had given serious thought to the historic importance of that first step on the moon—the man who took it would be another Columbus, thought Christopher Columbus Kraft Jr. He believed that man should be Neil Armstrong, who, unlike Aldrin, had never petitioned for the privilege. Kraft voiced his concerns first to Slayton, then to George Low, who both agreed with him. Sometime in mid-March, they met with Gilruth in his office. Everyone was aware that Buzz had been lobbying for the honor, and no one was critical of him for that. After some discussion, they made a unanimous decision. Neil Armstrong was their choice. He was the right kind of man to be the first to walk on the moon.

* * *

Next up was the dry run for the lunar landing, and it had just as many new dangers as the previous flight. Apollo 10's crewmen—Tom Stafford, John Young, and Gene Cernan—were all seasoned; each had flown at least one Gemini mission, and Stafford and Young had flown two. They had plenty of rendezvous and docking experience, which would be needed for this most ambitious Apollo effort to date. And unlike the last time, the simulators didn't keep breaking down, so the crew members were well rested at launch. But they almost didn't get to fly it.

Early in 1969, NASA's new administrator, Thomas Paine, had suggested that the Apollo 10 mission should be an unmanned flight. He felt it might be the best way to get to a manned lunar landing with Apollo 11. A quick study was conducted, and it concluded that it would take six months to a year to mechanize the mission—far too long if they were aiming for a 1969 landing. There were too many functions that needed to be handled by the astronaut.

Administrators had briefly thought about allowing Apollo 10 to attempt a landing, but that hadn't been considered long. There were still too many unknowns that had to be investigated and tested—combined operations with two spacecraft orbiting the moon, communications with the LM at lunar distances, and rendezvous and docking near the moon for starters, to say nothing of the LM descent, landing, EVA, and ascent.

Apollo 10's launch into space on May 18, 1969, went off without a hitch, if you didn't count some bone-rattling pogo vibrations from both the second and third stages. Three days later, the crew fired the hefty SPS engine and moved into an orbit sixty-nine miles above the lunar surface. On the twelfth revolution, Stafford and Cernan transferred to the LM, separated from the command-service module, powered their descent engine, and arced down toward the moon to an altitude of forty-seven thousand feet. They skimmed over its craters and *mares*—and even closer to its mountains, some of which reached twenty thousand feet. They would go no closer, since Young in the command-service module couldn't descend any lower than that to rescue them in case of emergency. Standing at the controls and tethered in place, the crew flew over the Sea of Tranquility, the planned landing site for the next flight, and the awed men took plenty of photographs and film while observing boulders as large as forty-story buildings. The landing radar, essential to any complete descent, worked in its debut. When Stafford began preparing to activate the ascent engine to push them toward a rendezvous with the command-service module, Cernan—anticipating his commander's procedures—flicked a navigational control switch to help him. A moment later, an unsuspecting Stafford flicked the same switch, setting it back to where it had been.

The LM immediately started gyrating as its rendezvous radar tried to lock in on the command-service module somewhere below the horizon and its thrusters fired in an effort to move toward it. "Son of a bitch!" yelled Cernan as he and Stafford were thrown around and the craft began to roll end over end at about sixty degrees a second. Stafford said, "Let's go to PGNS," referring to their main guidance system, and flipped another switch, but that didn't help, and he said, "Goddamn!" The LM's rendezvous radar that should have locked onto the command-service module above had instead locked onto the moon below. Stafford said, "We're in trouble," but hundreds of hours of failure simulations had prepared them for this scenario, and he switched over to the manual system, jettisoned the descent stage, and coolly got the spacecraft under control using his attitudinal thrusters. Then they fired the ascent engine, which propelled them toward the correct rendezvous point with Young. A couple of orbits later, the two vehicles were successfully docked. When Stafford and Cernan were safely transferred to the command-service module, the LM was released to eventually enter orbit around the sun. Three days later, the command module hit the Earth's atmosphere at a record speed for a manned vehicle, 24,791 miles per hour, and splashed down just two miles from the recovery ship.

There was a public furor over the men's language during the descent's live broadcast—"Air Turns Blue as Astronauts Blow 'Cool' Image," read one headline—and one high-profile minister demanded a public apology. NASA ordered Cernan to offer one, and he did. The astronauts had also used even stronger profanity, but the avalanche of letters received by NASA were generally supportive of the crew's blue language; the opinions ran twenty-five to one in favor of their expressing themselves freely. In spite of the language controversy and the underlying reason for its use, the flight was a success. It also contributed valuable data on the mascons, especially since the LM had followed the exact trajectory outlined for Apollo 11—which would now attempt a landing on the moon.

If, that is, all the elements of the mission were ready. There were grave doubts among the upper management of NASA that they would be, and the pacing item—that is, the thing that would take the longest to complete—was the astronauts' training. Because of the landing, the Apollo 11 crew needed more preparation than any of the previous crews, and it wasn't clear that the requirements could be met in time.

On June 12, a flight-readiness review meeting was held to determine if the mission would proceed on July 16. They would need to begin loading some of the spacecraft's hypergolic fuels on June 16. A dozen Apollo managers met in General Sam Phillips's office at NASA headquarters in Washington, DC. Several more directors participated via conference call from Houston, Huntsville, and Cape Kennedy. But the decision would be made by Phillips.

After hearing from everyone, Phillips asked Deke Slayton about the crew.

"Our situation hasn't changed appreciably. Training is scheduled up to the sixteenth. We've had to compromise in the CSM-LM area," Slayton said, referring to the lunar module. "We should have one hundred more hours, but we'll have to fit the training in only half of that. I think we're comfortable with what we've got. It will be a ten-hour day, six days a week for the crew from now on. They shouldn't have to work past eight o'clock at night. The LLTV is an open area. Neil will fly the LLTV Saturday, Sunday, and Monday, and maybe Tuesday and the following weekend." There had been another accident with the LLTV in December 1968, and it had been out of commission for four months after that. But Armstrong was adamant about its importance in training for the landing.

They discussed the LLTV some more, then moved on to other parts of the astronauts' schedule.

Phillips said, "Deke, we don't want to wind up with the crew beat down two days before the launch."

"We're about where we were this time before Apollo 8," Slayton said. "We're in better shape than with 9, but not so good as 10. I have no reservations about the crew being adequately trained." He had asked the crew if they needed another month. Armstrong had told him they'd be ready in July.

Phillips asked Slayton what it would mean if they delayed until August.

"Honest to say," Slayton told him, "I don't think we'd be all that much better off."

The last man to give his opinion was Dr. Charles Berry, the surgeon in charge of the astronauts and a man disliked by most of them. He appeared frequently on televised press conferences, where he billed himself as their personal doctor, but they rarely saw him. He told Phillips that he had a "gut feeling" that the crew would not be in the best condition for a July launch, though his reasons were vague.

That didn't sit well with Phillips. He polled the room again and received a *Go* from everyone. He gave a short summary of the issues. "I came into this discussion fully prepared to hold up for an August launch—if there were grave questions," he said. "I have not seen that that is the case. My summary is that nothing is now apparent to prevent our proceeding toward a July launch.

"So, Rocco," he said to Rocco Petrone, the launch director at Cape Kennedy, "go ahead with your hypergolic loading."

Apollo 11 was on for July 16.

IV.

DOWN

CHAPTER FOURTEEN

"YOU'RE GO"

If we get those first three guys back alive, we're going to be damn lucky.

ROBERT GILRUTH

A MONTH BEFORE THE launch, the Apollo 11 crew moved to Cape Kennedy, where they would live and train in semi-isolation. The idea was to keep them focused and healthy by reducing their contact with—well, almost everyone. That included friends, family, and the unrelenting media. Finishing their training requirements in time for their mission would be challenging enough, and the fewer distractions the better. Despite these measures, there was still plenty of stress to go around.

They spent most of their time in simulators, Collins in the command module, Armstrong and Aldrin in the LM, though occasionally an integrated sim was run with all three connected to a fully staffed Mission Control in Houston. As the days went by, these sessions became increasingly difficult. One of them occasioned a rare argument between Armstrong and Aldrin.

After a late dinner in their crew quarters on the third floor of the assembly and test building, just a block away from the simulator building, Armstrong retired to his bedroom. Aldrin and Collins stayed up in the common lounge area, Collins with his favorite libation, a martini, Aldrin with his Scotch. Buzz was annoyed with

Neil's decision earlier, during one particularly hairy descent, to allow the LM to crash into the lunar surface. Armstrong, who had spent countless hours in simulators as a test pilot before becoming an astronaut, had considered it an important learning experience, not just for the pilots but also for Mission Control. Every so often, when Armstrong thought things were slow during a LM sim, he'd reach over and pull a switch or flip a circuit breaker just to keep the flight controllers on their toes. That didn't go over well with Buzz, who had never been a test pilot; he saw the simulation as a game to be won, and deliberately crashing into the moon instead of aborting the flight didn't make sense to him. It would, he believed, reflect badly on his and the crew's ability to perform.

Aldrin insisted on telling Collins all about his annoyance, and fueled by irritation and Scotch, he became louder; finally, Armstrong, clearly irritated himself, walked out of the bedroom in his pajamas. "You guys are making too much noise," he said. "I'm trying to sleep."

Collins quickly excused himself and went to bed; Aldrin and Armstrong argued late into the night. By the next morning, they had ironed out their differences—or at least had come to an understanding about them. But if their relationship had been strained by the first-man brouhaha, it was even more so now.

Since the day the crew of Apollo 11 had been announced, the astronauts had been inundated with interview requests from media outlets around the world. NASA's public-relations department handled the demands and sent some requests to Slayton, who nixed the vast majority. But this mission was one of the most newsworthy stories of the century, and a certain amount of media cooperation was necessary, even at Cape Kennedy. The press wanted time with the three astronauts, and they wanted revelations from them—especially the mission commander. The private and reticent Armstrong rarely satisfied them.

In interviews, he was so guarded, he often seemed like a defendant under cross-examination at a murder trial. Friends always mentioned his dry wit, but it was rarely in view before the press. Aldrin answered questions more quickly, though he wasn't much livelier; he lived up to his reputation as an egghead, the Mechanical Man. But the media loved the easygoing Collins, who was erudite and charming with a self-deprecating wit: "I hate geology," he said. "Maybe that's why they won't let me get out on the Moon." He made no secret of his distaste for machines and computers, and his disarming honesty and insouciance beguiled the press corps. If only he were the commander, this man with his laid-back attitude and easy smile. But he wasn't, Armstrong was, and so he received the lion's share of the questions, and his measured responses were rarely helpful to a reporter looking for a glib sound bite.

It wasn't the luck of the draw that got Gene Kranz the plum assignment of flight director for the lunar descent, and it wasn't the fact that he was a good friend of Cliff Charlesworth, the man who made the decision. It was his previous experience with the LM.

The prime flight directors, Kranz, Charlesworth, and Glynn Lunney, had been taking turns as lead flight. The Apollo 11 mission was Charlesworth's turn, and as lead, he was the one who assigned the flight directors for the various phases of the mission. It would have three new phases—lunar landing, surface EVA, and lunar ascent. Everyone wanted to work the landing, including Charlesworth, but after weighing the strength and experience factors, Charlesworth told Kranz the landing was his. Lunney would handle the lunar ascent, he himself would take the EVA and launch, and another flight director, Milt Windler, would take the reentry and fill in on the occasional graveyard shift while the crew slept.

Kranz had just gotten the assignment of his life, and he knew it. After bouncing off the walls for a while and calling his wife to tell her, he began picking his landing team. Flight shifts were formed on a mission-by-mission basis. "The branch chiefs carefully matched the personalities and strengths of controllers to those of the individual flight directors and their capabilities to handle mission events," remembered Kranz, and each team would focus on training for its specific phase. His White team included experienced controllers Don Puddy, Jay Greene, Bob Carlton, Chuck Deiterich, Granville Paules, and Steve Bales.

If astronauts were the stars in this spaceflight production, and Kraft and his flight directors and controllers in Mission Control their supporting cast, then the hundreds of thousands of other NASA and contractor employees were the technical crew behind the scenes, all performing important and often unrecognized jobs. The men who ran the simulators were just such unsung heroes. Tasked with training the astronauts—either separately with the LM or the command module or together in integrated simulations with or without Mission Control—simulation supervisors (SimSups) and their teams honed the skills of both astronauts and flight controllers to a fine edge, sometimes acting like Marine drill instructors turning raw recruits into fighting machines. With lives in the balance, it was not enough for teams to know what to do when everything worked, a daunting task in itself given the myriad systems involved—they needed to know what to do when things went wrong. To that end, the sim teams ran their charges through every problem or abort situation conceivable and some that were not. The variations were endless and included a host of different failures in virtually every system: fuel, engine, computer, communications, spacesuit, navigation, thrusters, oxygen, and more, even the sudden illness of one of the astronauts. One sim included a controller's faked heart attack; another, a blown fuse on Bales's console. The SimSups' job was to make sure the two groups were prepared for every eventuality, and they took that job seriously. Often, an especially difficult simulation—such as one that ended in an abort that the crew shouldn't have called but did, or one that wasn't called but should have been—would result in a change to the mission plan. But one tenet of the simulations was strictly obeyed: there was always a way out that did not involve an abort.

Jack Garman and Bales had started working together while preparing for Apollo 10. Bales had been guidance officer for that flight's LM activation phase, and Garman had been the software-support expert, working from the GUIDO staff support room (SSR) outside and down the hallway from Mission Control. Now, a few weeks before the Apollo 11 flight and toward the end of Mission Control's sim training, Garman was picked to play both sides of the fence.

SimSup Dick Koos, a former army sergeant who looked and sounded more like a college professor, was in charge of the descent-stage simulations. Koos asked Garman if he could provide some kind of computer glitch, error codes he could run past the flight controllers. Garman did. Who knows? he thought. The knowledge might come in handy.

July 5, eleven days before the flight, was to be the final day of simulations for Gene Kranz's White team, which had been assigned the lunar landing; the team members still had plenty of other things to do in preparation, but he felt good about their progress. They were ready for the mission. Traditionally, the last few sims were nominal, with no complications, a kind of confidence-boosting graduation day.

After lunch, from the glass-fronted sim room on the right side of the MOCR, Koos looked down at the White team as his instructors sprang the computer-program alarms on Kranz's unsuspecting controllers. In the back room, Garman played along, acting as if he knew nothing, and Bales, the GUIDO in the Trench and the man responsible for the LM computer, called an abort at ten thousand feet. The puzzled crew in the LM—Dave Scott and Jim Irwin, the backups for Apollo 12, which was scheduled for late in the fall—had heard an alarm, seen it on their onboard computer, and asked for assistance; now they confirmed the abort, throttled down the descent engine, and ignited the ascent engine to rendezvous with the command-service module. Kranz was furious with Koos; he felt they should have landed. In the debriefing afterward—there was always a debriefing, during which every controller took a turn in the hot seat and defended the calls he had made—he started in on the SimSup. Then Bales said they didn't have a rule for alarms programmed to be used in testing because that would never happen during an actual flight. And since it should never have occurred, he assumed it meant something serious had gone wrong—and so he'd called the abort.

Koos was unwavering. "No," he said. "You should not have aborted for those computer alarms. What you should have done is taken a look at all of the functions. Was the guidance still working? Was the navigation still working? Were you still firing your jets?"

Bales was devastated. At the end of the day, Kranz called his Trench together. There were no mission rules for the alarms, and he wanted some before the launch. Bales said he'd pull a team together and get some answers. Later that night, Bales called Kranz. Koos was right, he told him. They should have continued. He and his team would come up with rules that night and go through some extra training runs in the morning with Koos just to make sure they were on top of the problem. They did go through some extra runs—four hours' worth.

That wasn't good enough for Kranz. He told Garman, "I want you to study and write down every possible program alarm whether they can happen or not." Garman met with his MIT software people to go over the various alarms. He found that there were alarms in the computer just for programming purposes—they'd been put there to ensure that the computer's cycling time was adequate to handle all the guidance and control, and the programmers just hadn't taken them out. The MIT group kept telling him they'd never come up in an actual flight, but they finally categorized them into actionable types. Garman made a handwritten list of all twenty-nine, along with the correct response to each. He put it under the Plexiglas on his SSR console.

The mission rules were constantly evolving—and increasing. Kranz had begun recording the what-ifs during Mercury, and since then, the list had grown from a booklet to a book to a very large book. It was vital for both the crew and controllers to know what could go wrong during a flight and what the responses should be to each problem so they would not have to figure out what to do when seconds might mean the difference between life and death. For Apollo 11, the book ran to more than 330 pages. Since its initial publication on May 16, it had been revised three times, and the changes kept coming. On launch day, seven last-minute changes would be handwritten into each copy. Every flight controller had a copy, and of course there was a copy aboard the command module. No one could memorize every rule and every response, but it was at hand, and since it was organized by system and subsystem, it wouldn't take someone more than a moment or two to find the solution to virtually every problem—or at least, that was the idea.

Not every rule was black or white. There were some gray areas, especially in rules pertaining to the landing. This would become a point of contention between Mission Control and Armstrong.

In more than one discussion about the LM's abort rules, Armstrong made it clear that he wouldn't feel bound by them. After one long debate about insufficient data during the landing, he shook his head. "You must think I'm going to land with the window shades down."

In a planning session shortly before the mission, Kranz—whose White team was training for the descent exclusively—went over landing strategy with the crew. As they discussed various abort situations, Armstrong remained silent, occasionally smiling and nodding.

Kranz saw something in Armstrong's manner that led him to believe that the Apollo 11 commander had "set his own rules for the landing," the flight director wrote later—that he could "press on accepting any risk as long as there was even a remote chance to land." That was okay with Kranz, who, like Armstrong, had flown in Korea: "I had a similar set of rules. I would let the crew continue as long as there was a chance"—even if that meant overruling a mission rule.

Kranz's boss, Chris Kraft, got the same impression during other descent discussions. He and Armstrong butted heads over the rules concerning a landing radar failure. Armstrong didn't want some nervous flight controller aborting the descent based on questionable information. "I'll be in a better position to know what's happening than the people back in Houston," he told Kraft. "And I'm not going to tolerate any unnecessary risks," Kraft fired back. "That's why we have mission rules." They finally agreed that the rules would remain as written, though Kraft too had doubts: "I could tell from Neil's frown he wasn't convinced. I wondered then if he'd overrule all of us in lunar orbit and try to land without a radar system." Unlike Kranz, Kraft wasn't fine with Armstrong—or any astronaut—flouting a mission rule. As far as Kraft was concerned, that was insubordination, and he had made it abundantly clear over twenty manned missions that he wouldn't put up with it.

Even NASA administrator Thomas Paine, likely alerted by Kraft, got involved. Six days before the launch, he shared a quiet dinner with the crew in their quarters and asked them not to take any risks. He promised them that if they didn't have a chance to land this time, they'd get another one. "If you want to abort," he said, "I'll see that you fly the next moon landing flight. Just don't get killed."

They nodded their heads and responded properly. But Armstrong still reserved the right as commander to make the final decision.

Just as important as the mission's training was its timing. Armstrong and Aldrin would be heading from the east, away from the dazzling sun. They would land at dawn, when the angle of sunlight on the moon's surface was at ten to twelve degrees, meaning the shadows of features—hills, craters, boulders, mountains—were long but not too long and the light not too bright or high, which would cause a washout and make visibility difficult. There was only one day in the lunar month that fit those requirements, July 20, which meant a departure date four days earlier, July 16. If they missed this window, the mission would be scrubbed until the same time next month.

In all, the final mission plan for this first landing attempt had taken six years to create. The moonwalk itself—the two hours and twenty minutes Armstrong and Aldrin were scheduled to spend outside the LM—had taken two years to choreograph. They spent dozens of hours tromping around fully suited on a realistic facsimile of a lunar landscape, complete with sand, practicing the deployment of several experiments devised by NASA's scientists. If they landed, they wouldn't have time for much of that in their brief sortie out of the LM, though later missions would. Get out, grab a bunch of rocks and soil, salute the flag, get back in, and get home safely—that would be more than enough for this mission.

Even the astronauts' postflight activities were planned, at least for the first few weeks after they got back. Years earlier, the scientific community had begun to express alarm at the idea of back contamination—the notion that moon bacteria brought back to

Earth might prove dangerous. Although the chances of that were minimal, complex safeguards and protocols were put into effect to keep any alien occupants of the command module, and its potentially deadly human occupants, from destroying life on Earth. Upon the astronauts' return, after they splashed down in the Pacific, a frogman would toss three biological isolation garments (BIGs) into the command module. After donning them, the astronauts would scrub themselves with disinfectant in their life raft, then they'd be hoisted into a copter and taken to their recovery carrier, where they would be whisked into a biologically sealed, modified Airstream trailer and, eventually, to an eighty-three-thousand-square-foot, state-of-the-art facility at MSC called the Lunar Receiving Laboratory. There, they and any lunar samples would be isolated, along with a few doctors, a cook, several technicians, a PR person, and a group of mice who would also be exposed to the lunar soil. The astronauts would be monitored for a total of twenty-one days, starting from Armstrong and Aldrin's lunar-surface EVA. At that point, if there were no health issues—for the men or the mice—they would be allowed to leave the facility. If something went amiss, they would remain in the LRL until it was deemed safe to release them. No one really knew what would happen if the astronauts weren't given the green light—theoretically, they might stay there for the rest of their lives.

The public's fears of such a catastrophe went back at least as far as 1898 and H. G. Wells's classic *War of the Worlds*, in which he raised the issue of the potentially dangerous consequences of microbes crossing alien biology, though in his story, it saved rather than harmed Earth. The fears were further stoked by a few novels published during the run-up to the flight. One was a July 1968 paperback by Harry Harrison titled *Plague from Space:* "THE SPACE PROBE RETURNED TO EARTH CARRYING A CARGO OF WRITHING DEATH," screamed the copy on its bloodred cover above an illustration of a suited, helmetless astronaut whose face was covered with hideous boils. *The Andromeda Strain*, a highly effective thriller published in May 1969, made even more of an impact, and it quickly became a bestseller. Written by a young doctor named Michael Crichton, it featured a deadly, mutating microorganism that comes to Earth aboard a military satellite and wreaks havoc on the planet. Three days before the Apollo 11 launch, on Sunday, July 13, the book ranked number five on the *New York Times* fiction bestseller list.

In addition to the long hours of training the astronauts endured, they also had endless lists of nontechnical tasks to take care of. Each man was permitted a half-pound personal preference kit (PPK), a small bag in which he could carry items for friends, family, and co-workers—coins, medallions, miniature flags, jewelry, and any number of other mementos that could, upon their return from the moon, have significant monetary and historic value—and deciding what to take was surprisingly time-consuming. Armstrong requested and received permission to take two small pieces of the original 1903 Wright Flyer. They also needed to design a mission emblem, and they had to come up with call signs for their two ships, for communication purposes. After much deliberation, Jim Lovell suggested an American eagle for the emblem, which was quickly adopted, and Collins supplied a rough design that was refined and finally approved. That led to the landing craft's name: *Eagle.* For the command module, a NASA public affairs officer offered *Columbia*, and that stuck.

There was also a stainless-steel plaque that would be left on the moon, bolted to one of the LM's legs. Below an image of two hemispheres of the Earth was this legend:

HERE MEN FROM THE PLANET EARTH
FIRST SET FOOT UPON THE MOON
JULY 1969, A.D.
WE CAME IN PEACE FOR ALL MANKIND

Below the inscription were the signatures and names of the three crewmen and, beneath that, of President Nixon. When the LM lifted off, leaving the four-legged descent stage, the plaque would remain on the moon unchanged in the vacuum of space for all eternity. Other items to be left on the surface included an American flag with a telescoping support rod along its top length, an Apollo 1 mission patch to honor its crew, and medals commemorating Soviet cosmonauts killed in action.

The crew spent the long Fourth of July weekend in Houston with their families. Then they flew back to Cape Kennedy, where they would remain semi-quarantined to protect themselves from catching any last-minute viruses or other bugs. While there, they continued to train on the simulators.

Armstrong and Aldrin had been training on the LM simulator for almost two years, first as backups, then as part of a prime crew, but it was only after the Apollo 10 launch in May that they got top priority. For the first time, crew training, not the command-service module or the perpetually tardy LM, was the pacing item in the schedule. Mike Collins spent four hundred hours in the command-module simulator, mostly by himself, though occasionally his crewmates practiced with him; both had spent just as much time in the LM simulator. When they teamed up with Collins to do an integrated sim while connected to Mission Control, the runs could be surprisingly close to the actual thing, and all involved were generally wrung out when they were done. The crew would finish up the day reviewing the 240-page flight plan.

Like the command-module simulators, the two LM simulators were state of the art, the world's most realistic arcade game, though this game was deadly serious. Each was a life-size replica of the LM, accurate in almost every way. TV cameras linked to computers projected whatever the astronauts would really be seeing at that point in the mission and moved along with the LM's progress, even casting a simulated shadow on the simulated moon. Most of the descent was somewhat crude, since no previous mission had produced images to approximate the last several thousand feet, but as the LM neared the simulated surface, the scene became more accurate; the moving image was actually a stationary camera focused on a revolving, realistically sculpted plaster-of-Paris moon hanging above it. Combined with an exact-replica cabin featuring accurate instrumentation and computer readouts, these proved highly valuable exercises. The men could participate in full spacesuits or in shirtsleeves, and they could even be tethered to the deck with elastic cords as they would be during an actual mission.

The LM-sim guys threw every abort and emergency situation at them. Days usually began at eight a.m. and ended about five p.m. Other LM crews—the Apollo 11 backups of Lovell and Haise, the Apollo 12 crew of Conrad and Bean and their backups, Dave Scott and Jim Irwin—also were training on the two LM simulators. There were hundreds of episodes with much repetition of both basic procedures and those involving complications. After a while, even the aborts and emergencies became second nature. Overtraining was a fundamental tenet of Apollo mission preparation, just as it had been for Mercury and Gemini; this overtraining enabled an astronaut "to perform better in the presence of stress in an actual mission," according to an early NASA report on Apollo training.

A direct result of this was the reduction of fear. Astronauts hated the word and avoided using it. (Only one, the ill-fated Gus Grissom, had admitted he'd been scared during a mission.) When pressed on the subject, the men preferred the euphemism *apprehension*, and they used even that term sparingly. "We literally trained out fear," said Wally Schirra. "You're not born with the ability [not to be afraid], you sort of develop it over the years. That's why they selected people like us, because we could do it," said Bean. Fear led to panic, which led to mistakes. Train the fear out, and you lowered the chance of a mistake in a life-or-death situation to zero, or close to it. And since there was no better equivalent to high-stress situations involving spacecraft than similar situations involving high-performance aircraft, it was no coincidence that Deke Slayton would pick six test pilots (and one fighter pilot) to land the LMs in seven Apollo missions.

Saturday, July 5, 1969, was media day for the three astronauts. It started with a morning press conference in the MSC visitors' center, followed by smaller sessions with wire services, magazine writers, and the three national TV networks. The men walked onto the stage wearing masks and kept them on until they were seated behind a clear plastic shield with a large fan behind them that was presumably blowing any reporter germs away. The crew found this somewhat perplexing, since every day they came into contact with dozens of NASA personnel who went home to their families and friends at night and came back to work the next morning. Using the same unclear reasoning, Dr. Charles Berry had decided that President Nixon's request to dine with the crew the night before the launch should be denied, and that decision invited the same puzzlement, as there would be almost a dozen non-crew people at the dinner.

After a long day of interrogation by the media, the crew spent the rest of the weekend with their families and returned to the Cape on Monday.

The last week or so of training, their workdays were limited to ten hours, not the fourteen to sixteen hours they'd been working for six months straight. If they were going to be in tiptop shape for the flight, they'd need all the rest they could get. Two nights before the launch, they did a nationally televised thirty-minute interview from their crew quarters; a panel of reporters questioned them from a NASA news facility ten miles away. The crew appeared subdued and withdrawn. When asked if he had any fears regarding the mission, Armstrong said: "Fear is not an unknown emotion to us. But we have no fear of launching out on this expedition."

On July 15, the day before the departure, Armstrong and Aldrin spent some time in the morning in the LM simulator, but most of the day was spent relaxing, swimming, and taking it easy at a cottage on the coast that NASA had bought for the occasional use of its astronauts. Aldrin borrowed a metal detector from their personal chef, Lew Hartzell, and spent a good part of the afternoon searching for treasure he never found. Collins also took the day off, but he hung around the crew quarters. That evening, each one called home for a final talk with his family.

Dinner would be early, and simple. Hartzell, a former tugboat cook, had learned his trade in the Marines; he liked his beer and had the belly to show it. He loved cooking for the astronauts. Al Shepard had hired him before Gemini 4. After a dozen other candidates hadn't sufficiently impressed him, the meal Hartzell prepared for him did the trick. Shepard told him the astronauts were meat-and-potatoes men, and basic fare was what he always gave them. In between missions, Hartzell worked the occasional yacht run up and down the East Coast, but when he cooked for the astronauts, he avoided the fancy touches he learned on those luxury cruises. He would always return to the Cape before a flight, and he had been providing victuals for the Apollo 11 astronauts since they'd moved into the crew quarters a month ago; he even prepared them overstuffed sandwiches that they wolfed down during their quick lunch breaks. Tonight, it was an easy dinner for less than a dozen people: Apollo 11's prime, backup, and support crews, and Slayton. The only exception was backup LM pilot Fred Haise, who would be in the command module hours before the crew, making sure it was shipshape and that every switch and control was in the right position.

Hartzell cooked the traditional prelaunch dinner, which continued the low-residue diet the crew had been eating for almost a week: salad, broiled sirloin steak, mashed potatoes, tomato puree, asparagus, cottage cheese, a fruit bowl, and bread and butter. The prime crew partook of almost everything, it was noted, except the salad, asparagus, and fruit.

Fifty-fifty—that's what Armstrong, Aldrin, and Collins estimated the chances were of Apollo 11 making a successful landing. They thought there were still too many unknowns in the descent from lunar orbit to the surface. Something they didn't understand was bound to go wrong. Most of the other astronauts felt the same way.

Armstrong actually liked working with unknowns. He found the process fascinating in a problem-solving way, and that was a big reason he had enjoyed test-piloting so much. This was just another test flight—only with much higher stakes.

He had decided that his crew had a 90 percent chance of making it back to Earth alive—certainly better odds than playing Russian roulette with a six-gun, but still risky. But he and his crewmates all concurred—it was a risk worth taking. And it looked like they'd be taking that risk in the morning; dark clouds had hovered overhead most of the past two days, but the forecast for the next day was broken cloud cover at fifteen thousand feet—satisfactory for a launch.

Armstrong might not have been so optimistic if he'd seen the mission risk assessment that had been submitted to Paine in a briefing on July 10. One chart, entitled "Certain Loss of Crew," listed the many possible catastrophic developments at each crucial phase of the mission, from "space vehicle breakup" and "failure to jettison launch escape tower" during launch to "overturning on lunar surface" and "descent propulsion system failure" during powered descent, and several others. Another chart, "Possible Loss of Crew," listed many more, most of them under "Powered Descent."

Armstrong and his crewmates might not have been dwelling on such things, but there were plenty of others at NASA who were.

Bill Tindall, for instance, was still so worried about the mascons that just two days before the launch, he'd sent out a memo about them; Apollo 10 had helped map the dozen they knew about, but no one could be sure that there weren't more, and if there were, that might affect the LM's descent or ascent. Cosmic radiation was another concern. The Earth's atmosphere provided adequate protection, and the continuous low dose received beyond it would be tolerable, particularly within the fairly thick walls of the command module. But solar flares were something else; at any moment, the sun might blast massive clouds of atomic particles that could reach the moon in eight to ten hours, and spacesuits or the paper-thin walls of the LM would be no defense against such a radioactive tsunami. The result would be radiation poisoning and possibly even death. Such flares were impossible to predict, though they did occur in eleven-year cycles, and this year was a peak year during this cycle. Fortunately, NASA had a network of seven observatories to detect and report dangerous flares, which meant that if Armstrong and Aldrin were on the lunar surface, they would have time to halt their EVA and blast off to rendezvous and dock with Collins.

But Armstrong avoided worrying about circumstances he couldn't control. He believed his crew was sufficiently prepared. It had been tight, and he and Aldrin particularly had barely gotten in the amount of training deemed necessary, but they'd done it—although Riley McCafferty, simulator chief at the Cape, believed they were unprepared for anything but a nominal mission and that they would run into trouble if anything unexpected occurred. But the three astronauts knew the machines and systems well and were confident that the hundreds of thousands of people involved in readying the three million parts of the Saturn V, the two million parts of the command-service module, and the one million parts of the LM could not have been more conscientious if their own children were going to be aboard. There was still tremendous pressure on the three men, and each prayed that he wouldn't be the one to screw up and possibly endanger the mission or his shipmates. They were all aware that any kind of failure during the next eight days would tarnish America's image.

Collins, especially, worried about the many things that could go wrong when the LM separated from the command-service module and descended to the lunar surface and, later, when the astronauts rendezvoused. There were eighteen variations on the procedure in case of emergency, and some of them required complicated operations performed flawlessly. One of the mission phases he had to practice in the simulator was making the burn that would send him back to Earth if his two crewmates had to be left on the moon to die. He had developed tics in both eyes at the thought of being unable to retrieve his two comrades and having to turn homeward and leave them on the moon or circle it until they perished...("Of such possibilities are nightmares bred," he later wrote.) He would carry a small notebook listing the eighteen variations on the procedure and keep it clipped to the front of his spacesuit. All three men knew that the nation's hopes rested on their faultless performance. Despite the pressure, there was a certain relief just before the launch—not just because they had finished the many months of nonstop training and studying, but because they were finally getting on with it. Beyond doing their duty for the country, they were looking forward with great anticipation to returning to space. It was an exhilarating experience that each man—and every astronaut—had relished.

They might have been amiable strangers, but they were the perfect crew for the first attempt at a moon landing. Armstrong—with his wealth of experience in the Korean War, as a test pilot, and on a treacherous Gemini 8 mission—would get the LM down to the surface if anyone could. He could handle any unknowns that arose with his skill, knowledge, and nerve, and if that didn't work, there was always his luck. Aldrin, Dr. Rendezvous, would do a damn good job as LM navigator, and if communications with Mission Control broke down or if the LM's onboard computer failed, he'd use his slide rule to get them back to the command-service module. And Collins, as good and knowledgeable a command-module pilot as there was in the astronaut corps, would get them home if it was humanly possible.

And if it wasn't possible and they didn't come back, there was a plan for that too. Frank Borman, who had retired from the astronaut corps to become NASA's liaison with the White House, suggested to Nixon's speechwriter William Safire that he prepare a speech for the president in case the mission went badly and Armstrong and Aldrin were left stranded on the moon. Safire composed a short address that began, "Fate has ordained that the men who went to the moon to explore in peace will stay on the moon to rest in peace"; it ended with an homage to lines written by Rupert Brooke, a British poet who fought in World War I: "For every human being who looks up at the moon in the nights to come will know that there is some corner of another world that is forever mankind."

The president would telephone each of the new widows before the statement. Afterward, a clergyman would treat their deaths as a burial at sea and commend their souls to "the deepest of the deep."

In his room at the Cocoa Beach Holiday Inn, Wernher von Braun, now gray-haired and fleshier than the Peenemünde wunderkind of thirty years ago but still handsome, sat on the floor, cross-legged and in his shirtsleeves, looking over the launch schedule one more time. On Sunday, he and his wife, Maria, had flown down from Huntsville with his deputy director Eberhard Rees, and he had spent the previous few days in endless management meetings and press conferences. Earlier in the evening, he had been the keynote speaker at a gala dinner for industry bigwigs, astronauts, and other luminaries. Despite the host mentioning von Braun's Nazi affiliation in his introduction, the crowd had given him a standing ovation. Now von Braun called his old friend Kurt Debus, the director of Kennedy Space Center, to wish him luck the next day. The launch would be a reunion of sorts for the Peenemünde group; in addition to the many German engineers at the Cape and in Huntsville, their old boss Walter Dornberger, recently retired after many profitable years with Bell Aerospace, would be there, as would Hermann Oberth, the rocketry pioneer who had inspired them all. After calling Debus, von Braun said some prayers for the next day's flight and then went to bed, though he didn't sleep well. His thoughts were on the launch tomorrow of his Saturn V, which would be making the most important flight of its—and his—life.

It seemed as if almost all of the million people outside Cape Kennedy's gates were partying (the Cocoa Beach bars, which usually closed at three a.m., had received special permission to stay open till five a.m.), but the Apollo 11 crew retired to their individual rooms by nine p.m. By ten o'clock, all three were asleep. While they slumbered, a light rain fell, and flashes of lightning could be seen far to the north.

Earlier in the evening, Collins had managed a quick look at the Saturn V, which was "suspended," he later wrote, "by a crossfire of searchlights which made it sparkle like a delicate opal and silver necklace against the black sky." Launchpad crews cooled down the fuel systems and began transferring the cryogenic propellants—liquid hydrogen and oxygen—into the Saturn V and preparing the spacecraft. Mission specialists entered the command and lunar modules to ink in last-minute updates to the checklists and flight plans. Guenter Wendt and his crew were everywhere with their clipboards, reviewing every last detail before the skyscraper-size rocket hurled its human payload into the airless void of space.

At 4:15 a.m., on the third floor of the operations and checkout building eight miles away, Deke Slayton walked down the crew-quarters hall and knocked on each astronaut's door. "It's a beautiful day," he said. "You're GO."

CHAPTER FIFTEEN

THE TRANSLUNAR EXPRESS

I am far from certain that we will be able to fly the mission as planned. I think we will escape with our skins, or at least I will escape with mine.

MICHAEL COLLINS

THE DAY OF THE trip to the moon started like any other day, with a quick shave and shower. To ensure the three astronauts were healthy and record their vitals one last time, nurse Dee O'Hara, every astronaut's favorite pulse-taker, gave each one a quick final physical. She'd been the personal nurse to all astronauts since before the first Mercury flight, and when she'd moved with them to Houston in 1962, she extended her services to their families as well. After all these years, O'Hara had an easy rapport with the men, and they appreciated her warmth, cheerfulness, and professionalism. A devout Catholic, she prayed the rosary on her beads during every mission, and she would for this one also.

Next was the traditional launch breakfast—steak, eggs, toast, juice, and coffee, as prepared by Lew Hartzell—with Deke Slayton and Bill Anders, the backup LM pilot. Paul Calle, part of NASA's art program, sat quietly in a corner sketching them. Then it was back to their rooms to brush their teeth and pack their belongings, after which they headed upstairs to the suit room. There they donned long johns and urine-collection devices before being helped into their spacesuits by Joe Schmitt's four-man team. Schmitt had handled these duties for Alan Shepard, Gus Grissom, John Glenn, and virtually every NASA manned flight since. Each astronaut carefully snapped his plastic bubble helmet down into the neck ring and locked it in place, then he plugged into a portable oxygen ventilator that he carried like a briefcase. This purged his body of any nitrogen, which prevented the bends when the cabin pressure was reduced soon after liftoff. The watch each man wore on his suit's wrist, an Omega Speedmaster, was set to Houston time.

Slayton, Joe Schmitt, and the three astronauts passed well-wishers as they walked down the corridors and out of the building, where a crowd of TV crews and photographers waited. They all got in the transfer van together, but Slayton was dropped off at the control center, where he wished the astronauts, who were fully suited and unable to hear anything but their ventilator fans, good luck. The crew was driven the last three miles to the launchpad. The sun was about to rise as they arrived at pad 39A. Sheets of frost slid off the Saturn where air had condensed and frozen against the super-cold fuel tanks. The astronauts and Schmitt and his assistant took an elevator to the base of the launch tower and then another to level nine, where they were met by Pad Führer Guenter Wendt. The three seats in the craft were positioned side by side below the hatch opening; since Aldrin would sit in the center seat, he would be the last one to enter the command module, so he was let off on the level below, where he waited near the elevator. Armstrong and Collins walked over the creaking thirty-foot swing arm into the small White Room surrounding the side hatch, where technicians began the elaborate process of buckling them in and connecting them to the ship's life-support system.

Wendt and his closeout crew had been busy for hours, and in the White Room there was now a brief pause while he and the astronauts exchanged gag gifts, a tradition that had developed over the years in an attempt to ease some of the pressure. He gave the crew a large key to the moon made of Styrofoam and aluminum foil. From his watchband, Armstrong pulled out a card that read *Space Taxi. Good Between Any Two Planets* and gave it to Wendt. While Schmitt helped Armstrong into the cabin, Collins, who had been carrying a brown paper bag, pulled out a wooden plaque with Wendt's name on it and the inscription TROPHY TROUT. To the plaque was nailed a frozen (but rapidly thawing) trout all of seven inches long. Wendt was an avid fisherman, and Collins had been out on the water with him many times.

While he waited, Aldrin drank in the view: the reddish sun just rising above the azure-blue ocean half a mile to his left, the crowded highways and beaches in the distance, Cape Kennedy and the Vehicle Assembly Building to his right. Thirty feet away was the booster, frost dropping from it and liquid oxygen boiling off. "I could see the massiveness of the Saturn V rocket below and the magnificent precision of Apollo above," he remembered with typical scientific appreciation. Then one of Schmitt's assistants tapped Aldrin on the shoulder, and it was his turn. Gifts were usually lighthearted, but Aldrin gave Wendt, a fellow Presbyterian, a copy of *Good News for Modern Man*, a condensed version of the Bible. Schmitt helped him into the cabin and got him secured.

Fred Haise was still in the command module, in the space below the cloth-and-canvas seats called the equipment bay. He had been busy for hours reviewing switch positions and running through a checklist 417 steps long so that when the crew entered, they would have little to do besides throw a few more switches on the wraparound instrument panel. Haise finished up, wriggled out of the hatch behind the center couch, then reached back in and shook each man's hand. When it was time, Wendt tapped Aldrin on the helmet and wished the crew luck, and the hatch was closed and locked. A few minutes before eight a.m., the Pad Führer and his team descended to the ground, and the swing arm was pulled away.

The crew was alone atop the thirty-six-story rocket with its six million pounds of propellants.

It was about then when Collins looked over and noticed that the abort handle at Armstrong's left side, now powered up and ready, was dangerously close to a large pocket on Armstrong's suit. Just one counterclockwise twist would fire the three rockets of the escape tower above them and jerk the command module up and away from the stack below it. A slight movement of his left leg could snag the handle, so Collins pointed it out to Armstrong, who quickly pulled the pocket as far to the right as he could.

Unlike Apollo 9 and Apollo 10, Apollo 11 fascinated the entire country—actually, the entire world. People from every state in the Union and many countries outside it had begun descending on the area a week before the launch, and by the morning of July 16, there were almost a million of them. There were no motel rooms available within a fifty-mile radius of Cape Kennedy, so some motels allowed extra guests to sleep in the lobby or on deck chairs around the pool. The rest of the visitors congregated along U.S. Highway 1 and the beaches that ran parallel to it, setting up camp in their tents, crude shelters, vans, trailers, and cars. Every public park in the area was converted into a campground. There seemed to be a lot more kids around than usual for a launch, likely because parents wanted their children to witness history in the making.

Sequestered three miles away near the VAB were the six thousand special guests NASA had invited: nineteen governors, forty mayors, sixty-nine foreign ambassadors, thirty-three senators, two hundred congressmen, untold numbers of senior NASA employees and Apollo contractors, and plenty of other dignitaries and celebrities, including former President Johnson and his wife, Lady Bird. Jim Webb was also there, ready to witness his first rocket launch. Charles Lindbergh, who had inspired so many of the astronauts and engineers, was there, sitting next to an old friend named Claude Ryan. In 1927, Ryan's small California aviation company had built a plane to Lindbergh's specifications for an ocean flight the young pilot had planned. The *Spirit of St. Louis* had flown from New York to Paris in thirty-three and a half hours. Forty years later, Ryan's firm had built the landing radar for Apollo 11's LM.

That morning, a few buses had dropped off about a hundred more special guests, among them the Reverend Ralph David Abernathy, a civil rights leader and close friend of the late Martin Luther King Jr.; the two had led the Poor People's Campaign to DC the previous summer to focus the nation's attention on problems of hunger and poverty. Abernathy had not originally been invited, but the day before the launch, he had arrived at the Cape with two hundred and fifty followers and two mule-drawn farm wagons to protest the billions of dollars spent on space exploration while those same problems existed. At Cape Kennedy's visitors' center, they were met by Thomas Paine, a Democrat whom President Nixon had decided to keep on as NASA's acting administrator after a fruitless search for a qualified Republican. Abernathy tried to make clear that he was proud of the space program's accomplishments. Paine told him that he was a member of the NAACP and sympathetic to their cause. "If we could solve the problems of poverty by not pushing the button to launch men to the moon tomorrow, then we would not push the button," he said. He invited the reverend and a delegation of his protesters to watch the liftoff the next morning from the VIP viewing area. The group spent the night at a nearby trailer park. Early in the morning, buses had picked up Abernathy and a hundred of his people. Now they were here to watch the launch with NASA's other invited guests.

Nearby, at the press site on the shore of the barge canal, there were 3,493 American journalists and 812 reporters from fifty-five countries. Three thousand boats of varying sizes floated in the

Banana River to the east and the Indian River to the west. On one of them, a motor cruiser owned by North American Aviation that was moored three miles from the launch site, Jan Armstrong would watch the launch with her boys, Rick, twelve, and Mark, six. The Aldrin and Collins families had elected to remain in Houston to avoid the mad crush at the Cape.

About a hundred and seventy-five miles off the coast were uninvited guests with a different agenda. While the vessels the Soviets usually deployed to monitor launches from Cape Kennedy were fishing trawlers bristling with radar and antennas, now they'd sent a flotilla that included a guided missile cruiser, two destroyers, two submarines, and a sub tender, all of them spread out over an area of twelve square miles, their heavy instrumentation a sure sign they were ready to track Apollo 11 from liftoff to orbit.

For a long time, NASA had suspected that the Soviets were jamming communications during missions. Just in case they tried this time, a massive surveillance dish was hoisted onto the roof of the VAB. If a disrupting signal was detected, the dish could pinpoint the location. The countdown would be halted and corrective action taken—hopefully, a simple change in frequency would do the job.

The Soviets had all but conceded the prize hanging in the sky 240,000 miles away. But they had one more trick up their sleeves.

After Premier Nikita Khrushchev had been deposed and the Gemini program began to pull the Americans ahead in the space race, Sergei Korolev was able to convince the USSR's new leader, Leonid Brezhnev, of the importance of a lunar landing. But the Soviet moon program had started late and never caught up. It faced many obstacles, including severe underfunding, inefficiency, insufficient testing, the Marxist state's distrust of science, and the military's penchant for siphoning off funds from politically motivated space projects toward newer and better strategic systems.

Korolev's death, in January 1966, had dealt a severe blow to the project.

But Vasily Mishin and his team had plowed ahead, still hopeful that they could beat the Americans. The Soviet cadre of cosmonauts picked for the lunar missions hadn't begun training until January 1968. Their insufficient funding precluded proper simulators, so they practiced landings with jury-rigged copters. Spacewalker Alexei Leonov was the front-runner for the honor of commanding the first mission.

They were aiming for a late 1969 or early 1970 lunar landing and had believed it unlikely that the U.S. program could recover from the Apollo 1 tragedy quickly and then pull off the series of highly complicated step missions required to make a landing by the end of the decade. But their own frequent failures—due in large part to poor quality control and a singular lack of strong ground-testing—colored their opinion; they didn't know how well engineered and how thoroughly ground-tested Apollo was.

In the summer of 1969, the West had little information about the Soviet plans for space other than what scientists could gather from the photographs of Webb's Giant. Earlier in the year, there had been statements from academicians and cosmonauts, including Leonov, indicating they were working on a moon landing. No one outside the USSR knew that the Soviets had already had a trial run of the huge moon rocket; the CIA's spy-plane surveillance missed it. Although the N1 had never been ground-tested, on February 21, Mishin's team was confident it was ready, and an unmanned stack had been launched at Baikonur. The rocket lifted off, but sixty-nine seconds later, severe pogo problems resulted in a fuel leak that started a fire, causing all thirty of the N1's first-stage engines to shut off. At seventeen miles up, it turned earthward and crashed thirty-one miles from the launch site, though the launch escape system pulled the spacecraft mock-up away to safety.

The problems were considered fixable, and a second N1 was prepared for launch. Almost five months later, at 11:18 p.m. on July 3, they tried again. This time, the rocket barely cleared the gantry before it collapsed onto the launchpad and exploded into a sun-bright fireball that quickly became a purple-black mushroom cloud. Large white-hot pieces of debris rained down as far as six miles away, and windows twenty-five miles away shattered in the most powerful non-nuclear explosion in the history of rocketry. Miraculously, no one was killed.

The CIA had missed the February failure and would not discover the July 3 disaster until mid-August, when spy-plane photos revealed that a large launch complex had been destroyed. It didn't matter. With the N1's failure died any Soviet hope of beating the Americans to a manned lunar landing.

On Sunday, July 13, three days before the Apollo 11 liftoff, the USSR launched another of its Luna probes toward the moon. With characteristic reticence, TASS, the Soviet news agency, noted only that its mission was to conduct experiments around the moon and near it. The probe was widely seen as a final desperate attempt to upstage the Americans; possibly the Soviets meant it to land, scoop up some soil, and return to Earth, thus achieving a couple more lunar firsts. Others postulated a more sinister purpose: maybe the Soviets would try to observe Apollo 11...or interfere with it, or even shoot it down.

By the morning of the Apollo launch, the three-ton Luna 15 was nearing the moon on a minimum-energy trajectory, and Chris Kraft was incensed. The possibility of a collision with Apollo 11 was remote, but he was convinced that the Soviets had more than once deliberately operated their communications at or near American radio frequencies during missions. He mentioned the Russian probe during a press conference to put some pressure on them, but that hadn't seemed to do any good. He couldn't exactly ask the Soviets for their mission trajectory and communication details. But he knew someone who could.

Frank Borman and his family had just returned from a nine-day visit to the USSR, where he'd been welcomed warmly, met some cosmonauts and scientists, and drunk many vodka toasts. It was the first time an American astronaut had been allowed in the country. Kraft called Borman, still working as NASA's White House liaison, and explained the situation. Borman put a call through to Dr. Mstislav Keldysh, president of the august Soviet Academy of Sciences. The two had hit it off and had discussed cooperation between their space programs.

He'd left Keldysh a message asking for the orbital parameters of the Luna probe and looking for reassurance that it wouldn't interfere with Apollo 11. He hadn't heard back by the day of the launch, but maybe the Russian would get in touch soon. Keldysh had seemed like an intelligent and reasonable man.

The morning was already sweltering, with a bright sun and the temperature near ninety degrees.

The countdown went smoothly save for a couple of minor problems that were fixed without causing a delay. At T minus nine seconds, the five massive F-1 engines ignited, and at zero, as they reached their full thrust of 7.6 million pounds, the launch tower's swing arms pulled back, the twenty-ton hold-down clamps at the booster's base sprang free, and hundreds of thousands of spectators watched, transfixed, as the 6.5-million-pound spacecraft began to rise, sluggishly at first, its thrust-to-weight ratio so close that it appeared to ascend in slow motion until it finally cleared the launch tower, sheets of ice breaking off into thousands of shards. At 9:32 a.m., exactly on schedule, Apollo 11 blasted off toward the moon with three men in a small nose cone atop the largest rocket ever sent into space.

The sound took fourteen seconds to reach the closest observers, three miles away, and when it did, the roar was deafening. The ground shook as Apollo 11 continued to soar upward on a long pillar of fire and then arced out over the Atlantic to begin its five-hundred-thousand-mile round-trip journey.

In firing room 1 in the launch control center, beside the VAB, Wernher von Braun stood next to George Mueller in the mission managers' row. While the last twenty seconds of the countdown ticked away, he put down his binoculars and stared out the large blast-proof windows, then bowed his head and began to recite the Lord's Prayer to himself. As the rocket climbed, the four-hundred-man countdown team started cheering, and von Braun lifted his head and joined them.

It was surprisingly quiet in the command module—the F-1s were a distant rumble, not unlike a commercial airliner's engines at takeoff. The spacecraft shook as it lifted off, and the astronauts were jostled against their straps as the four outer engines swiveled back and forth, adjusting to stay balanced and straight—if the rocket came into contact with the launch tower, it would mean catastrophe. The men felt only 1.25 g's, slightly more than normal, but far less than they had endured in training. The force steadily increased as the rocket picked up speed, and two minutes and forty seconds into the flight, at the end of the first-stage burn, the men felt four times their normal weight.

When the rocket was forty-five miles up, the astronauts were jerked forward as the spent first stage fell away into the sea. The five smaller J-2 engines of the second stage took over, and the ride smoothed out. A few seconds later, at sixty miles altitude, the launch escape tower was jettisoned, and at a hundred and ten miles, the single J-2 of the third stage ignited, this one causing a rougher, shakier ride as the second stage released. The ascent to orbit had taken twelve minutes.

At least fourteen danger points would occur during their flight. NASA called them go/no-go decision points, those critical events that involved complex mechanisms and split-second timing that had to transpire flawlessly for a successful mission. (There might even be a few more than fourteen, depending on how many midcourse trajectory corrections were needed.) At every point, the flight director would ask each controller whether the system he monitored was functioning properly and ready to go. The launch had been the first danger point—every rocket launch was dangerous, considering the massive amounts of propellants involved. The next point, which would occur during the second orbit at a hundred and ten miles altitude, was translunar injection (TLI). Restarting the third stage's single J-2 engine for a five-minute-and-forty-seven-second burn would, if perfectly timed, increase the spacecraft's speed to 24,258 miles per hour, fast enough to pull it out of Earth orbit and propel it toward the moon—or, rather, a specific point where the moon would be when Apollo 11 reached it in three days. And restarting the J-2 was risky, given the nature of its propellants; liquid oxygen at negative 297 degrees and liquid hydrogen at negative 423 degrees had to be handled extremely carefully. But Collins had practiced the burn many times. Midway through the second orbit, after checking out their spacecraft to make sure all systems were nominal, the crew donned their helmets and gloves again, in case the TLI went badly, then waited for word from Mission Control.

It finally came as they passed a hundred miles over Australia. "Apollo 11, this is Houston. You are go for TLI."

The burn went off without a hitch. The cabin shook and the thrust pushed the astronauts back in their seats at one g. Then the engine shut down automatically and they were on their way to the moon. Von Braun's rocket had performed flawlessly once again. "Hey, Houston, Apollo 11," said Armstrong after shutdown. "That Saturn gave us a magnificent ride."

A short while later, Collins separated the command-service module from the third stage, turned around, then docked with the LM, whose four-panel protective shroud had peeled away like large silver petals. Then one of the astronauts threw a switch on the control panel that released the LM from the Saturn third stage, and the odd-looking spacecraft—the command-service module secured nose to nose with the LM—continued moonward. The almost empty third stage's trajectory would be changed to send it into an orbit around the sun.

Besides the attitudinal thrusters, they had one large engine left, the big service-module engine attached to their rear that resembled a large bell. Its thrust of 19,500 pounds would be used on several occasions during the next few days. It had to work every time. Now they started it for a three-second burn to get safely away from the third stage and fine-tune their trajectory. Its light kick—one-fifth of a g—was reassuring.

The passive thermal control was next. In the vacuum of space, with no atmospheric protection against the sun's rays, the temperature was over 280 degrees on the side facing the sun and negative 280 degrees on the other side. Fuel-tank pressures could rise to dangerous levels, and radiators and other parts could freeze, so to distribute the sun's heat evenly, Collins used his thrusters to position the spaceship broadside to the sun and then induced it to rotate slowly on its long axis—one full turn every twenty minutes, hence the nickname "barbecue mode." Apollo 11 was now slowly spinning at an angle as it moved toward the moon.

Two delicate maneuvers were behind them with no more scheduled until they reached the moon's vicinity in three days. The crew appeared to relax, although it was hard to tell with Collins, who tended to mask anxiety with humor anyway. Armstrong joshed with CapCom Jim Lovell, his backup, who had been teasing him about taking his place. Then Lovell and Aldrin chatted; the two had been crewmates on Gemini 12 three years before.

After their course was set, the crew changed out of their bulky spacesuits into two-piece nylon jumpsuits, a difficult chore in zero gravity. They took turns; as one man bounced around the cabin, his shipmates helped him. The suits were folded, bagged, and stowed under the center couch. No one had shown any signs of space-sickness or any other illness, and no major problems had cropped up, although there were some minor ones, like an oxygen-flow indicator that malfunctioned. It was deemed useless, irreparable, and unnecessary, since Mission Control could monitor oxygen. Everything appeared shipshape. After the crew took care of various chores, it was time for dinner. The food, packets rehydrated with a hot-water gun and eaten with a spoon, was good—their first meal of chicken salad, shrimp cocktail, and applesauce eaten with a spoon was a major improvement over the tubes they'd squirted into their mouths on Gemini flights. Each man had meals color-coordinated for him and planned out for every day, and there was a well-stocked snack pantry they could partake of. (The kitchen "cupboards" were on the left side of the cabin, making the right side, with its waste-management systems, the bathroom area.) And since the sun was always on them, there was no sunrise or sunset, so they operated on the time their watches were set to, Houston's central daylight time. At 10:30 p.m. CDT, fourteen hours after liftoff, they turned the radio down, fastened covers over the windows, and closed their eyes.

They soon settled into a routine. They would wake up, have breakfast—including dehydrated coffee, predictably disappointing even with hot water—and clean up. The CapCom relayed the flight-plan updates, relevant telemetry readings, the news from Earth, including the baseball scores, and the latest on Luna 15 (which by Thursday had entered orbit around the moon). They would troubleshoot minor problems, and Collins would take star sightings to check location and trajectory; on the ground, Mission Control would continue to monitor telemetry and relay any necessary information.

The three Gemini vets enjoyed the experience of weightlessness, particularly in a spacious cabin with room to move about. *Spacious* was a relative term, though; it was roomier than either the Mercury or the Gemini spacecraft but still not much larger than the inside of a station wagon. But the lack of gravity made the area above them usable also, so the short connecting tunnel leading to the LM could serve as a cubbyhole to relax in. Below the couches in the equipment bay, one man could stand up fully, and two men could stand there after the center seat was folded up. There was room for two men to stretch out and float weightless in their zipped-up, lightweight nylon sleeping bags slung fore to aft under the left and right couches. The third crewman slept above them in the left-hand seat with its seat folded flat, loosely buckled in to keep from floating away and with his headset volume lowered—a call would come in only in an emergency. Though there was no true day or night, their bodies' circadian rhythms kept them on the same sleep/wake schedule they had always known. With the craft's shades pulled down over the windows and its interior lights dimmed, only the soft whirring of ventilators interrupted the unearthly quiet of cislunar space that lay just a few inches outside the gray alloy walls.

The gravitational pull of the Earth extended more than halfway to the moon, and their escape velocity of almost 25,000 miles per hour gradually diminished to a tenth of that speed. Not that the crew could tell; since there was nothing to compare their speed to—no trees or buildings or asteroids whizzing by—they had no sense of movement. Only the barbecue roll, the gauges, and the slowly shrinking Earth gave the men any indication that they were moving. Their destination couldn't be seen yet because of its proximity to the blinding sun.

Besides sleeping, daily chores, and the near-constant chasing after floating items of all sizes—from pens and sunglasses to crumbs and cameras, each of which had to be stowed away or anchored to one of the dozens of Velcro panels affixed to the cabin walls—the astronauts did not have much to do in cislunar space once the spacecraft was set on the proper trajectory, and the computer and Mission Control were in charge. Most other flights included a long list of experiments; not this one. They could be distractions, and this crew had to be as rested and prepared as possible for their Sunday excursion, although they would set up a few experiments on the surface. But daily telecasts had been planned, and on Thursday evening, the second day out, the crew aimed a color-TV camera out the window and shot the Earth for fifteen minutes while delivering a running commentary on what they were seeing. "Hey, Houston," Buzz said, "do you suppose you could turn the Earth a bit so we could get a little bit more than just water?" Then twenty-one minutes were spent showing the inside of the command module.

When the camera was turned off, there wasn't a lot of conversation among the three men, especially compared to their Apollo 10 predecessors. "It's all dead air and static," said one NASA official. Armstrong and Aldrin were laconic by nature, and they had the descent flight plans and checklists to review and ninety-two large-scale lunar-surface photos to study. Collins was more loquacious, although not to the point that he talked to himself. But their moods had definitely lightened up; even Aldrin had shaken off his not-the-first-man disappointment and seemed to be in a fine mood, validating Slayton's faith in him. He wasn't even trying to communicate telepathically.

Collins took care of most of the housekeeping; he charged batteries, purged fuel cells, dumped wastewater, chlorinated drinking water, and prepared most of the meals, the solar system's only combination pilot/cook/housekeeper/handyman. Sometimes they played music on a small cassette player, mostly easy listening and classical, including Dvořák's Ninth Symphony, "From the New World," and a favorite of Armstrong's entitled *Music Out of the Moon*, a 1947 album of jazz instrumentals featuring the otherworldly sounds of a theremin. One day, Collins ran in place for a while, his hands above his head against the bulkhead. So did Armstrong, briefly—a mild shock to anyone aware of his avoidance-of-exercise philosophy.

On the distant blue sphere behind them, the crew's families watched telecasts with friends and the occasional astronaut. This wasn't their first space rodeo; each of the three women had previously endured what the wives had come to call a "death watch"—waiting and watching at home to see if their husbands would survive a mission. But the media attention for this one was off the charts. Between *Life* reporters and photographers in their houses, and TV crews and newspeople outside on their front lawns, there weren't many places to avoid the spotlight. They were dutiful and gave interviews; everyone in the astro-community knew it was expected. But they tried to balance the attention with some sense of normalcy, so the three families alternated between watching the TV broadcasts and going on with ordinary activities. They ate at their favorite restaurants and snuck out, hidden in their neighbors' cars, to shop and get their hair done (they had to look good for the cameras), though one female reporter tracked a wife to the beauty parlor and had her hair done too. On Friday, the Aldrin family, which now included Buzz's uncle Bob Moon and his wife, hosted an afternoon pool party. The other two wives attended, and all three talked to the press on the front lawn.

* * *

On Thursday evening, Frank Borman had received a phone call from Russia. It was Keldysh's assistant saying that the academy president had gotten his message. A little while later, Borman received a cablegram from the USSR with the precise trajectory of the Luna probe's orbit and the assurance that it would not intersect with that of Apollo 11; Borman would be notified if there were any changes. It was signed *Keldysh*. For the first time, the Soviets had revealed mission details to their Cold War rivals while the mission was still in progress. There was no mention of the probe's purpose or its radio frequencies, but Borman and Kraft took it on good faith that there would be no radio interference. Late the next morning, the two held a short press conference to announce the news; that information was relayed to Apollo 11, and everyone breathed a sigh of relief.

Friday afternoon, the astronauts took the TV camera out again for a twenty-minute shot of Earth, then shots of the crew and the cabin, and then, for the first time, a shot of the LM. After Collins removed the probe and drogue assembly from the connecting tunnel between the LM and the command module, Aldrin and Armstrong floated up and squeezed through the yard-wide opening into the LM to begin preparing for the separation on Sunday. They followed that with a ninety-six-minute telecast during which they roamed through both modules and displayed an Earth no larger than a silver dollar and becoming even smaller.

At 11:12 p.m., traveling at a snail's pace of 2,040 miles per hour and just 38,000 miles from their destination, the spacecraft entered the moon's gravitational field. The craft began picking up speed, and the men woke up early on Saturday to find themselves only ten thousand miles from the moon. While they ate breakfast, they entered the moon's shadow, and a few hours later, they stopped the barbecue mode and swung the joined spacecraft around. For the first time in almost twenty-four hours, they had the opportunity to see the moon clearly. It was no longer a distant, flat-looking, grayish-white orb the size of an extended thumb but a massive, darker, and fully three-dimensional world that filled the window and brightly reflected the earthshine—four times as bright as moonshine—from their home planet behind them. The light lent a cold bluish tint to the moon's features, adding shadows to its craters and mountains. It looked close enough to touch and cold and ominous. "It's a view worth the price of the trip," Armstrong said.

They could also see stars again; the dazzling sunlight had prevented that for days. "The sky is full of stars, just like the nights on Earth," reported Collins.

The seventy-three-hour voyage—the translunar coast—had been, in the best sense of the word, uneventful. Only one three-second burn on Thursday had been necessary to fine-tune their trajectory, and that had been so accurate that two others had been canceled. Now, as they neared the moon, another danger point approached: lunar-orbit insertion (LOI). They had to turn their ship around and apply a braking burn that would slow their speed from 5,225 miles per hour to 3,248 miles per hour, allowing the craft to be captured by the moon's gravity and drop down to an elliptical orbit. If they didn't get it right, Apollo 11 would sail around the moon in a huge arc and then head back toward the Earth in a free-return trajectory—or it would be carried the other way, toward the sun. If the burn was too long by just a few seconds, their reduced speed would send them crashing to the lunar surface. They had to fire up the service-module engine again for six minutes and three seconds—and they had to do it eight minutes after they disappeared behind the moon, where they'd be unable to communicate with Mission Control if a problem occurred.

The burn would be handled by the onboard Apollo Guidance Computer (AGC). After checking their numbers several times, they would have to enter the directions and figures manually. One wrong digit could mean a catastrophic change to their trajectory.

Both the LM and the command module carried an AGC, and it was up to the task. Designed by the MIT Instrumentation Lab and built by Raytheon, it was a wondrous machine. Most capable computers of the day took up a large room, like the IBM System/360 Model 75J mainframes on the first floor of the Mission Control Center in Houston, and a mini-computer was the size of a phone booth. The two carried on Apollo 11 were each the size of a briefcase—not quite portable, but the first embedded computer, and a micro-computer for its time. Computers used large, fragile disks and tapes, and they broke down frequently; this one absolutely, positively could not. To increase reliability, the AGC employed a fixed core rope memory: thin copper-wire ropes woven by a staff of Massachusetts women whose ancestors had worked in the New England weaving industry and who would maintain their expertise between mission-software jobs by knitting and getting paid for it. (Some at NASA called this long, painstaking process the "little old lady" method.) Using looms, they would thread hundreds of these wires around and through many rows of magnetic ceramic-ferrite cores, or microchips, affixed to a board, and each one would take six weeks to build and test. It was one of the first computers to use these new miniaturized, integrated circuits. The result was a computer with a 36,864-word fixed memory and 2,040 words of erasable memory—the equivalent of seventy-two kilobytes—with a processing speed of roughly one megahertz. That was a limited amount of memory, but it was enough to perform the tasks assigned to it, such as measuring velocity changes, determining rendezvous and course corrections, and making minute adjustments to the trajectory during descent. Moreover, the AGC was nearly indestructible and would immediately recover from a crash or overload and continue right where it left off. The AGC would also prioritize jobs in case of memory overload and drop those not considered essential.

The computer's keypad interface was an eight-inch-by-eight-inch DSKY (display and keyboard) on the control panel of each spacecraft (there were two in the command module), with a calculator-type, nineteen-pushbutton keyboard, ten indicator lights, and a digital display. The computer was capable of solving real-time problems, and though initiating tasks could be cumbersome, requiring many key presses, they were performed quickly. They had to be.

As the flight progressed, shifts came and went in Mission Control. Cliff Charlesworth's Green team had been on console for the liftoff; six hours later, they yielded to Kranz's White team; eight hours after that, Glynn Lunney's Black team took over, and then it was the Green team's turn again. The aroma of stale sandwiches, pizza, Mexican food, and burnt coffee mingled with the blue-gray haze of cigarette smoke in the subdued lighting. Kraft, Gilruth, Paine, and Mueller often watched the proceedings from the top row. An ever-changing cast of astronauts moved in and out of the room to sit and stand near the CapCom console; others watched in the first row of the glass-fronted VIP area, where people came, lingered for a while, and left. Many off-shift controllers on their way home made a stop at the Singing Wheel just a mile down the road to decompress and talk shop. Others remained at Mission Control Center, spending the night in the flight controllers' dormitory on the floor above the MOCR.

Just before noon on Saturday morning, Charlesworth polled his flight controllers one by one on lunar-orbit insertion—the burn that would slow down the spacecraft enough to drop it into orbit—then he gave astronaut Bruce McCandless, the CapCom, the okay. At 11:58 a.m., McCandless told the crew, "You're go for LOI."

Aldrin said, "Roger. Go for LOI."

"All your systems are looking good going around the corner," said the CapCom, "and we'll see you on the other side."

Fifteen minutes later, the spacecraft disappeared behind the moon, just 309 nautical miles from its surface, and Mission Control became noticeably quieter. A few groups of men stood chatting in low voices. Flight controllers stared at their consoles or the big TV monitors in the front of the room. On the large lunar map, the radar dot representing the spaceship had moved to the left edge and then vanished. If all went well, it would reappear on the right side of the map after about thirty-five minutes.

In their craft behind the moon, the crew entered the numbers, and then Collins punched in the command to start the service-module engine again. It fired right up and burned five minutes and fifty-seven seconds, placing the spacecraft into an elliptical orbit that ranged from sixty to one hundred and seventy nautical miles above the moon's pockmarked far side—a near-perfect burn and orbit insertion.

Using the thruster jets, Collins maneuvered the spacecraft so the moon was visible in the front windows, and they spent the next twenty minutes gazing down at it and taking photos. The rugged surface, completely covered with craters of all sizes, was even more forbidding than the near side. The satellite's cold, stark visage contrasted with the vibrant blues and greens of the Earth they'd left behind. Like the crew of Apollo 8, they argued over its color and came to the same conclusion: various shades of gray, with some browns here and there, depending on the angle and the light. Three days after leaving Earth, they finally experienced a sense of movement as they made a complete circle of the moon every two hours.

The crew was mesmerized by what they saw. Though Collins had taken far fewer geology lessons than his shipmates, he became excited over one area below them. "Oh, boy, you could spend a lifetime just geologizing that one crater alone, you know that?"

Armstrong said, with little enthusiasm, "You could..."

"That's not how I'd like to spend my lifetime," Collins said, "but picture that. Beautiful!"

Aldrin had been quiet for a while. He said, "Yes, there's a big mother over here too."

"Come on now, Buzz," said Collins. "Don't refer to them as big mothers. Give them some scientific name."

A few minutes later, they edged around the moon's right side.

"There it is," Aldrin said. "It's coming up!"

Collins said, "What?"

"The Earth. See it?"

"Yes," said Collins. "Beautiful."

As they took it all in, there had been nothing but static from Mission Control. Then, right on schedule, they heard McCandless's voice.

"Apollo 11, Apollo 11, do you read? Over."

"Yes, we sure do, Houston," replied Aldrin. "The LOI burn just nominal as all get-out, and everything's looking good."

It took a minute or two before communications were fully restored. Then McCandless inquired about how the burn figures looked.

With a grin, Collins said, "It was like—it was like perfect!"

In Mission Control, the tension dissipated, the volume increased, and many onlookers began to filter out.

A couple of hours later, at 2:56 p.m., the crew began a thirty-five-minute TV broadcast. From a side window, they focused the camera on the moon below, pointing out the LM's planned flight path and the landmarks along it highlighted by the sun's long shadows. The crew took turns commenting on what they saw. Then they got their first view of their landing site in the Sea of Tranquility. They swiveled the camera to keep it in the frame as they ended the broadcast.

"And as the moon sinks slowly in the west," said Collins, "Apollo 11 bids good day to you."

About an hour later, they made a second burn on the far side—this one only seventeen seconds—to drop Apollo 11 down to a slightly lower and almost perfectly circular orbit of fifty-four by sixty-six nautical miles. Two more danger points had passed—now six in all. There were several more to come.

Armstrong and Aldrin floated up into the LM to give it another once-over and prepare for the next day. They spent two and a half hours powering it up, presetting switches, and working a long communications checklist. Then the three gathered in the command-module cabin to eat dinner as soft music played on the small tape recorder. Afterward, Collins tended to the usual housekeeping while his crewmates got their suits and equipment ready. They prepared to retire. Collins would sleep top-deck on the left seat, with his headset on in case of emergency. Armstrong and Aldrin would be in their sleeping bags below; they both needed a good night's rest. The temperature inside was sixty-nine degrees, a few hundred degrees cooler than outside in the sunlight. It was almost midnight on Sunday—the day they would attempt what they had come all this way for.

Collins felt the need to say something. "Well, I thought today went pretty well," he said. "If tomorrow and the next day are like today, we'll be safe."

A half an hour later, as their spacecraft continued to circle the lifeless moon sixty miles below, they were asleep.

CHAPTER SIXTEEN

DESCENT TO LUNA

The unknowns were rampant.

NEIL ARMSTRONG

"APOLLO 11. APOLLO 11. Good morning from the Black team."

It was six a.m. Collins, loosely belted and floating over the left seat with his headset on, took a while to wake up. "Good morning, Houston... you guys wake up early."

"It looks like you guys were really sawing them away," said Cap-Com Ron Evans.

A few minutes later, the spacecraft disappeared around the moon's left side for the ninth time. None of them had slept long—six hours at most. Aldrin had had a fitful night, but the other two had slept soundly. All three ate breakfast. In Mission Control, when Evans saw their craft reappear around the right edge of the moon, he read them the day's headlines. He reported that Miss Philippines had won the Miss Universe crown the night before, that congregations in churches around the world were including the Apollo 11 crew in their prayers, and that the Russian Luna probe was still circling the moon, though its purpose remained a mystery and its orbit far from theirs. After the crew heard the latest updates on their families, it was time to get down to business.

In Mission Control, there was about to be a changing of the

guard. Glynn Lunney's Black team had been there most of the night while the crew slept, and they were about to hand over the console to Gene Kranz's White team. They wouldn't go far—Black was also the ascent team and had to be present during the descent in case of an emergency liftoff, so most of them would stick around, sitting in chairs behind their replacements. The two teams compared notes, and the White team began computing numbers for the upcoming maneuvers.

Kranz arrived carrying two bags. Others in NASA, especially the younger flight controllers, had let their hair grow out, but not him; he still sported a crew cut. His sharp features and intense gaze gave his face the look of a bird of prey. The larger bag held his lunch, several snacks, and candy bars—he liked to eat his way through a shift. In the other bag was his new mission vest, made by his wife, Marta—a tradition started back in Gemini. Each one was slightly different and always a surprise for the controllers. This one, which he called his landing vest—the first—was white-and-silver brocade. Kranz made his way to his console on the second row—there were four rows in all, with preassigned consoles—and put the food in a drawer and his vest on. When Chris Kraft walked in and passed by Kranz, he patted him on the shoulder and said, "Good luck, young man." He took a seat on the level above, management row, where he was soon joined by most of NASA's top brass: Bob Gilruth, Robert Seamans, George Low, and Sam Phillips. Kraft and Low had determined that the overall probability of success for the entire mission was roughly 56 percent. Now Low turned to him and said, "I've never seen this place so tense."

The glass-fronted viewing room overlooking the MOCR seated seventy-four. On this occasion, there were many more than that standing in the aisles, on the sides, on the steps. The gathering was an illustrious collection of NASA officials and other dignitaries, among them Wernher von Braun, Eberhard Rees, Kurt Debus,

Rocco Petrone, John Houbolt, Thomas Paine, Bill Tindall, Guy Thibodaux, and several astronauts, including the first American in orbit, John Glenn. On the right side of the MOCR was the simulation-control area, where instructors lined the windows to watch. In staff support rooms throughout the building, systems experts and contractor representatives sat, ready to aid in any way they could.

As Kranz settled in at his console, he looked up into the viewing room and saw Tindall there. He waved him down into the MOCR. Tindall declined, but Kranz insisted. If anyone deserved to be there, it was Tindall; he had been instrumental in making this moment a reality. Tindall made his way in and Kranz cleared off a chair right next to him.

The evening before, after their short four-hour shift had ended at about nine, Steve Bales, Jay Greene, and a couple of other controllers on the descent team had gone out to Perusina's, a good steak house on the Gulf Freeway in Dickinson. They'd returned to the Mission Control Center building and gone to sleep in the controller dormitory above the MOCR. It was large enough to sleep twenty and had its own showers. Bales had woken up refreshed, eaten at the temporary cafeteria set up in the lounge next door, and walked into the MOCR at 7:30 a.m. feeling good. It would be eight hours before the landing attempt, and he had a roomful of computer experts, including Jack Garman, standing by in a staff support room down the hall, as well as a bunch of MIT engineers, the men who had created the computer, sitting in a Cambridge, Massachusetts, classroom listening in on a dedicated line. Every flight controller had a similar support network.

But Bales's good mood was short-lived. As his team began their shift, the pressure they felt was evident. No one was smiling. He took his seat at the far right on the bottom row, near the rear-projected screens in front, with the rest of the flight dynamics team.

At his right sat Granville Paules, a tall, blond guidance officer who would support Bales on the guidance systems, including the on-board computer. To their left was flight guidance officer (FIDO) Jay Greene, a pipe-smoker from Brooklyn, who would handle trajectory—where the spacecraft was and where it was going. He was not shy in making his opinions known. To his left, retrofire officer (RETRO) Chuck Deiterich, a native Texan, would manage its return, whether nominal or an emergency; he would be ready with possible abort options if necessary. Like most members of the Trench, he had an attitude that said, *We're the guys really running the show—the rest of you are just plumbers.* These men thought of themselves as astronauts on the ground, the ones who really did the job of running the spacecraft.

In the next row up were the flight systems controllers. Right behind Bales, on the last console on the right, was dry, imperturbable Bob Carlton, the veteran controller in charge of the LM's guidance, navigation, and control. At thirty-nine, he was considered the old man of the group; the average age of the rest of the White team was twenty-six. They called him the Silver Fox, for his gray hair. On the extreme left side of the row sat the flight surgeon. Next to him was the CapCom, astronaut Charlie Duke, who spoke with a light North Carolina drawl. He hadn't been assigned to a flight yet, but he'd done such a good job in the same role for Apollo 10 that Neil Armstrong had requested him for the descent. Duke knew the LM propulsion systems well, and that wouldn't hurt. His job was important, and difficult, for he would have to sort through several voice loops to get the information the astronauts would need and then deliver it to them on the ground-to-air loop—the only one they would communicate on. Jim Lovell and Fred Haise sat on his left, and behind Duke was Deke Slayton, drawing on a cigarillo. Pete Conrad and Dave Scott, preparing for Apollo 12, sat nearby. Three other systems controllers sat between the CapCom and Carlton.

At about seven forty-five, Kranz and his controllers donned their headsets, plugged into their consoles, and began punching up various controllers' voice loops to monitor; depressing a foot switch or activating a control on a headset would allow them to speak on a loop. Most of them were smokers—cigarettes, cigarillos, or pipes—and one of their first orders of business was to empty the amber ashtrays, still full from the evening before, and light up.

Some members of the Black team left, wishing the White team good luck on their way out. John Hodge, the former flight director who had overseen Neil Armstrong and Dave Scott's Gemini 8 near catastrophe three years before and then moved into management, stopped to talk to Kranz.

Apollo 11 was still completing an orbit of the moon every two hours. In the command module, all three suited up—Armstrong and Aldrin in their liquid-cooled undergarments that would keep their temperature down on the hot lunar surface, Collins in his own suit—and then attended to various housekeeping chores. After that, Aldrin pulled himself through the tunnel and did more systems checks to *Eagle* while Armstrong, with Collins's help, struggled into his EVA suit in the navigation bay. Then Aldrin and Armstrong switched places, and Aldrin put on his suit. It took each man a half an hour to fully suit up, with their helmets and gloves locked into place, though fortunately their suits were only half pressurized for now, so they weren't as stiff as usual. About nine thirty a.m., Armstrong entered the LM. Once Collins had reinstalled the LM drogue and the command-module probe, Armstrong sealed the upper hatch.

An hour later, Duke said, "Apollo 11, Houston. We're go for undocking."

Three minutes later, *Columbia*, the command-service module, and *Eagle*, the lunar module, disappeared around the moon. When they reappeared at the right edge of the large map of the moon on a front screen in Mission Control, there would be two radar dots, not one. Kranz told his team to take five, and there was a mad rush to the restroom. It would be the last break before descent.

When everyone returned, Kranz directed his flight controllers to a private communications loop.

"We are about to make history," he told them. "We have trained and prepared for this moment and we are ready. But I want you to know this—whatever happens when we walk out of this room, we walk out together as a team." Then he switched back to the flight director's loop, ordered the doors to the MOCR locked until the landing attempt was over, and lit up another Kent.

"You cats take it easy on the lunar surface," said Collins on the other side of the moon. "If I hear you huffing and puffing, I'm going to start bitching at you." He was as aware as anyone of how strenuous an EVA could be—and how dangerous.

"Okay, Mike," said Aldrin.

Collins retracted the probe, vented the air in the tunnel connecting the two spacecraft, and, at 12:46 p.m., threw a switch that opened the capture latches. The release of the remaining air in the tunnel caused the two vehicles to gradually drift fifty feet apart. Armstrong maneuvered with his thrusters to regain the precise orbital parameters. Then he detonated explosive bolts that released *Eagle*'s spring-loaded landing gear. There was a shudder and a click as the four legs, including the spindly, sixty-seven-inch landing probes extending from three of them, swung out and locked into place.

They flew formation together—station-keeping—as Armstrong rotated *Eagle* one full revolution so that Collins in *Columbia* could inspect it for damage and make sure the legs and probes were in position. The twenty-three-foot-high LM consisted of two parts. The descent stage was essentially four large propellant tanks and a gimballed rocket engine encased in an octagonal metal chassis with four legs sprouting from it. Above that was the ascent stage, a hodgepodge of tanks, boxes, antennas, and radar surrounding a pressurized cabin just large enough for two astronauts in spacesuits, with another rocket engine underneath it. The cabin's walls were an aluminum alloy skin, about twelve-thousandths of an inch thick—the width of three layers of tinfoil. Most of the bottom half and the four slender legs were wrapped with gold Mylar insulation, and the top half was black, silver, and gray. It might have been mistaken for a large alien insect from another galaxy.

Collins made sure the landing gear was down and locked. A few minutes later, he said, "I think you've got a fine-looking flying machine there, *Eagle*, despite the fact that you're upside down."

"Somebody's upside down," Armstrong said.

"There you go...you guys take care."

"See you later," Armstrong said, as if he and Aldrin were going out to get a bite to eat and would be back soon.

So far, things had gone as planned. Collins fired the service-module engine to move two miles away. In *Eagle*, more system activations and reviews followed; they checked on their spacesuits, the guidance system, the control thrusters, the descent propulsion system, and its propellants. The radio signal was weak and staticky, and the air-to-ground from *Eagle* was dropping out, so they adjusted the antennas to improve it. Though the rendezvous radar wouldn't be needed for the descent, they powered it up so it would be ready in case they had to make a quick abort.

There were no seats in the LM, so both men stood shoulder to shoulder in the cramped, gray-painted cabin. Armstrong was on the left and Aldrin on the right, their boots secured and waists tethered, an array of controls, gauges, switches, and displays before them. Each man had a hand controller on either side of him, at waist level, and a small, upside-down triangular window sixteen inches in front of his face.

In Mission Control, fifteen minutes after separation, a radar dot appeared on the right side of the large moon map. It was *Columbia*, in a higher orbit now than *Eagle* and visible first.

A few minutes later, *Eagle* emerged from behind the moon, moving at about four thousand miles an hour, and reestablished radio contact.

"How does it look, Neil?" said Duke.

"The *Eagle* has wings," Armstrong said.

After Kranz polled his controllers, Duke said, "You're go for DOI"—descent-orbit insertion. Seven minutes later, *Eagle* disappeared behind the moon's left side again, followed by *Columbia.* Both would be out of contact until they reappeared thirty minutes later.

In the middle of *Eagle*'s passage across the far side, Armstrong maneuvered the LM so that its rocket engine was aimed forward, in the direction it was traveling. That meant he and Aldrin were looking down, so they could track their progress by landmarks they had studied for months. He fired the descent engine just about opposite the landing area on the near side. The burn, like the others, had to be precise, twenty-eight and a half seconds; a few seconds too long could cause the LM to crash into the moon. This one went perfectly and slowed *Eagle* enough to allow the moon's gravity to pull it down to fifty thousand feet. *Columbia* remained sixty miles above them and behind. Armstrong and Aldrin could feel a slight gravitational pull—maybe a third of a g. The glycol pumps circulating coolant were loud and constant.

The altitude of fifty thousand feet was not arbitrarily chosen. Since the moon has no atmosphere and little gravity, an orbit could theoretically be set for any altitude, as close to the ground as possible—but that was theoretically. An orbit was never perfectly circular, and the moon was not a perfect, smooth sphere; lunar terrain altitudes could vary by twenty thousand feet or so, and uncertainties in the guidance system by another fifteen thousand. To be on the safe side, fifty thousand feet had been chosen for the LM's beginning orbit altitude.

From that altitude, the descent to the lunar surface two hundred and sixty miles downrange—powered-descent initiation (PDI)—was planned to take twelve minutes. The braking phase would come first, a steady burn of eight and a half minutes to drop them down to about seven thousand feet and slow them to four hundred miles per hour while covering all but four and a half miles. Then the short final approach of one minute and forty seconds would begin, during which they would travel the remaining distance, drop to five hundred feet, and further slow their speed, to about twenty miles per hour. The last minute or so, mostly a vertical drop to the surface two thousand feet downrange, would be the landing phase. The onboard computer was programmed to land the LM, but the crew could take control at any time. It was expected that Armstrong would assume control as *Eagle* came within a few hundred feet of the surface.

Collins in *Columbia* came around the left edge of the moon first. He had been visually monitoring *Eagle* continually since separation in case there was an emergency.

"How did it go?" said Duke, the CapCom in Mission Control.

"Listen, babe. Everything's going swimmingly. Beautiful," Collins said. It was the first report of the LM burn.

"Great. We're standing by for *Eagle*," said Duke.

"Okay. He's coming along."

Two minutes later, *Eagle* acquired signal—sixteen minutes to PDI. In Mission Control, telemetry began to pour in, and flight controllers scanned through the constantly moving green numbers on their CRT screens to make sure it was safe to begin powered descent. Tindall scooted his chair closer to Kranz. The flight director realized his palms were sweating, but he said to his crew, "We're off to a good start. Play it cool."

Slayton, sitting next to Duke, told the astronauts hanging around the CapCom station to find other consoles for the rest of the flight—it was too crowded, and Duke didn't need any distractions. Conrad moved down next to Greene at the FIDO console in the first row and plugged in his headset. Scott sat on a step over on the left.

As *Eagle* came into view and radio communications were restored, Aldrin confirmed a quality burn. A minute later, the signal was lost, and both voice and telemetry were glitchy. *Eagle*'s high-gain antenna must have been pointed toward Earth, but it was out of position, and the spacecraft itself was interfering. As *Eagle* descended, the LM system back room scrambled to find an alternate antenna. Without the exact altitude, Mission Control would be unable to precisely compute the descent engine ignition point.

Communications improved briefly, just enough for the LM crew to provide the data. Duke struggled to maintain radio contact and suggested Armstrong change *Eagle*'s attitude slightly to improve reception but couldn't tell if *Eagle* had received his suggestion. Some telemetry data managed to get through, just enough for Kranz to decide to go forward—to try a landing. He polled his team for PDI, five minutes away. Each one responded with a "Go." "You're go for powered descent," Duke said, but he was unable to speak directly to *Eagle*, so he relayed the go order through Collins in *Columbia*, fifty-five miles above his crewmates. The signal was lost again, then reacquired after *Eagle* made the change Duke had suggested.

The landing radar locked onto the surface and began transmitting info, and Bales knew right away that something was wrong.

"Flight," Bales said, "we're out on our radial velocity—we're halfway to our abort limits. I don't know what's caused it, but I'm going to keep watching it." (Bales would later find out that the tunnel's unvented air had combined with thruster burns and other variables to add a bit of movement to *Eagle* at undocking and increased its velocity—"like popping a cork," as Kranz would put it later—just enough to make a difference down the flight path.) Bales was worried, and with good reason—what if it was a guidance-system problem? It might get worse. His nightmare might even happen—he would have to call an abort. He had quit smoking a few months ago after an X-ray had shown a shadow on his lung, and now he fleetingly wished he hadn't.

But after thirty seconds, he determined that it wasn't a guidance-system issue but a navigation problem. It was holding steady, though that meant a new problem: *Eagle* would be landing about three miles farther downrange than planned. But at least he wouldn't have to call an abort.

Bales relaxed a bit. It looked like his problem was over. He turned to Greene on his left, busy recomputing potential abort modes, and said, "We're in great shape."

Four minutes later, at 3:05 p.m., *Eagle*'s descent engine roared to life at 10 percent power for twenty-six seconds to settle the fuel in the tanks, then powered up close to full throttle. Facedown and feet forward, the two astronauts could feel the brake through their boots as their speed and their altitude began decreasing. While Aldrin kept his eyes glued to the computer display and the other gauges, Armstrong watched the features below as they sped over the brownish-gray lunar surface—craters, hills, ridges, and cracks, their long shadows stretching westward before the rising sun. When they passed one called Boot Hill, he said, "Our position checks downrange show us to be a little long." That corroborated what Bales had told Kranz.

"We confirm that," Bales said.

"Rog," said Kranz. The far western end of the targeted landing area, he knew, was rougher terrain. Landing would be more difficult for the fragile LM.

A minute later, he went around the room polling on whether

to continue powered descent. When he got to his guidance officer, Bales shouted, "Go!"

Kranz chuckled, but it relieved some of the tension.

Duke said, "You are go to continue powered descent."

At forty thousand feet, there was lots of static on the radio as Armstrong used his right-hand controller to roll over to windows-up, the astronauts' feet still forward. The Earth came into view, and the landing radar antenna, now pointing down at the moon, began acquiring the velocity and altitude the computer needed to calculate rate of descent. As it continued to descend, *Eagle* swayed left and right every few seconds as the engine's propellants sloshed back and forth.

Five seconds later, a warbling alarm sounded in their headsets, and both Armstrong and Aldrin looked down at the computer display between them to see a yellow caution light.

"Program alarm," Armstrong said with some urgency, and Aldrin punched in a command to ask the computer to define the problem. They both looked down and saw *1202* on the digital readout.

"It's a twelve-oh-two," said Armstrong. He and Aldrin looked at each other. Neither of them had heard of this alarm. They could probably find it in the book they'd brought along, but there wasn't a spare second for that. They were trying to land a strange spacecraft that had never landed before on an alien world no human had ever touched, and in a low-gravity vacuum. Was the computer going to die on them?

Duke, flustered, said, "Twelve—twelve-oh-two alarm," and looked toward Bales. A silence of several seconds followed. "It's the same one we had in training," Duke said on the flight director's loop.

In Mission Control, more than one heart skipped a beat. This was the alarm that had caused Bales to call an abort two weeks before. It meant that the computer was being overloaded with data—the unnecessary rendezvous radar that had been turned on added just enough to kick its computational load over 100 percent—and it was going to ignore that secondary task, reboot, and restart selected programs where they had been before the reboot. In other words, it would continue with more important tasks, as it had been programmed to do. The simple but sturdy rope-core memory assembled by the little-old-ladies method was doing its job, but no one in Mission Control knew that. The radar had been on during simulations with no problems.

Bales had been busy assessing the fresh radar telemetry when the 1202 sounded. He said, "Stand by," and reached for his notebook. Garman had made him a copy of his cheat sheet, and Bales had stuck it in there. Next to him, Paules said, "That's like the one we had in the sim."

To Garman, on the dedicated GUIDO loop that only the two of them shared, Bales said, "What's that?"

Garman had seen the alarm appear on his screen seconds after it sounded—the time it took the LM telemetry to reach Earth—and he was already looking at the cheat sheet under the Plexiglas on his console. His back room was on a dedicated, open line to a roomful of engineers at MIT, but he wasn't sure they knew, and he didn't need them anyway. He told Bales, slowly and calmly, "It's executive overflow. If it does not occur again, we're fine...that has not occurred again—okay, we're go. Continue."

On the air-to-ground loop, Armstrong asked, as insistently as anyone had ever heard him, "Give us a reading on that twelve-oh-two program alarm." Usually Aldrin talked to Mission Control about such things, since Armstrong's attention was on the window in front of him and the ground coming toward them. But Armstrong's eyes were now fixed on the instrument panel. To the right of Armstrong's window lay the red ABORT and ABORT STAGE buttons.

Bales said, "We're...we're go on that, Flight." *Go* in this case meant "nothing to worry about."

Kranz said, "We're go on that alarm?"

Kranz was about to accept Bales's decision, but Duke had heard it and he didn't wait for Flight to confirm, as CapComs usually did. As a pilot, he knew the men in the LM were thinking abort, and they needed an answer fast, so twenty-one seconds after Armstrong first mentioned the alarm, he said, "Roger. We got you—we're go on that alarm."

"It's—if it doesn't recur, we're go," said Bales.

Garman, on the GUIDO loop, reassured him. "If it's continuous, that makes it a no-go. If it reoccurs"—he meant with at least several seconds in between—"we're fine." Sitting next to him was thirty-seven-year-old Russ Larson, MIT's representative to the LM crews, almost a foot shorter than Garman. He didn't know what the alarms were and was too scared to speak. He could only give Garman a weak thumbs-up.

Someone in a backroom loop said, "Hey, this is just like a simulation," and for some reason, everyone relaxed a bit. The computer was still doing its job of firing thruster jets and navigating, and it appeared the problem was gone.

Fifteen seconds later, the same 1202 alarm sounded with another yellow light.

On the GUIDO loop, Garman said, "Tell 'em to leave it alone and we'll monitor it, okay?"

Bales passed the word along, as did Duke.

Aldrin was worried about further overloading the computer, so he stopped asking it for landing radar data. He had begun calling out critical info: velocity rate and descent speed in feet per second, and altitude. The engine throttled down—a good sign; it meant the computer was still working—and at twenty-one thousand feet and seven minutes into the descent, *Eagle* began to pitch over so the astronauts faced forward, standing again. Forty seconds later, when the craft was at sixteen thousand feet, the lunar surface began to creep into the bottom of Armstrong's window.

Kranz said, "Everyone hang tight. Seven and a half minutes."

Bales told him, "The landing radar has fixed everything; the LM velocity is beautiful." More than a minute had gone by since the last alarm, and they were receiving a clear signal. To Greene he said, "Jay, we're in good shape now, babe."

The LM was dropping quickly, at a hundred feet per second, and at almost nine minutes, it was five thousand feet above the surface. To test his manual controls, Armstrong checked the right-hand controller, which adjusted attitude; he moved the toggle switch from auto to attitude hold, tested yaw and pitch, and pushed it back to auto. He knew he'd need it soon.

Kranz polled the room for the last time, now asking whether or not to land. Bales gave another resounding "Go!" At three thousand feet, Duke gave them a go for landing, and ten seconds later, there was a familiar sound.

Aldrin said, "Program alarm," paused, then said, "Twelve-oh-one."

Both Garman and Bales were all over it; it was a similar alarm. "Same type," said Bales. "We're go, Flight."

Duke didn't bother waiting for confirmation. "We're go. Same type, we're go."

The alarms had caused Armstrong and Aldrin to spend too much time monitoring gauges. When Armstrong was finally able to devote his full attention to the triangular window in front of him, he realized he'd lost track of the landmarks he had studied at such great length. At two thousand feet, he could see a crater coming up. He asked Aldrin for a reading for the landing point.

Aldrin punched the question into the computer, then read the answer. "Forty-seven degrees."

Both panes of Armstrong's window were inscribed with a grid that was calibrated to his height and eye level in degrees, and it showed where the computer, which didn't know the LM was going long, planned to land them. He moved his head to align the grids and looked out, about a mile or so away. "That's not a bad-looking area," he said at a thousand feet.

He changed his mind a few seconds later. The computer was about to land them on the side of a crater about the size of a football stadium with boulders as big as cars surrounding it. For a second, Armstrong considered trying to land there—maybe he could find an open, level space just short of the crater where the slope wasn't more than fifteen degrees, roughly the maximum safe angle estimated for a successful ascent. A much sharper angle and the LM might tip over; even in the one-sixth gravity of the moon, it would be impossible to right. But they were moving too fast to try landing.

"Twelve-oh-two alarm," said Aldrin.

"Roger, no sweat," Garman said to Bales.

"Roger, twelve-oh-two, we copy it," said Duke.

Aldrin, the Mechanical Man, began to coolly call out the critical data Armstrong needed: rate of descent, in feet per second, and altitude. Seconds later they were at six hundred feet. Another 1202 alarm. Everyone ignored it.

There was an old saying among pilots: When in doubt, land long—what was up ahead of you was easier to see than what was under you. Armstrong decided to do that.

He said, "I'm going to..." but he didn't need to finish. Everyone knew he was taking over manual control. He quickly slowed the descent to almost zero by switching the autopilot from auto to attitude hold—that would also ease the computational burden and prevent further program alarms. The computer was still involved, but he would be telling it what to do; he, not the computer, would now control the landing. Armstrong pitched *Eagle* forward to almost level and guided it over the rim of the large crater and its boulder field, bending the trajectory slightly left to avoid a particularly large rock and then resuming his original course. The LM flew better than he had expected; all those hours in the unforgiving LLTV had been worth it.

The ground was getting closer—five hundred feet below them. They were moving faster than they ever had at this point in a sim and still slowly descending over what looked like the lifeless bed of an ancient sea.

In Mission Control, a controller reported, “We’re on attitude hold,” and Duke said, “Attitude hold.”

Slayton slapped his arm and said, “Shut up, Charlie, let ’em land.”

Duke said, “Yes, sir,” then said to Kranz, “I think we better be quiet, Flight.”

“Rog,” said Kranz. He told his team, “Okay, the only callouts from now on will be fuel.”

Aldrin mentioned the forward velocity for the first time: “Fifty-eight forward”—about forty miles an hour.

“No problem,” Armstrong said calmly, though his heartbeat had doubled to a hundred and fifty. With his right hand, he carefully tilted *Eagle* back to allow the rocket engine to slow their velocity, and with his left, he toggled the controller switch, one click at a time, adjusting his rate of descent by one foot per second. Between the noise of the rocket engine below them and the frequent cracks made by the sixteen small thrusters arrayed around the exterior, it was loud in the cabin. Up ahead, he could see a relatively smooth area between some large craters and a boulder field. It wasn’t perfect, but it would have to do.

He said, “How’s the fuel?” The LM descent engine didn’t carry more than about twelve minutes’ worth.

“Eight percent,” said Aldrin.

"Okay. Here's a—looks like a good area here."

Aldrin shot a quick glance out his window and saw the LM's shadow on the ground beneath them and ahead—the sun was low behind them. "I got the shadow out there," he said.

They were at two hundred feet and had slowed to thirteen miles an hour. They scooted over a small crater.

By this time in every simulation, the LM had touched down, crashed, or aborted. Their new landing site wasn't even in view, and fuel was getting low. Propellant slosh in the tank meant that an accurate fuel-level reading was impossible. Like seafaring explorers centuries before, they were now in terra incognita. They continued skimming over the brightly lit gray surface a hundred feet below them. Another 1202 alarm sounded.

Duke said, "You're GO," though they knew that already.

Duke knew Armstrong had no intention of aborting; he was going to try to land, alarms or not, just like he'd told them he would in all those planning sessions. No one on the ground knew why *Eagle* hadn't landed yet—it should have—but they all knew there had to be a good reason. A display on one of the front screens showed the plot of the expected trajectory; *Eagle* had followed this graph closely at first, but not anymore. They were four miles downrange.

Aldrin looked at the fuel reading—a red light had just come on. "Five percent," he said. "Quantity light." If they were still aloft in ninety seconds, they were supposed to make a decision: land within twenty seconds or abort. Aldrin was concerned, but there was nothing he could do. The last thing he wanted to do was disturb his commander's concentration. He continued to read out numbers—altitude and rate-of-descent speed and forward velocity in feet per second. "Sixty feet, down two and a half...two forward... two forward...that's good."

A red low-fuel-level light had flashed on Bob Carlton's console also. He had started a stopwatch and was now counting down the seconds before the *Eagle* ran out of propellants. "Sixty-five, sixty-four, sixty-three, sixty-two, sixty-one, sixty..."

Duke said, "Sixty seconds"—one minute until the mandatory abort-or-land decision.

The ground was close now. They had entered the dead man's zone, an old helicopter term meaning that if they tried to abort at this point, their downward velocity would crash them into the ground before they could ignite the ascent engine. Armstrong wanted to maintain a slight forward speed so they wouldn't fall into a hole that he couldn't see under or behind them.

Aldrin said, "Forty feet, down two and a half. Picking up some dust...thirty feet, two and a half down. Faint shadow."

Armstrong had noticed it too. A sheet of dust dislodged by the engine exhaust flared up from the surface and made it almost impossible for him to judge his rate of descent or determine where the ground was. He could see some rocks ahead through the slowly moving dust, and he tried to base his velocity decisions on that.

Everyone in Mission Control was silent except for Carlton, who called out, "Thirty seconds," and Duke said, "Thirty seconds," but Armstrong was coming down and fuel was not a factor for him anymore; if the engine quit, they would just fall to the ground. He figured the compressible landing gear could absorb a fall from thirty or forty feet.

"Drifting forward just a little bit—that's good," said Aldrin.

They were moving left when a blue light flashed on the control panel: LUNAR CONTACT. One of the probes had touched the surface.

Aldrin said, "Contact light."

Armstrong said, "Shutdown," and quickly turned off the engine—the back pressure could cause an explosion.

Aldrin said, "Okay. Engine stop."

Neither of them felt the touchdown, but they had stopped moving. Armstrong watched, fascinated, as dust particles continued to race out to the horizon, which appeared surprisingly close, and then disappeared over it. Except for the glycol pumps, it was silent.

At 3:17:39 p.m. Houston time, Armstrong and Aldrin grinned at each other and shook hands firmly, then checked to make sure that everything, especially the engine arming control, was off.

Duke half collapsed onto his console in relief and said, "We copy you down, *Eagle*."

"Engine arm is off," said Armstrong. "Houston, uh..." He waited a couple of seconds, and then with added spirit, he said, "Tranquility Base here. The *Eagle* has landed."

In Mission Control, when Armstrong shut down the engine, everyone started to breathe again. Carlton looked at his stopwatch. It said eighteen seconds—meaning eighteen seconds of fuel left, roughly, though no one could know for sure. Later it would be determined that they had as much as forty-five seconds left, but with fuel slosh, it was hard to tell at the moment.

Steve Bales had just started his abort procedures when he heard "Contact light." Now he was puzzled—Tranquility Base? In every one of the hundreds of sims they'd done, he'd never heard that call sign. Then he thought: *What a wonderful name.*

The only person Armstrong and Aldrin had mentioned the base name to was Charlie Duke. Now he said, "Roger, Twan... Tranquility. We copy you on the ground. You got a bunch of guys about to turn blue. We're breathing again. Thanks a lot."

Aldrin said, "Thank you."

In Mission Control, a few arms were thrust in the air, but the

White team stayed in their chairs; there was still work to do. On the top row, Gilruth wiped his eyes and shook hands with Kraft, who then walked down to the GUIDO console and clapped Bales on the shoulder. The viewing room erupted in cheers as people stood or drummed their feet. The instructors did the same in the simulation control area on the right. In the back rooms, men jumped to their feet, screaming and shouting. Von Braun, with tears in his eyes, turned to Houbolt, the man who had crusaded so tirelessly for the lunar-orbit rendezvous. He gave him an okay sign and said, "Thank you, John."

CHAPTER SEVENTEEN

MOONDUST

We were lucky.

FLIGHT CONTROLLER GLYNN LUNNEY

SOMEWHERE ON THE SOUTHWESTERN shores of the Sea of Tranquility, on a level, rock-strewn plain, two men in a small spaceship clapped each other on the shoulders and shook hands. Then they got back to work.

There was no time to waste. Several steps needed to be taken in case the LM had been damaged during the landing and they had to take off. Kranz's White team would decide that, but the spacecraft had to be ready.

In Mission Control, the unflappable Gene Kranz found himself overwhelmed with emotion and unable to speak. He knew he needed to do something to get himself and his team focused. With the exception of his flight controllers, all the people in the building, even the ones in the staff support rooms, were cheering and celebrating the moment. But decisions had to be made, and quickly, so he slammed his right forearm down on his console, flipping a pen in the air in the process. It did the trick.

Kranz said, "Okay, everybody, T one. Stand by for T one." Time-1, or T1, was the first opportunity for an emergency liftoff, and it would occur immediately. A minute later, he said, "Okay, all flight controllers, about forty-five seconds to T one—stay/no-stay." It was Bob Tindall who had pointed out that *go* and *no-go* didn't make sense when the spacecraft was sitting on another world. *Go* could be interpreted in different ways—continue the mission and stay on the moon, or leave immediately. So stay/no-stay it was.

If the crew had to make an emergency liftoff due to a major problem—a leaking fuel tank, a damaged engine, a compromised environmental system, the sun's heat affecting the LM's fluids adversely, or a footpad sinking into that deep layer of moondust—*Eagle* could launch using its ascent engine in two minutes and rendezvous with *Columbia* before it got too far away.

Each controller scrambled to assess the telemetry on his screen and the state of his system. Thirty seconds later, when Kranz polled them for T1, each one answered, "Stay." Then they started evaluating more thoroughly for T2, eight minutes after landing. When Kranz polled them, there was another unanimous round of *stays*. *Eagle* would remain on the surface for now. It would be another two hours before *Columbia* was overhead again.

Just sixteen minutes later, a problem was detected by Bob Carlton's backroom team. The pressure in the descent engine's fuel line was rising. In the subzero temperatures in the shade—about negative 250 degrees—some of the fuel had frozen into a solid plug after the engine shutdown. The other end of the line was blocked by the engine-shutoff valve, so the pressure could cause the fuel to explode or a relief disk to blow. It could be catastrophic either way.

"Flight, the descent engine helium tank is rising rapidly," said Carlton. "The back room expects the burst disk to rupture. We want the crew to vent the system."

They decided to ask the crew to "burp" the engine—that is, flick the engine on and then off again quickly to relieve the pressure. CapCom Charlie Duke was just about to relay the order to *Eagle* when the ice plug melted by itself, and the pressure dropped to zero.

From his position sixty miles overhead and two hundred miles west of the landing site, Mike Collins had listened intently to the air-to-ground transmission, hanging on every word. He had stood in *Columbia*'s lower equipment bay with his right eye at the sextant, trying to keep the LM in view as it slowly dwindled to a dot and then disappeared. At the first 1202 alarm, he grabbed his checklist and began looking through it, but before he could find the right page he heard Duke's "Go." He froze when he heard him say, "Thirty seconds," and exulted a moment later when he heard Armstrong announce, "The *Eagle* has landed."

Although TV commentators emphasized how lonely Collins must be, especially when he disappeared behind the far side of the moon, Collins didn't feel that way at all—he was quite content despite his unprecedented solitude. He only wished he could sight the LM whenever he passed over it. Knowing where *Eagle* was would supply valuable information to the Apollo Guidance Computer to calculate rendezvous maneuvers for the next day. He continued to listen in as the *Eagle*'s crew described the sights around them while preparing for their return in less than twenty-four hours.

At the Armstrong home in El Lago, Jan Armstrong retreated to her bedroom during the descent, leaving family, neighbors, and guests, including Bill Anders, one of the backup crew members, in the living room watching CBS's coverage with Walter Cronkite and his on-air partner that day, Wally Schirra. One of the Armstrongs' guests was a Catholic priest, in case the landing went bad, though neither of the Armstrongs was Catholic or especially religious.

She sat at the foot of the bed and studied a lunar map. Her son Rick came in and sat on the floor, and they listened to one of the two squawk boxes installed in her house. Every astronaut home had a speaker transmitting the air-to-ground loop. Jan was fairly knowledgeable about the descent, more so than most of the other wives, and she was frustrated by the freewheeling speculation, much of it pessimistic, by the commentators. Since there was no live camera feed of the landing, models were used to illustrate it to television viewers. Anders came into the bedroom and sat on the edge of the bed with Jan next to the moon map and provided technical clarification. As her husband guided the LM down the last few hundred feet, Jan sank to the floor and hugged Rick, murmuring, "Good...good...good."

In the living room of the Aldrin home in Nassau Bay, Joan Aldrin stood leaning against a door frame, tears in her eyes, unable to look at the TV screen. When *Eagle* landed, she collapsed onto the floor in relief while everyone else applauded. She got up to throw kisses at the TV screen and pass around a box of cigars, then she went into her bedroom to collect herself. In the Collins home one street over, Pat sat on her living-room couch, nervously smoking, with friends and family. A reporter and photographer from *Life* magazine were there to capture the personal side of the story, as was usually the case during a flight; a team was at each of the other houses as well. At touchdown, Pat smiled for the first time in a long while and watched as the usually eloquent Cronkite became inarticulate. He took off his glasses and said, "Man on the moon! Whew...boy!" Next to him, an overwhelmed Schirra just wiped his eyes.

After the landing, each of the wives gave an interview in her front yard, all of them offering variations on the "proud, thrilled, happy" speech every veteran astronaut wife was accustomed to giving. But they knew there was another critical go/no-go point—a big one—coming up the next day. None of them would sleep well that night.

* * *

Once they received the okay to stay, Armstrong and Aldrin went through a simulated countdown for the next day's liftoff; they hadn't practiced it for a week, and they wanted to make sure the launch-prep procedures worked in the real world. That went fine, and after they both gave detailed descriptions of their views of the lunar surface—and, through a small window above them, the gibbous Earth, hanging in the velvet-black sky like a blue-and-white moon—they told Houston what controllers there had half expected to hear: they wanted to go outside sooner than planned.

Even if the astronauts had taken off immediately after landing, John F. Kennedy's goal would have been met. But a full day's stay, including a two-hour-and-twenty-minute EVA, had been scheduled, chiefly to gather moon rocks, set up several experiments—and see what it was like to walk on the moon. Science had not been the prime factor behind the moon landing or, indeed, behind the entire Apollo program, and this mission was about getting there and getting back, but NASA had agreed to a few experiments for the science guys. Just in case the crew was tired from the descent, the timeline called for a meal and then four hours of rest.

But the two astronauts weren't tired—they were pumped and probably wouldn't have been able to sleep so soon after landing on the moon. They had discussed it with Chris Kraft and Deke Slayton, and all had agreed beforehand that they might forgo the rest period. So, just after four p.m. Houston time, they requested official permission to skip it. Mission Control gave them a go.

First, they quickly prepared a meal. Just before they ate, Aldrin pulled out from his PPK a tiny vial of wine no larger than the tip of his pinkie, a silver chalice about the same size, and a wafer, all given to him by his Presbyterian minister. He radioed to Houston and, with the whole world listening, asked for a few moments of silence. He said, "I'd like to take this opportunity to ask every person listening in, whoever and wherever they may be, to pause for a moment and contemplate the events of the last few hours and to give thanks in his or her own way." Then he poured the wine into the chalice, ate the wafer, and drank the wine in a private Communion ceremony no one but Slayton and Armstrong had known about beforehand.

It took a few hours to prepare for the EVA. They were especially careful about securing their suits, helmets, gloves, and various connections; one less-than-perfect joining could mean a quick death. Their heavy-duty spacesuits were twenty-one layers thick and, like the Apollo Guidance Computer, had been fabricated using the "little old lady" method. Each suit would be pressurized to 3.5 pounds per square inch, making it bulky and not very flexible. Each clear bubble helmet had a special outer helmet equipped with a gold-plated visor to reflect the glare of the blinding sun, unfiltered by an atmosphere. Gloves were heavy gauntlets and, like the helmets, were locked and double-locked. The self-contained backpack each man would wear—the personal life-support system, or PLSS—could provide enough oxygen, cooling water, and electric power to keep him alive in the moon's vacuum and extreme temperatures for four hours. The hundred-and-ninety-pound suits weighed only thirty pounds on the moon's surface, but the tight confines of the LM complicated any movement in a fully pressurized suit, as did *Eagle*'s thin skin, which could be pierced by a pen or any other pointed object and lead to a major pressure leak. A good part of the astronauts' prep time was spent storing the dozens of items that were scattered around the cabin, from food packages to checklists. Finally, at 9:39 p.m., they finished depressurizing the LM and pulled open the hatch at their feet, which hinged on the right. It was clear that it would have been nearly impossible for the lunar pilot on the right side to get around the commander and go out first.

Armstrong got down as low as he could. With Aldrin guiding him, he began to slowly back out of the thirty-two-inch-square opening. It was a tight squeeze with the bulky PLSS and its communications system antenna. He reached the small porch outside, grabbed its side rails, then climbed slowly down the nine-rung ladder attached to the forward left leg. Halfway down, he pulled on a lanyard that deployed a desk-like storage unit to the left of the ladder and activated a Mylar-wrapped TV camera that would be trained on him. Aldrin powered it up, and it began transmitting ghostly black-and-white images to the estimated 530 million people watching—one of every six Earthlings, the largest TV audience in history.

The landing had been so soft, the LM legs hadn't compressed more than an inch or two, so the end of the ladder was still three feet from the ground. Armstrong dropped onto the saucer-like pad at the bottom of the leg. "I'm at the foot of the ladder," he said. "The LM footpads are only depressed in the surface about one or two inches, although the surface appears to be very, very fine-grained as you get close to it. It's almost like a powder. The ground mass is very fine.

"I'm going to step off the LM now," he said, still holding on to the rail with his right hand.

At 9:56:15 p.m. Houston time, he reached out with his booted left foot, hesitated a moment—scientist Thomas Gold's insistence that the moon was covered with a deep layer of dust, though obviously discounted by the landing, was still in the back of his mind—then gingerly stepped onto the surface of the moon.

He said, "That's one small step for man...one giant leap for mankind."

Armstrong hadn't spent much time before or during the flight thinking about what his first words would be—only after the landing did he decide. Several people, including his crewmates on the way to the moon, had asked him if he knew what he'd say, but he'd deflected the question. "Not yet, I'm thinking it over" was his usual reply. He'd meant to say "one small step for *a* man," since without the indefinite article, the line didn't make sense—*man* and *mankind* meant essentially the same thing. But he either forgot or misspoke. In any case, his statement would be both praised for its elegance and criticized for its blandness. Armstrong would later express the hope that "history would grant me leeway for dropping the syllable and understand that it was certainly intended," since without it, the statement was, he said, "inane."

On CBS's telecast, where the words FIRST STEP ON THE MOON hung above Armstrong's shadowy TV image, Cronkite didn't seem to notice the omission. "Well," he said, "for thousands of years now, it's been man's dream to walk on the moon. Right now, after seeing it happen—knowing that it happened—it still seems like a dream."

All over the world, people stopped what they were doing and watched the images from space, vicariously experiencing the adventure. In casinos in Las Vegas, Monte Carlo, and elsewhere, gamblers and dealers at blackjack tables gazed at TVs set up just for the broadcast. At airports throughout the world and many train stations, hurrying commuters paused. In New York's Central Park, ten thousand watched on giant screens; bars and restaurants throughout the United States and in much of the free world showed the broadcast. In Warsaw, several hundred Poles crammed into the lobby of the U.S. embassy to see it. Even the pope, at his summer villa, sat mesmerized in front of a TV. Despite the turmoil of the time, for one day, the billions of inhabitants of Earth shared the same sense of yearning and wonder as a human walked on the satellite above them, so far away.

Though several Communist countries aired the live telecast, including Poland, Romania, Czechoslovakia, Hungary, Bulgaria, and Yugoslavia, it wasn't shown publicly in the USSR, China, North

Korea, or North Vietnam. But in a Soviet military center in Moscow, ten cosmonauts gathered early in the morning to listen to radio transmissions, watch it on TV, and, with a mixture of envy and admiration, applaud Armstrong as he stepped onto the surface. A few days later, they would drink a toast to the safe return of the Apollo 11 crew.

In the harsh light of the lunar dawn, Armstrong moved his right foot onto the surface and stood fully on the ground. "The surface is fine and powdery," he said, "and I can kick it up loosely with my toe. It adheres in fine layers like powdered charcoal to the sole and sides of my boots."

He bounced up and down a few times, then finally let go of the LM's handrail and began walking, tentatively. He assured Mission Control that there was no problem adjusting to the moon's low gravity; no one had been sure of what it would be like to get around in the cumbersome suits. He found that the best method of moving was a slow lope. Visibility was not a problem; with the low, blinding sun to the east and the bright Earth above, the lighting was "like being on a sandy athletic field at night that is very illuminated with flood lights," Armstrong would write later.

Using a "Brooklyn clothesline" pulley, Aldrin sent down a Hasselblad color camera to Armstrong, who began taking photos. Then Armstrong used a handled scooper with a bag on the end to collect some rocks and soil, a contingency sample in case the EVA had to be cut short.

Twenty minutes after Armstrong's first step, Aldrin descended the ladder and joined him. "Beautiful view!" he said, standing on the footpad. The former pole vaulter jumped back up to the last rung just to show how easy it was.

"Isn't that something? Magnificent sight out here," Armstrong said.

Aldrin looked around and drank in the alien landscape. "Magnificent

On May 6, 1968, Armstrong narrowly escaped death when the LLRV trainer he was piloting malfunctioned and pitched sideways two hundred feet above the ground. Armstrong's ejection seat blasted him away (bottom left) before the LLRV crashed and exploded.

Armstrong and Aldrin in the LM simulator at Cape Kennedy on July 11, 1969.

The LM simulator in Houston, with technicians at left.

The "Great Train Wreck": the Apollo command-module simulator at the Cape.

The traditional steak-and-eggs breakfast on the morning of the launch of Apollo 11 (left to right): Bill Anders, Armstrong, Collins, Aldrin, and Deke Slayton.

Armstrong (foreground) and Collins, followed by a technician, cross the access walkway to the command module.

Apollo 11 launches at 9:32 a.m. EDT on July 16, 1969, from pad A, launch complex 39.

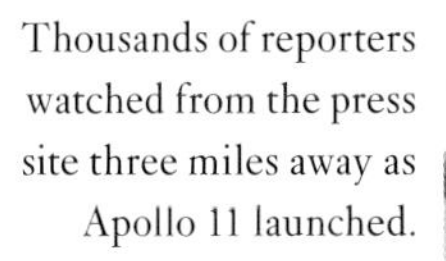

Thousands of reporters watched from the press site three miles away as Apollo 11 launched.

Soon after Apollo 11 cleared the tower, the launch team in the Firing Room listened to congratulatory remarks from Vice President Spiro T. Agnew.

At a July 18 pool party at the Aldrin house, (left to right) Jan Armstrong, Pat Collins, and Joan Aldrin meet the press on the front lawn. *(AP)*

Mission Control during the Apollo 11 mission.

The flight dynamics staff-support room during Apollo 11. Jack Garman, wearing a dark jacket, sits second from left in the front row.

Garman (shown here on the right receiving an award from Chris Kraft after Apollo 8) was a computer whiz kid, only twenty-four during Apollo 11.

Though it had the equivalent of only 72 kilobytes of memory, the rope-wired Apollo guidance computer was extremely reliable.

APPLICABLE TO: IN DESCENT, AVERAGE-G ON

ALARM CODE	TYPE	PRE-MANUAL CAPABILITY	MANUAL CAPABILITY
00105 MK ROUT. BUSY 00430 CAN'T INTG. S.V. 01103 CCSHOLE-PROG. BUG 01204 NEG. WAITLIST 01206 DSKY, TWO USERS 01302 NEG. SQ. ROOT 01501 DSKY, PROG. BAD 01502 DSKY, PROG. BUG 00607 LAHB, NO SOLN	POODOO " " " " " " " "	* PGNCS GUID. LOST; * PGNCS/AGS ABRT/ABRT STG * (decision how on current rules) * (NO LR DATA)	PGNCS GUIDANCE NO/GO (PGNCS GO for TAPE METERS, CROSS-POINTERS, CONTROL, ABORTING) (NO LR DATA)
"O.F." = Overflow, to many. CONTINUING ⟸ OCCURRENCE OF: 01104 DELAY ROUT. O.V. 01201 EXECT. O.F (VAC) 01202 EXECT. O.F. (JOBS) 01203 EXECT. O.F. (TASKS) 01207 EXECT. O.F. (HRKS) 01210 TWO USERS 01211 MRK ROUT. INTRPT 02000 DAP O.F.	BAILOUT " " " " " " "	DUTY CYCLE MAY DEGRADE PGNCS (AGS CONTROL MAY HELP—SEE BELOW) [WATCH FOR OTHER CUES, PGNCS CONDITION UNKNOWN, DSKY MAY BE LOCKED UP, DUTY CYCLE MAY BE UP TO POINT OF MISSING SOME FUNCTIONS (NAV. LAST TO DIE) SWITCH TO AGS (FOLLOW ERR NEEDLES) MAY HELP (REDUCES PGNCS DUTY CYCLE SIGNIF.)]	SAME AS LEFT (except "other cues" which would otherwise be cause for ABORT PROBABLY AREN'T, INSTEAD IT WOULD BE PGNCS GUIDANCE NO/GO — COMPLETE MANUAL LANDING IN AGS.)
ISS WARNING WITH: 00777 PIPA FAIL 03777 CDU FAIL 04777 PIPA, CDU FAIL 07777 IMU FAIL 10777 PIPA, IMU FAIL 13777 CDU, IMU FAIL 14777 PIPA, CDU, IMU FL	LIGHT ONLY " " " " " "	PIPA/CDU/IMU FAIL DISCRETES PRESENT (Other mission rules suffice; alarm may help point to what rule will be broken)	same as left
00214 IMU TURNED OFF	LIGHT ONLY	* AGS ABRT/ABRT STAGE	SWITCH TO AGS PGNS NO/GO on GandC (poss. NO/GO on NAV.)
01107 E-Mem. Destroyed	FRESH STRT	* AGS ABRT/ABRT STAGE	SWITCH TO AGS PGNCS NO/GO! (IMU as ref. okay)
CONTINUING ⟸ 00402 BAD GUID. CMDS	LIGHT ONLY	* IF ALARM DOESN'T STOP: same as "POODO's" (ABRT!)	IF ALARM DOESN'T STOP: Same as "POODO's"
CONTINUING ⟸ 01406 GUID. NO SOLN 01410 GUID O.V.	LIGHT ONLY	PGNCS GUID. NO/GO AS LONG AS ALARM OCCURRING (ATT. HOLD, CONST. GTC, CONT. OK) (ABRT WILL PROB. COME FROM CURRENT RULES e.g. GTC vs. V) WATCH GTC ⟵	same as left (except prob. no abort.)

Jack Garman's cheat sheet, which he consulted to make sure the 1201 and 1202 alarms would not impede the landing. *(Courtesy of Jenny Arkinson and Mary Garman)*

Flight controller Steve Bales would be on the hot seat when alarms began going off during the Apollo 11 LM descent.

(left to right) CapCom Charlie Duke and backup crew members Jim Lovell and Fred Haise during the lunar landing on July 20, 1969.

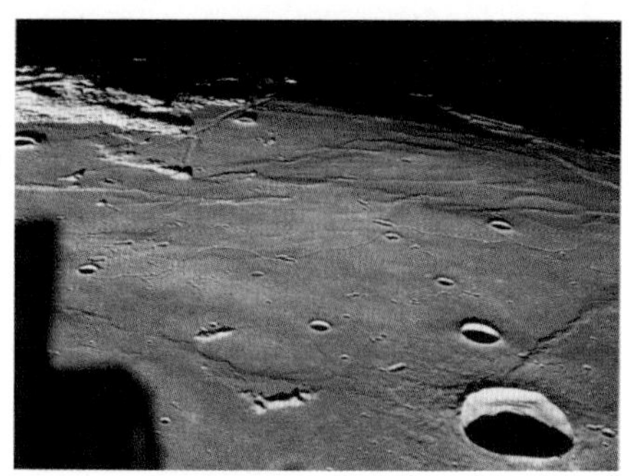

The Apollo 11 landing site on the Sea of Tranquility, moments before the LM landing, in a photo taken from an LM window. The landing site is just this side of the edge of darkness, about a third of the way from the right side of the photo.

Armstrong takes the first step onto the moon's surface.

Aldrin during the lunar surface EVA, standing near the LM.

Gene Kranz (left) behind flight directors Glynn Lunney (center) and Cliff Charlesworth during the Apollo 11 EVA.

Aldrin took this photo of a tired but happy Armstrong in the LM after their moonwalk.

The LM ascent stage rises from the lunar surface to dock with the command-service module, with the Earth in the background.

After the *Apollo 11*'s successful splashdown at 11:49 a.m. CDT on July 24, the Mission Control staff celebrated. In the back row are (from left) Max Faget and NASA officials George Trimble, Chris Kraft, Julian Scheer (behind the others), George Low, Bob Gilruth, and Charles Matthews.

President Richard Nixon greets the three Apollo 11 astronauts on the USS *Hornet,* the prime recovery ship. The crew is in the Mobile Quarantine Facility to avoid the potential spread of moon germs.

New Yorkers welcomed the Apollo 11 crew in a record-tonnage ticker-tape parade on August 13, 1969.

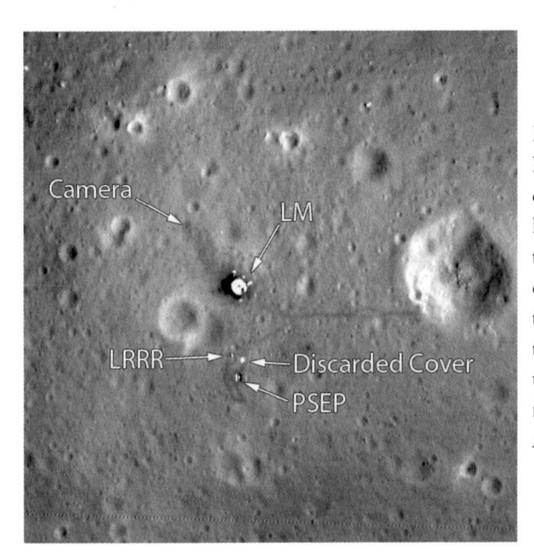

Forty-three years later, in 2012, NASA's Lunar Reconnaissance Orbiter captured this photo of the Apollo 11 landing site from fifteen miles above the surface. The LM descent stage is clearly visible, and the dark lines are the tracks made by the astronauts as they walked around the area setting up experiments and visiting a crater fifty yards to the east. *(NASA/GSFC/Arizona State)*

desolation," he said. Then he peed in his pants—a lunar first, he would claim later. Fortunately, he was wearing a urine-collection device.

While Armstrong set up the TV camera on a tripod sixty feet from the LM, Aldrin began experimenting with various kinds of gaits: a two-legged kangaroo hop, a long, stiff-legged skip, and the easy lope Armstrong had settled on. They both examined the LM for any fuel leakage or damage but saw none. Armstrong read aloud the words on the plaque attached to the LM leg, then they got to work. They had several jobs scheduled but not much time to do them—and Mission Control didn't want them to tax themselves; the memory of those exhausting Gemini space walks was still vivid. While Armstrong set up the TV camera on a tripod sixty feet away from the LM, Aldrin began experimenting with various kinds of gaits: a two-legged kangaroo hop, a long, stiff-legged skip, and the easy lope his commander had settled on. Armstrong gathered more rocks while Aldrin set up one of the three main experiments, a solar-wind collector.

Next up was an American flag, placed thirty feet in front of the LM. They pounded in the metal staff about six inches—just enough to keep it standing—then pulled out the telescoping crossbar on top to keep the banner extended. They couldn't get it out all the way, which lent the flag a wavy appearance. Buzz stood back and saluted it. Five minutes later, there was an interruption to their schedule—a surprise.

CapCom Bruce McCandless said, "The president of the United States is in his office now and would like to say a few words to you."

Richard Nixon had been watching the TV broadcast with Frank Borman. He'd had little to do with the Apollo program, which had come about because of two of his bitter rivals, Presidents Kennedy and Johnson, though he'd been gracious enough to phone and congratulate Johnson and Mamie Eisenhower, the late president's widow. NASA had asked him if he would make a call to the astronauts if they could work it out, and no politician would have refused this stage and this audience. "Hello, Neil and Buzz," Nixon said. "I'm talking to you by telephone from the Oval Room at the White House, and this certainly has to be the most historic telephone call ever made." The president kept it short; he congratulated them and ended with "For one priceless moment in the whole history of man, all the people on this Earth are truly one—one in their pride in what you have done, and one in our prayers that you will return safely to Earth."

Armstrong thanked him, and both men saluted the flag. Then it was back to the checklist sewn onto each man's left gauntlet. Armstrong continued to photograph while Aldrin set up the other two experiments: a passive seismometer, which would detect and transmit any lunar tremors, and a set of panels that would reflect lasers aimed from Earth and, when received back, would measure the exact distance from the Earth to the moon. Armstrong decided to visit a crater about sixty-five yards east, and he picked up some interesting-looking rocks around its edge. They would prove to be the most geologically fascinating moon material the mission would return to Earth.

Then it was time to return to the LM. Aldrin was halfway up the ladder when Armstrong called up to him. "Buzz? How about that package out of your sleeve? Get that?"

"No."

"Okay, I'll get it. When I get up there."

"Want it now?"

"Guess so."

Aldrin pulled out a small white cloth pouch and tossed it down. They didn't want to make a big deal out of its contents, but there was one last thing they needed to do. Inside the pouch was an Apollo 1 patch that had belonged to Scott Grissom, Gus's oldest son; a metal case containing a half-dollar-size silicon disk with goodwill messages from seventy-three countries etched into it; two medals commemorating Vladimir Komarov and Yuri Gagarin, a nod of respect to their rivals; and a tiny gold olive branch.

The pouch landed in the lunar dust near Armstrong. With his right boot, Armstrong nudged it under the ladder. "Okay?"

"Okay," Aldrin said, thinking about Ed White, his squadron mate in Germany who had talked to him about becoming an astronaut. Then Aldrin clambered up to the porch and went through the hatch. After shuttling up the cameras and two airtight rock boxes, Armstrong followed him. Eleven minutes after midnight, he closed and secured the hatch. They pressurized the cabin and stowed the moon rocks—forty-eight pounds of them—then depressurized, reopened the hatch, and tossed out everything they wouldn't need again: backpacks, heavy boots, a spare Hasselblad, and any refuse they could find, though they couldn't do much about all the lunar dust they had dragged in on their suits. It looked like charcoal and smelled like wet ashes or spent gunpowder. For what they hoped would be the last time, they secured the hatch and repressurized.

Then it was time for dinner—cocktail sausages and fruit punch—followed by bedtime. Mission Control signed off at 3:24 a.m.

The lunar module was designed for many purposes, but sleeping wasn't one of them. The two astronauts kept their helmets on to avoid breathing in the moondust, which seemed to be everywhere. Aldrin curled up on the floor, his head on the right side. In the light gravity, he was fairly comfortable, though he couldn't stretch out fully. Armstrong sat on the ascent-engine cover behind their stations a few feet above Aldrin, leaned his head back on a shelf, and rigged a sling for his feet from a waist tether that hung from the instrument panel. The position was tolerable, and he thought he could sleep there. They were both tired. It had been a long day, but a good one.

Neither of them slept well. They might have if it weren't for all the warning and caution lights, the many illuminated switches, the light seeping in through the covered windows, and a small telescope that focused a bright light on Armstrong's face. It was too cold, somewhere around sixty-one degrees, and adjusting the temperature controls only seemed to make it colder. And somewhere near Armstrong's head there was a loud water glycol pump.

While they drifted in and out of sleep, their thoughts were on the next day. There were several more go/no-go decision points to come in the flight, all of them dangerous, but only one hadn't been attempted on a previous mission: the liftoff from the lunar surface. And unlike virtually every other system in the LM, there was no redundancy, no backup, for the ascent engine. Its thirty-five hundred pounds of thrust had to work tomorrow or they would be stuck on the moon and die when their oxygen ran out—about another twenty-four hours. The sixteen thrusters, clustered in four groups of four, could muster only one hundred pounds of thrust apiece, and only four of them provided lift; four hundred pounds was not nearly enough to get the five-ton LM off the lunar surface, even in the moon's minimal gravity. But the engine had been designed to work as simply and reliably as possible. There were only four moving parts and no ignition system or fuel pumps; small explosive charges would open the valves to the tanks, and a pressurized helium system would force the hypergolic propellants together, at which point they would ignite spontaneously, even in the vacuum of space. In tests, this particular engine had never failed—but that was on Earth.

If it failed...Armstrong and Aldrin hadn't discussed that turn of events, but if it happened, Aldrin had already decided that he'd work on the problem until the lack of oxygen caused him to fall asleep. There were other, much quicker, ways to die: they could depressurize the cabin or go outside, then unplug their spacesuits' oxygen-inlet hoses or unlock their helmets. In the airless vacuum, they would lose consciousness in fifteen seconds at most, and death would follow a few minutes after as the gases in their bodies quickly expanded and liquids vaporized, leaving shriveled, bone-dry corpses.

Another problem—one that could have been calamitous—was being addressed. As Aldrin had prepared to retire, he'd noticed a small piece of black plastic on the floor. He looked at the rows of circuit breakers on his side and found the broken one. It was part of the ascent-engine arming circuit breaker, which would supply electricity to the engine before they pushed the final button to start it. If this circuit breaker was not pushed in and engaged—and right now it wasn't—the engine could not be activated. Aldrin realized he must have hit it with his backpack at some point. He told Mission Control. When he was informed that they'd find a work-around for it, he told them he had a plastic felt-tip pen he could use to push the remaining part of the breaker in; it would fit right into the slot. Mission Control told him to hold off until morning.

After the landing the previous day, Kranz had asked Steve Bales and a few other flight controllers to join him in the regular post-shift press conference. Bales almost nodded off in it. He got out as fast as he could and went up to the flight controller lounge, sat down in a chair, and didn't move for thirty minutes. Then the rest of the descent team started showing up. A few hours later, they all watched the EVA on TV. Once the flag was planted, Bales headed across the hall to the bunk room and quickly fell asleep.

He was on console again for the ascent—he was the only controller on both the descent and ascent phases—and after breakfast in the cafeteria, he walked down to Mission Control to join Glynn Lunney's Black team. The MOCR was already filling up; so was the VIP viewing room. No one had yet figured out the cause of the 1201 and 1202 alarms, which meant that they might recur during ascent through docking—the onboard computer would be even busier than yesterday. Bales and his team had prepared backup procedures for several computer-failure scenarios. He hoped they wouldn't be needed.

Twenty minutes before liftoff, Jack Garman, in the staff support room, got on the GUIDO loop. A large MIT team in Boston had studied the problem all night and thought they had the answer—the rendezvous radar. To avoid risking a similar alarm scenario during the ascent, the crew would need to put the radar in the manual position, not the automatic position it had been in during the descent.

All flight controllers hated last-minute changes, but Bales hated them more than most. He said, "What the hell, Jack? Now? Here we are twenty minutes before liftoff, this is our procedure. And we're gonna change a thing now?"

Garman said, "Steve, they don't have time to discuss the details. They feel this is absolutely the best way to do it."

Bales said, "Well... I gotta believe them." He passed the message on to Lunney. The flight director wasn't pleased about the late input, but he told the CapCom to share the message with the crew.

At 9:32 a.m., six hours after Mission Control had signed off for the night, the shift CapCom, Ron Evans, gave them a wakeup call. Liftoff was scheduled for a little after noon, when *Columbia* would be close overhead. *Eagle*'s crew spent the next couple of hours preparing, going over checklists, and verifying that the many circuit breakers and switches were in the correct positions. The first change to the schedule was an important one: Evans told them to turn off the rendezvous radar—they didn't want to take the chance that it would overload the computer again. Then Aldrin used his felt-tip pen to push in the ascent-engine arming circuit breaker. Moments later, Mission Control confirmed that the circuitry was fine.

An hour and fifty-seven minutes before *Eagle*'s liftoff, Luna 15, the Russian probe, crashed about seven hundred miles northeast of

Tranquility Base into the Sea of Crises, under which a mascon lay. According to Soviet plans, a lone cosmonaut was supposed to have landed in the vicinity in the not-too-distant future on a mission that would never happen. Though TASS quickly announced that the research probe had been successful, few in the West believed it. Whether the crash was due to faulty hardware, incorrect or out-of-date lunar-altitude information, or the mascon, no one—not even the Soviets—would ever know.

At 12:37 p.m., after some adjustments had been made to make sure the fuel tanks were properly pressurized, Evans told *Eagle*'s crew, "You're cleared for takeoff."

"Roger, understand," said Aldrin. "We're number one on the runway."

Seventeen minutes later, at 12:54 p.m., Aldrin said, "Nine, eight, seven, six, five"—and then, as both he and Armstrong began to push buttons and flick switches—"Abort stage, engine arm, ascent, proceed."

Explosive devices separated the ascent stage, the top half of the lunar module, from the descent stage, and the engine fired and lifted them away as dust, debris, and shredded Mylar flew in every direction. Aldrin looked out his window just long enough to see the flag fall to the ground. The liftoff was smooth and swift, and a few seconds later, Armstrong said, "The *Eagle* has wings," and the LM pitched over forty-five degrees to begin moving horizontally. "We're going right down U.S. One," he said as they passed over crater after crater.

Seven minutes after liftoff, the engine cut off. It had done its job, which was to get *Eagle* up high enough and going fast enough to overcome the moon's meager gravity and reach orbit.

Above them, Collins caught sight in his sextant of a small blinking light in the darkness of space: *Eagle*. Just before Apollo 11 had launched on July 16, he'd received a telegram from a friend: BEST WISHES FOR A SAFE JOURNEY. DON'T FORGET TO WAIT FOR YOUR PASSENGERS WHILE THEY ARE OUT WALKING. He hadn't forgotten. The thought of having to leave his shipmates and return home alone had terrified him for the past six months. He'd been waiting almost a full day for their return, and he'd been trying not to think about the thousand things that could go wrong with the LM. Fortunately, he'd had little time to worry; there were eight hundred and fifty separate computer keystrokes necessary to effect rendezvous with *Eagle*, and he'd been busy since breakfast. He'd hardly breathed during their seven-minute ascent. Now, a couple of hours later, he was still on edge. Just in case, he kept his notebook containing the eighteen options available if it failed close at hand. He prayed he wouldn't have to consult it.

Rendezvous had been successfully performed several times in Gemini and Apollo, but it still demanded several tricky maneuvers and perfectly timed acceleration and braking burns. But Armstrong had achieved the very first docking in space, in Gemini 8, and he had practiced rendezvous and docking many more times for this mission. Over the next three hours, both he and Collins carefully maneuvered their spacecraft to allow *Eagle* to catch up with *Columbia*.

As they approached, their relief at the successful launch—and their confidence in the rest of the flight—was evident.

"One of those two bright spots is bound to be Mike," Armstrong said.

"How about picking the closest one?" said Aldrin.

"Good idea," said Armstrong.

The alignment looked good, so he engaged the three small capture latches and flipped a switch to draw the two spacecraft together. But the alignment had been off, maybe by fifteen degrees, and the LM started yawing dangerously to Collins's right. The automatic retraction cycle took six to eight seconds, and Collins couldn't stop it or release the LM. If *Eagle* continued to twist around, the docking equipment might be damaged, and they'd have to try a tricky EVA of both crewmates from *Eagle* to *Columbia.* Collins worked his right-hand controller and managed to swing *Columbia* around with *Eagle* until they were aligned and he heard a bang and the docking latches slammed shut. They were safely docked.

Almost two hours later, after Armstrong and Aldrin had disabled several of the LM's systems and prepared *Eagle* for its jettison, they doffed their helmets and gloves and crawled through the pressurized tunnel into the command-module cabin. Aldrin came through first. Collins floated up to meet him, and when Aldrin emerged with a big smile on his face, Collins grabbed his head and barely resisted the urge to kiss his forehead. He shook his hand, then Armstrong's. All three were almost giddy.

They transferred the rock boxes and camera-film cartridges into *Columbia* and used a small vacuum to clean the moondust they had tracked in. Then *Eagle*'s crew bade farewell to their ship, Collins flipped a few switches, and *Columbia* separated from it. The LM would circle the moon until its orbit deteriorated and it smashed into the lunar surface.

At 11:10 p.m., just before *Columbia* disappeared around the far side of the moon for the last time, the White team's CapCom, Charlie Duke, said, "You're go for TEI."

Behind the moon again, on the command-service module's thirty-first orbit, it was time for the transearth injection burn of 2:28. It would increase their speed by 2,236 miles an hour, enough to free them from the moon's gravity and send them back to Earth. Collins keyed in the command, counted down, and pushed the button to ignite the engine. The burn went perfectly, and when the spacecraft came in sight of Earth twenty minutes later, Armstrong said, "Hey,

Charlie boy, looking good here. That was a beautiful burn. They don't come any finer."

Their course was set for home. Like the voyage out, the transearth coast was uneventful, almost routine. The spacecraft was put into its broadside rotisserie roll to distribute the sun's heat. Over the next three days, the crewmen caught up on their sleep, did the usual housekeeping chores, listened to the news from Houston, put on a couple of TV shows for the folks back home, and took turns photographing the dwindling moon and the approaching Earth. The charms of weightlessness and the wonder of space were wearing off, and the crew was ready to get back. In the background they heard the steady whirring, gurgling, and humming of the spacecraft and the music cassettes they played; they talked infrequently. Even after the adventure they had just shared, there were no deep or intimate discussions. Though they had worked superbly as a team, they were still not and would never be close friends.

After a brief midcourse correction burn of just eleven seconds, all that remained was to reenter Earth's atmosphere without burning to cinders; they would be traveling at almost 25,000 miles per hour and would have to slow significantly. That meant entering at the correct angle of attack in the command module—the service module had been jettisoned and would burn up in its own reentry.

At 11:50 a.m. Houston time on July 24, the ninth day of their journey, the eleven-thousand-pound Apollo 11 command module—all that was left of the spaceship's massive six-and-a-half-million-pound stack of three booster stages and three modules—plummeted through the atmosphere upside down with its heat shield ablating in orange-yellow cinders, deployed its two drogue and three main parachutes, then splashed down hard in the mid-Pacific, about eight hundred and twenty-five nautical miles southwest of Hawaii and just thirteen miles from the prime recovery ship, the carrier USS *Hornet*. A helicopter dropped an inflatable raft and three frogmen into the water, and one of them threw three biological isolation garments into the open hatch to begin the back-contamination precautions. The astronauts donned the BIGs, jumped into the raft, and, as rehearsed, began spraying and scrubbing themselves down with disinfectant. They were reeled into the chopper one by one and flown to the carrier, where they were immediately escorted into the Airstream trailer converted into a mobile quarantine facility.

In Houston, almost two hundred people packed into Mission Control for the splashdown. When one of the side display screens showed TV coverage of the astronauts on the carrier, the MOCR doors were opened, men poured in from the other two shifts and from all the SSRs, and the room erupted as the controllers jumped to their feet and everyone cheered, many waving small American flags and puffing on cigars. Then the center screen displayed the text of Kennedy's challenge—I BELIEVE THAT THIS NATION SHOULD COMMIT ITSELF TO ACHIEVING THE GOAL, BEFORE THIS DECADE IS OUT, OF LANDING A MAN ON THE MOON AND RETURNING HIM SAFELY TO THE EARTH—above the month and year of the speech, May 1961. On the screen to the right, the Apollo 11 emblem flashed below this legend: TASK ACCOMPLISHED...JULY 1969. Those two screens remained up long after everyone had left.

After each of them had a quick shower and shave, the three astronauts walked to the rear of the Airstream trailer, where the president was waiting outside to talk to them through a small window. "This is the greatest week in the history of the world since the Creation," he said, and on a hopeful note, he added, "As a result of what you have done, the world has never been closer together." Then it was on to Pearl Harbor, where a cargo plane waited to take the astronauts to Houston.

The second evening on the carrier, Collins entered the command module through the tunnel connecting it to their trailer, and with a ballpoint pen, above the sextant mount on the wall of the lower equipment bay, he scrawled this legend:

Spacecraft 107—alias Apollo 11
alias "Columbia"
The Best Ship to Come Down the Line
God Bless Her

Michael Collins
CMP

Back in Houston, Steve Bales had left after the ascent shift, his last one on Apollo 11. He went home and slept for a long time. The next day, he drove back to work and walked into the staff support room. Jack Garman was there. Bales walked over to Garman, shook his hand, and said, "Jack, thanks for everything."

Then the two engineers got back to their jobs. At that point, the Apollo 11 crew was still in space, and Garman was providing backroom support for the onboard computer. Bales had to start preparing for the next mission, Apollo 12, which was scheduled for a moon landing in November. But before he did, he called his parents in Iowa. They were enormously proud of the part he'd played in the flight, but his mother said what made her the happiest was that they'd finally been able to see him in some of the live TV shots of Mission Control. Until the landing, they'd never been able to spot their boy.

The night of the splashdown, there were parties up and down NASA Road 1 and at almost every bar and restaurant in the vicinity.

The flight controllers gathered at the Singing Wheel. At the Nassau Bay Resort, three thousand NASA workers gathered around the large swimming pool, drinking and feasting on barbecue while bikini-clad go-go dancers gyrated to the accompaniment of a rock band called the Astronauts. Many of the guests ended up in the pool, fully dressed; at some point late in the festivities, the hotel piano was thrown in. At four a.m., police officers gently rounded up the last of the revelers.

The Fagets, Gilruths, Lows, and Krafts had dinner together at one of their favorite restaurants, Mike's Rendezvous, in the nearby town of Algoa, southwest of the Manned Spacecraft Center. Along with other NASA folks there, they celebrated the successful mission—and the successful answer to President Kennedy's challenge issued eight years ago. America, and the NACA-nuts, had triumphed.

On Sunday, July 27, Armstrong, Aldrin, and Collins arrived in the large quarantine facility at the Manned Spacecraft Center to find a flood of goodwill letters and messages from around the world. One telegram of congratulations began "Dear Colleagues" and was signed by every living cosmonaut who had flown in space. The Cold War had been thawing for a while and would continue to do so.

The astronauts and a dozen or so others—doctors, chefs, technicians, a NASA PR officer, a journalist, even a janitor—spent the next two weeks there. Between daylong debriefs and endless postflight reports and reviews, they unwound in various ways, from making phone calls and having window visits with their families to watching recent Hollywood movies shown on a large screen. During their stay, several mice had been injected with lunar soil to test for negative reaction; when none of them died and the astronauts and the dozen or so other people in the facility with them didn't get sick, it was determined there was no risk of contamination. At nine p.m. on Sunday, August 10, they were all allowed to leave the facility and resume their lives.

After spending some quiet time reuniting with their families at home, the astronauts couldn't avoid the spotlight. There were parades galore, including the traditional ticker-tape celebration down Broadway, and countless parties, dinners, and celebrations, with interviews and press conferences sandwiched in. In mid-September, the crew addressed a joint session of Congress. Then it was a round-the-world goodwill tour with their wives: twenty-eight cities in twenty-five countries in thirty-eight days, meeting kings and queens, shahs and dictators, presidents and prime ministers. By the time the crew returned home, they understood that their lives would never be the same.

A few years after he walked on the moon, Neil Armstrong agreed to appear in a documentary. The filmmakers shot his scenes one afternoon at the U.S. Space and Rocket Center museum in Huntsville, Alabama, where von Braun, his Peenemünde team, and fifteen thousand other Americans had developed the Saturn V that took Armstrong and his shipmates to the moon. After the film crew had interviewed Armstrong and packed up and left, museum director Ed Buckbee, whom Armstrong had known from the days when Buckbee was a public affairs official with NASA, asked him what he'd like to do: Hold a press conference, sign some books, meet the press? Armstrong said he just wanted to look through the museum. Buckbee took him around.

When they got to the lunar-module simulator, which the museum had received from NASA—the same one Armstrong had trained on for Apollo 11 at the Manned Spacecraft Center—Armstrong stopped. He said, "Can I get in?"

"Sure," Buckbee said.

Armstrong stepped in, moved over to the commander's position, and looked over the control panel. He flicked on the power. "Let me see if I remember my procedures," he said, and for the next forty minutes, he flipped switches, pushed buttons, and maneuvered his hand controllers as he went through a few simulations. When he was done, he turned the power off, got out, thanked Buckbee, looked through the rest of the museum, and headed home.

EPILOGUE

Max, they're going to go back there one day, and when they do they're going to find out it's tough.

BOB GILRUTH TO MAX FAGET

IN THE THREE AND a half years after Apollo 11, NASA sent six more manned spaceflights to the moon. They all came back, though the crew of Apollo 13 never landed. In fact, they barely escaped with their lives after an oxygen tank exploded during the translunar leg and critically damaged the service module. It wound up being NASA's finest hour, as Mission Control came up with ingenious solutions to one life-threatening problem after another.

By December of 1972, when the last Apollo flight lifted off from the lunar surface, the American public had become bored with moon trips. Not only had the national goal—to beat the Soviets to the moon—been reached, but NASA's extraordinary preparation and readiness resulted in increasingly fewer problems and perceived dangers. They'd made each of the last few missions seem like a walk in the park. Funding for the agency had been falling for years, and after the Apollo 11 mission, it plummeted dramatically. From a high of 4.41 percent of the total U.S. budget in 1966, it dropped to under 1 percent in 1975, and down to under half a percent in 2013.

Despite its meager budgets, and though the agency's accomplishments never matched its dreams and formal proposals, NASA continued to launch men—and, eventually, women and minorities—into orbit to conduct research. So did the Soviets, who established the first space station, Salyut, in 1971; the United States answered with Skylab in 1973. The Apollo-Soyuz Test Project (1975) was the first joint U.S.-USSR space mission (and it allowed Deke Slayton, his heart problem resolved, to finally fly into space). Both nations, along with three other space agencies, united to launch the International Space Station in 1998. There was also NASA's space-shuttle program, consisting of five operational Orbiters that flew 135 flights from 1981 through 2011. (The *Challenger* and *Columbia* tragedies in 1986 and 2003, respectively, were grim reminders of the dangers of space flight.) In 1974, China launched its own space station, Tiangong, and beginning in 2003, it sent humans into space aboard several Shenzhou spacecraft.

Because it's safer and cheaper, unmanned space flight has become the standard mode of exploration beyond Earth. Robotic probes have flown by, around, and sometimes onto every planet in the solar system and some of its moons and asteroids. Launched in 1977, Voyager 1 and Voyager 2 made the grand tour of the four large outer planets with flybys of Jupiter, Saturn, Neptune, and Uranus. Voyager 1 entered interstellar space in 2012; its sibling will do the same soon.

At the turn of the twenty-first century, interest in space and human space flight began to increase. NASA announced a new launch system, plans for a crewed mission to orbit the moon in 2022, and eventually a lunar-orbiting space station from which it will send a manned mission to Mars sometime in the early 2030s. (NASA may partner with other nations to do this.) Several commercial space ventures have also been started, the most prominent being SpaceX, which has announced its own Mars expedition. The future of the space industry—including tourism, mining for rare metals on planets and asteroids, and other potentially profitable businesses—appears limitless.

"Sometimes it seems that Apollo came before its time," wrote Apollo 17 commander Gene Cernan in 1999. "President Kennedy reached far into the twenty-first century, grabbed a decade of time and slipped it neatly into the 1960s and 1970s." History has supported his observation; in the half a century since he climbed into his lunar module, no one has ventured beyond low Earth orbit. But a new spirit of space exploration is in the air. "Man has always gone where he has been able to go," Apollo 11 astronaut Mike Collins told Congress in 1969. As long as some part of us remains human, we always will.

ACKNOWLEDGMENTS

Thanks to Bill Barry and his superb group of historians at the NASA History Office, particularly Liz Suckow, who went above and beyond the call of duty; Sam Cavanaugh at SMU's Fondren Library; Brian McNerney at the Lyndon Baines Johnson Library and Museum; Jennifer Manger at the Rensselaer Polytechnic Institute; Lauren Rose Meyers at the University of Houston–Clear Lake; Fordyce Williams at the Clark University Archives; Sultana Vest, Crystal Brooks, and Sylvia Aguillon at the Dallas Public Library; and Aaron Purcell and his excellent Special Collections staff at Virginia Tech, particularly Laurel Rozema, for their help with the Robert R. Gilruth Papers and the Christopher Kraft Jr. Papers.

I'm especially indebted to all the people formerly of NASA and otherwise whom I interviewed either in person or on the phone; all were unfailingly gracious and helpful: John Aaron, Buzz Aldrin, Steve Bales, Alan Bean, Ed Buckbee, Bob Carlton, Jerry Carr, Maurice Carson, Mike Collins, Maddie Crowell, Walt Cunningham, Jerry Elliott, Chuck Friedlander, Jack Garman, Dick Gordon, Fred Haise, Bill Helms, Albert Jackson, Sy Liebergot, Jim Lovell, Ken Mattingly, Edgar Mitchell, Sam Ruiz, Joe Schmitt, Rusty Schweickart, Reuben Taylor, Tom Weichel, and Al Worden.

Individuals to whom I owe a special debt of gratitude are Jenny Arkinson and Mary Garman, Steve Bales, Robbie Davis-Floyd, Rebecca Wright, Asif Siddiqi, Carol Faget, Walt Cunningham, Bob

Carlton, Glynn Lunney, Bob Carlton, Albert Jackson, David Whitehouse, and Mike Collins. Brent Howard planted the idea for this book; when it refused to leave my head, I knew it was the right one.

Thanks to Alex Shultz and Rachel Donovan for interview transcriptions; Todd Hansen, Ellen Kadin, Kathleen and John Wainio, and Marcia Cunningham for reading and commenting on chapters, and Marcia for her excellent research; Matt Polomik and Dave Hamon, who read for accuracy—I can't think of anyone who knows more about NASA history; and Melissa Shultz for reading every word and helping to make this so much better.

My literary agent, B. J. Robbins, is as good as they get and a friend to boot. At Rain Management Group, Michael Prevett is a longtime friend and film-rights facilitator—no one in Hollywood is better. Publicist Nicole Dewey's team at Shreve Williams was a joy to work with.

At Little, Brown, thanks to publisher Reagan Arthur and editor John Parsley for believing in this from the start; editor Philip Marino for taking over and steering the book to completion and his assistant Anna Goodlett for her help; assistant publicity director Lena Little, marketing director Pamela Brown, and rights director Laura Mamelok and their respective staffs (especially Ira Boudah, Katharine Myers, and Nel Malikova) for all their invaluable work; copyeditor extraordinaire Tracy Roe, who put the final polish on the text; designer Gregg Kulick for the wonderful cover; Marie Mundaca for the fine interior design; proofreaders Katie Blatt and Holly Hartman; indexer Anne Holmes; and Ben Allen, production editor, who wrangled all these critters together. It's not an easy job, and Ben does it superbly.

NOTES

PROLOGUE

3 **"called a rocket":** Leonard, *Flight into Space*, 10–11.

ONE: COSSACKS IN SPACE

9 **"Our aim from the beginning":** Dornberger, *V-2*, 140.

10 **reach the moon:** Wernher von Braun appeared in three Walt Disney *Tomorrowland* specials: "Man in Space," March 1955; "Man in the Moon," December 1955; and "Mars and Beyond," December 1957.

10 **"almost alien":** Quoted in McDougall, *The Heavens and the Earth*, 141.

11 **"wars of liberation":** Soviet premier Nikita Khrushchev, January 6, 1961.

13 **"anyone could launch":** *New York Times*, October 5, 1957.

13 **"outer-space arms":** Ibid., October 7, 1957.

13 **"leap into space":** Ibid.

13 **"in the world":** Ibid.

13 **"democracy or slavery":** *Missiles and Rockets*, November 1957.

13 **back to Russia:** Ibid., August 1957.

13 **were "irrelevant":** Quoted in Wasser, "LBJ's Space Race: What We Didn't Know Then."

14 **October 1957:** *New York Times*, October 5, 1957.

14 **"slough away":** Quoted in Baker, *History of Manned Space Flight*, 87.

14 **part of the Cold War:** A half a century later, long after the height of the Cold War, this worry over prestige may appear extreme—a tempest in a teapot. At the time, however, it was a legitimate concern. After the Gagarin flight, the *New York Times* opined: "The neutral nations may come to believe the wave of the future is Russian; even our friends and allies could slough away." After John F. Kennedy's election, an October 27, 1960, article in the *New York Times* headlined "Post-Summit Trends in British and French Opinion of the US and the USSR" cited a study showing that "current confidence is low in America's capacity for leadership in dealing with present world problems." The study stated that both British and French opinion put the USSR "overwhelmingly ahead." A story in the October 29, 1960, *Washington Post* on world reaction to the U.S. and Soviet space programs reported, based on polls in Britain, France, West Germany, Italy, and Norway, that "in anticipation of future US-USSR standing, foreign public opinion...appears to have declining confidence in the US as the 'wave of the future' in a number of critical areas of competition." For more evidence that the U.S. national prestige was a major concern, see Callahan and Greenstein, "The Reluctant Racer," 30–31, and Lewis, *Appointment on the Moon*, 159. For a detailed argument confirming America's loss of prestige, see McDougall, *The Heavens and the Earth*, 240–41.

15 **damaged America's:** "Reaction to the Soviet Satellite—A Preliminary Evaluation," Eisenhowerarchives.gov.

15 **"North Atlantic Treaty Organization":** Quoted in Dethloff, *Suddenly, Tomorrow Came*, 1.

15 **asked *Time* magazine:** *Time*, January 19, 1959.

16 **immediately after Sputnik:** Launius and McCurdy, *Spaceflight*, 74.

16 **end of the war:** Quoted in Caidin, *Red Star in Space*, 87.

17 **Lutheran boys:** *Time*, February 17, 1958.

18 **"intellectual capabilities":** Neufeld, *Von Braun*, 97.

20 **"the *spaceship* has been born":** Quoted in Ward, *Dr. Space*, 29.

21 **Hitler had hoped for:** Two other miracle weapons in early development were the hundred-ton A-10 rocket (which

would have been able to reach New York and Washington, DC, and cause terrible destruction) and a "sun gun"—a giant, radio-controlled orbiting mirror that would focus the sun's rays on Earth and destroy cities.

21 **top scientists died:** Neufeld, *Von Braun,* 154–56; Bergaust, *Wernher von Braun,* 33; Dornberger, *V-2,* 168. Dornberger said the 735 killed included "178 of the 4000 inhabitants of the settlement"; his total is repeated by Bergaust, *Wernher von Braun,* 29, but he gives no source for it. Neufeld wrote that there were 600 workers and 135 Germans killed.

21 **disease, beatings, or execution:** Though von Braun was not directly responsible for the use of slave labor, he knew all about it, for by his own admission and the testimony of others, he visited the Mittelwerk many times. But he denied knowledge of the catastrophic living and working conditions. It would be decades before the full truth of his awareness of those conditions—and the fact that he'd visited the Buchenwald camp to pick skilled slave laborers—was revealed publicly. He would also misrepresent his involvement with the SS, insisting that he was forced to join in 1940 by SS Reichsführer Heinrich Himmler himself (which might have been true), that he wore his SS uniform only once (which was not true), and that his regular promotions (three; he was eventually promoted to major) were routine and that he was notified of them by mail. He and others who knew him—most of them friends—claimed that his refusal to join would have meant his abandoning the work of his life, an argument hard to refute.

23 **"on the winning side":** Daniel Lang, "A Reporter at Large: A Romantic Urge," *New Yorker,* April 21, 1951, 82.

24 **his British nanny:** Stuhlinger interview, December 8, 1997, JSC Oral History Project.

25 **von Braun and a hundred and twenty-six:** Von Braun and Ordway, *History of Rocketry and Space Travel,* 118.

25 **no entry permits:** Lang, "A Reporter at Large," 76; Stuhlinger interview.

28 **would fail:** *Time,* October 21, 1957.

30 **"no turning back":** Quoted in Schefter, *The Race,* 34.

TWO: OF MONKEYS AND MEN

31 **"It doesn't really require":** Quoted in Slayton and Cassutt, *Deke!*, 82.

31 **for military use:** Launius and McCurdy, *Spaceflight.*

34 **"gates of heaven":** *Missiles and Rockets*, July 1957.

36 **"what I should do":** Davis-Floyd et al., *Space Stories.*

37 **"kept him alive":** Kraft, *Flight*, 68.

38 **trancelike state:** *Newsweek*, April 20, 1959, 64; Hall, *Space Pioneers*, 318–19.

38 **"grapefruit-sized satellite":** Robert B. Voas interview, May 19, 2002, NASA Oral History Project.

40 **"chance for immortality":** Carpenter et al., *We Seven*, 60.

42 **"sadists to a man":** Cooper and Henderson, *Leap of Faith*, 13.

42 **"drive them crazy":** Caidin, *Man into Space*, 132.

42 **down the road:** Sources for this discussion of astronaut testing include Swenson et al., *This New Ocean*, 129–32, 159–64; Santy, *Choosing the Right Stuff;* Conrad and Klausner, *Rocketman;* Glenn, *John Glenn;* Lamb, "Aeronautical Evaluation for Space Pilots"; Link, *Space Medicine in Project Mercury;* and the best and most insightful writing on the subject, Carpenter et al., *We Seven*, 165–95.

43 **his dad's new job:** Grissom and Still, *Starfall*, 64.

44 **"cities into nothingness":** Quoted in Caidin, *Buck Rogers.*

44 **test-pilot mortality rate:** Wolfe: *The Right Stuff*, 14.

45 **at least 50 percent:** *Newsweek*, April 20, 1959, 65.

45 **"All-American football team":** *New York Times*, April 10, 1959.

45 **a man into space:** *Newsweek*, April 20, 1959.

THREE: "THE HOWLING INFINITE"

46 **"The Howling Infinite":** This phrase can be found in Herman Melville's *Moby-Dick*, 46 23: "Better is it to perish in that howling infinite" (though Melville was not referring to space).

47 **executive officer:** Carol Faget, phone interview with the author, April 11, 2017.

48 **"the success of NACA's mission":** "Reflections of Joseph 'Guy' Thibodaux Jr."

49 **serious heat damage:** Maxime Faget interview, June 18–19, 1997, JSC Oral History Project; Davis-Floyd et al., *Space Stories;* "Reflections of Joseph 'Guy' Thibodaux Jr."

50 **"You put it on":** Quoted in Schanche, "The Astronauts Get Their Prodigious Chariot," 112.

51 **lowest bidder:** Swenson et al., *This New Ocean*, 136, 150.

52 **astronaut might survive:** Landis, "Human Exposure to Vacuum"; Smith, "Can You Survive in Space Without a Spacesuit?"

53 **dead within minutes:** Carpenter et al., *We Seven*, 153.

54 **"99 percent":** Cortright, *Apollo Expeditions to the Moon*, 55.

54 **make demands:** Hersch, *Inventing the American Astronaut*, 1.

54 **"ain't got no wings":** Shepard, "The First Step to the Moon."

55 **meager reserve-fuel capacity:** Cooper, "Annals of Space: We Don't Have to Prove Ourselves."

56 **"herd shot 'round the world":** James K. Hinson interview, May 2, 2000, JSC Oral History Project.

57 **drawings and blueprints:** Burgess, *Sigma* 7, 35.

58 **toss his cookies:** Thompson, *Light This Candle*, 186–87; Baker, *History of Manned Space Flight*, 47.

58 **"take his turn":** Carpenter et al., *We Seven*, 153.

58 **the best player:** Glenn, *John Glenn*, 204.

60 **"out of the way":** Ibid., 274–75.

60 **attempts reached orbit:** Bilstein, *Orders of Magnitude*, 55.

60 **"fourth would make it":** *Rochester Democrat Chronicle*, December 20, 1959.

62 **not-too-distant future:** Ibid.

62 **American prestige:** Williams, "Trade Winds," in "Go!" Other sources mention the chimpanzee total as six. See Lewis, *Appointment on the Moon*, 113.

62 **"Spam in a can":** Voas, "Project Mercury," 1.

63 **Mercury Seven were receiving:** Apparently Yeager's feelings about NASA occasionally took the form of action. According to astronaut Bill Anders, Yeager was "upset" that Anders had applied to be an astronaut "and actually put some energy into their trying to get me kicked out of the program" (William Anders interview, October 8, 1997, JSC Oral History Project).

When Frank Borman told Yeager, then his boss, that he'd been selected to be an astronaut, Yeager said, "Well, Borman, you can kiss your Air Force career goodbye," and dismissed him without another word (Kluger, *Apollo 8*, 25).

63 **"After the chimp, the chump":** GWS Oral History Project, Robert Gilruth interview no. 4, NASA.

63 **"some of this Spam bullshit":** Quoted in Mindell, *Digital Apollo*, 80.

64 **to be exact:** *Mercury Project Summary*, 181.

65 **multiple openings:** *Missiles and Rockets*, October 13, 1958.

65 **"when you join NASA":** Ibid., May 28, 1962, and March 3, 1962.

66 **having fun with the new guy:** Frank E. Hughes interview, March 29, 2013, JSC Oral History Project.

67 **"two hundred miles away":** *Missiles and Rockets*, March 1957.

67 **around for long:** Neufeld, *Von Braun*, 345.

67 **"a rich uncle":** Ibid., 343.

68 **"for spaceflight's sake":** Ibid.

68 **blows at a party:** Kraft, *Flight*, 103–4.

68 **"what flag he fights for":** Neufeld, *Von Braun*, 337.

68 **"our damned Nazi":** Ibid., 368.

68 **go in his place:** Ibid., 338; Cooper and Henderson, *Leap of Faith*, 148, 160; Mitchell, *The Way of the Explorer*, 30–32; Warren North interview, September 30, 1998, JSC Oral History Project; and several other astronaut interviews and authored books. Wally Schirra in his book *Schirra's Space* described him as "that great man von Braun, who was a dear friend of ours."

68 **opening in St. Louis:** Cunningham, *The All-American Boys*, 272.

70 **"giant fraternity party":** Landwirth and Hendricks, *Gift of Life*, 103.

70 **Mercury Five and Two:** Williams, "Who Flies First?," in "Go!"

71 **"keep his pants zipped":** Glenn, *John Glenn*, 230.

71 **mind his own business:** Ibid.; Thompson, *Light This Candle*, 226–27. A mutual friend of Shepard and Glenn told writer Neal Thompson decades later that Glenn had told him the guilty party was Shepard.

71 **von Braun's Saturn booster:** Baker, *History of Manned Space Flight*, 58.

73 **"moon and the planets":** *New York Times*, July 30, 1960.

FOUR: MAN ON A MISSILE

74 **"With Mercury we are using":** Williams, "Dangers," in "Go!"

74 **"The man is crazy":** *Time*, March 2, 1962.

74 **the art of sniveling:** Ibid.; *Amarillo Globe-News*, October 25, 1998. See also Jack D. Woodul, "The Further Adventures of Youthly Puresome," *Hook* (Fall 2006): "It was Naval Aviator's heaven—sniveling a hop on a beautiful day." Michael Collins also uses the word in *Carrying the Fire*: "I had sniveled whatever simulator time I could" (330).

75 **"a little arrogance":** Glenn, *John Glenn*, 194.

75 **slightly higher:** Williams, "Who Flies First?," in "Go!"

76 **"He was really sharp":** Thomas Sanzone interview, July 26, 2011, JSC Oral History Project.

76 **a marvel of tautology:** Caidin, *Man into Space*, 37.

77 **"lost in his life":** *Life*, February 2, 1962.

77 **he pulled out of it:** *Amarillo Globe-News*, October 25, 1998. This account is supported by other contemporary accounts, particularly Williams, "Who Flies First?," in "Go!"

77 **the viewing area:** Neufeld, *Von Braun*, 355; Kraft, *Flight*, 118; Faget quote, Schirra and Billings, *Schirra's Space*, 72–73.

78 **"Talk to me, dammit":** Kranz, *Failure Is Not an Option*, 27, 29.

78 **"going to be damn lucky":** Quoted in John Hodge interview, April 19, 1999, JSC Oral History Project.

80 **had not improved morale:** Williams, "But You Can't Put a Man in the Can!," in "Go!"

81 **"Spokesman Says U.S. Asleep":** Shepard and Slayton, *Moon Shot*, 105–6.

81 **Soviets' goal was obvious:** Daniloff, *The Kremlin and the Cosmos*, 138.

82 **Gagarin echoed that:** *Albany Times Union*, April 13, 1961; *Buffalo Evening News*, April 13, 1961.

82 **nothing substantive:** Seamans, *Project Apollo*, 119.

83 **Soviet space program:** Siddiqi, *Challenge to Apollo*, 11–12.

83 **the Soviet people:** Hardesty and Eisman, *Epic Rivalry*, xii.

83 **"into the cosmos":** Boris Chertok, quoted in Reichardt, "The Luna 1 Hoax Hoax."

84 **"plenty of work for everybody":** Quoted in Glenn, *John Glenn*, 237.

85 **shared by many Americans:** Williams, "But You Can't Put a Man in the Can!," in "Go!"

85 **with hydrogen peroxide:** Reynolds, *Kennedy Space Center*, 71, 82.

85 **a faulty computer:** Sources for this account of Shepard's flight are Burgess, *Freedom 7;* Lewis, *Appointment on the Moon;* Thompson, *Light This Candle;* Shepard and Slayton, *Moon Shot;* Slayton and Cassutt, *Deke!;* Swenson et al., *This New Ocean;* Grimwood, *Project Mercury;* Kraft, *Flight;* Williams, "Go!"; and *Results of the First U.S. Manned Orbital Space Flight.*

86 **"in ze nose cone":** Alan Shepard interview, February 20, 1998, JSC Oral History Project.

87 **nicknamed "Ol' Reliable":** Kraft, *Flight*, 139.

87 **southeast of Cape Canaveral:** Williams, "Surfside 5," in "Go!"

89 **white shirt with a thin tie:** Claiborne R. Hicks interview, November 4, 2000, JSC Oral History Project.

89 **in medical distress:** Robert B. Voas interview, May 19, 2002, JSC Oral History Project.

91 **overall scientific achievement:** McDougall, *The Heavens and the Earth*, 246.

94 **"scientist or an engineer":** Bizony, *The Man Who Ran the Moon*, 21.

94 **Johnson would remember later:** Murray and Cox, *Apollo*, 69.

95 **defending NASA's budget:** Lambright, *Powering Apollo*, 89–94.

97 **"spend a little money":** Launius and McCurdy, *Spaceflight*, 72.

97 **watched Johnson work his contacts:** Murray and Cox, *Apollo*, 80–81.

98 **"I'm ready":** Thompson, *Light This Candle*, 265.

98 **"Let's go":** Neufeld, *Von Braun*, 364; Ward, *Dr. Space*, 128.

98 **"250,000 pounds into earth orbit":** Quoted in Watkins, *Apollo Moon Missions*, xx.

99 **"launch windows":** Cortright, *Apollo Expeditions to the Moon*, 19.

99 **"'New York City overnight'":** Grissom, *Gemini!*, 177.

FIVE: IN ORBIT

104 **"an administrator's discount on it":** Shepard and Slayton, *Moon Shot,* 41; Lambright, *Powering Apollo,* 101.

105 **where Shepard had:** Sources for this account of Grissom's flight are Burgess, *Liberty Bell 7;* Lewis, *Appointment on the Moon;* Shepard and Slayton, *Moon Shot;* Slayton and Cassutt, *Deke!;* Swenson et al., *This New Ocean;* Grimwood, *Project Mercury;* Kraft, *Flight;* Williams, "Go!"; Carpenter et al., *We Seven;* and *Results of the Second U.S. Manned Suborbital Space Flight.*

105 **going to drown:** *Results of the Second U.S. Manned Suborbital Space Flight;* Williams, "Go!," 322–24. For the most thorough, even-handed examination of the hatch controversy, see Leopold, *Calculated Risk,* chapter 9.

106 **"full of sea water":** *New York Times,* July 22, 1961.

106 **"possibly I did":** Boomhower, *Gus Grissom,* 199.

106 **banged into it:** Slayton and Cassutt, *Deke!,* 101.

106 **"good portion of the time":** Quoted in Swenson et al., *This New Ocean,* 370.

106 **"Scared, okay?":** Quoted in Boomhower, *Gus Grissom,* 211.

107 **"automatic devices":** *New York Times,* July 22, 1961.

107 **"disappeared almost entirely":** Ibid., August 5, 1962.

109 **rushing a message in:** Lunney et al., *From the Trench of Mission Control,* 317.

110 **almost five hours:** Ibid., February 21, 1962.

113 **Glenn might be the first:** On a Discovery Channel miniseries called *Rocket Science,* produced and aired in Canada in 2002–2003, flight controller Gene Kranz said, "Every one of us believed we would lose one, possibly even more, astronauts during the Mercury program."

113 **would remember thinking:** Williams, "The Flight of *Friendship Seven,*" in "Go!"

SIX: UNDER PRESSURE

117 **"We knew that human beings":** Robert B. Voas interview, May 19, 2002, JSC Oral History Project.

118 **"act up":** Quoted in Slayton and Cassutt, *Deke!,* 110.
118 **the Mercury program:** Lamb, *Inside the Space Race*, 108–11.
118 **"everyone in that room":** Shepard and Slayton, *Moon Shot,* 154.
118 **but without success:** *Desert Sun*, March 31, 1962; Loudon Wainwright, "Comes a Quiet Man to Ride Aurora 7," *Life*, May 18, 1962.
119 **think that was right:** Schirra and Billings, *Schirra's Space*, 77.
120 **meager jet time:** Carpenter and Stoever, *For Spacious Skies,* 175.
120 **"vague and detached":** Paul Haney, quoted in Hall, *Space Pioneers*, 388; Williams, "Go!"
121 **assistant flight director for Carpenter's flight:** This description of Kranz's early progress at NASA is primarily drawn from his book *Failure Is Not an Option.*
122 **"sunrise to worry about":** Quoted in Young et al., *Journey to Tranquility*, 148.
123 **sounded delirious:** Kranz, *Failure Is Not an Option,* 89.
123 **testing a new airplane:** Ibid., 91.
123 **wildly oscillating capsule:** Swenson et al., *This New Ocean*, 446–56; Slayton and Cassutt, *Deke!,* 114; Kranz, *Failure Is Not an Option.*
124 **"broke up his tranquility":** Robert F. Thompson interview, August 29, 2000, JSC Oral History Project.
124 **had also been concerned:** Williams, "The Flight of *Aurora Seven*," in "Go!"
124 **responded autocratically:** Carpenter and Stoever, *For Spacious Skies,* 301; Liebergot and Harland, *Apollo EECOM*, 110; Gerry Griffin interview, March 12, 1999, JSC Oral History Project.
124 **"they didn't either":** Kranz, *Failure Is Not an Option,* 91.
125 **"must work for me":** Ibid., 31.
125 **remote tracking station:** Ibid., 127.
127 **veiled reference to Carpenter:** Schirra interview with Francis French for CollectSPACE.com, February 22, 2002.
127 **told what to do:** Williams, "The Flight of *Sigma Seven*," in "Go!"
130 **Schirra's backup:** Sources for this account of Cooper's flight are Burgess, *Faith 7;* Lewis, *Appointment on the Moon;* Cooper and Henderson, *Leap of Faith;* Swenson et al., *This New Ocean;*

Grimwood, *Project Mercury;* Kraft, *Flight;* Williams, "Go!"; and *Mercury Project Summary.*

131 **switches on his control panel:** Grelsamer, *Into the Sky with Diamonds*, 110.

131 **"stop holding your breath":** *Time*, May 24, 1963.

131 **'band of bright blue":** Quoted in Barbour et al., *Footprints on the Moon*, 66–67.

132 **"made another revolution":** Robert Smylie interview, April 17, 1999, JSC Oral History Project.

132 **an amphetamine:** Williams, "The Flight of *Faith Seven*," in "Go!"

132 **floated by the military:** *Geneva (NY) Times*, May 17, 1963.

132 **"mission would have failed":** Swenson et al., *This New Ocean*, 494–501; Kraft, *Flight*, 183.

133 **repeated trial and error:** This statement is a recasting of a similar statement in the book written by the Mercury Seven themselves (though ghostwritten by professional writers). See Carpenter et al., *We Seven*, 210.

133 **to the president:** Stafford and Cassutt, *We Have Capture*, 109; Wilson, "Mercury Atlas 10."

134 **"Man is the deciding element":** Quoted in Young and Hansen, *Forever Young*, 160.

134 **"six men back":** Robert Gilruth R. C. Macalester College lecture, January 29, 1964, in Robert Gilruth Papers, box 4, folder 6, Special Collections, Virginia Tech.

134 **thought of them as his "boys":** Quoted in Launius, "Heroes in a Vacuum," 10.

134 **"things was launched":** Quoted in Chaikin, "Bob Gilruth."

135 **a kind of NACA university:** Burrows, *This New Ocean*, 346.

136 **Maybe he could again:** Siddiqi, "A Secret Uncovered"; Day, "The Moon in the Crosshairs."

SEVEN: THE GUSMOBILE

137 **"It was like the Blue Angels":** Agle, "Flying the Gusmobile."

137 **for good reason:** Walter Schirra interview, December 1, 1998, JSC Oral History Project.

139 **"never got solved":** Quoted in Murray and Cox, *Apollo*, 109.

140 **considered this suggestion:** Turnill, *The Moonlandings*, 89; *Missiles and Rockets*, June 25, 1962.

142 **"I vowed to dedicate":** Quoted in Hansen, *Enchanted Rendezvous*, 15.

142 **reserved and reticent:** Bill Causey quote in "The Lunar Module," an episode in the documentary miniseries *Moon Machines*.

143 **"very much like a motorcycle":** Owen G. Morris interview, May 20, 1999, JSC Oral History Project.

143 **"that's no good":** *Life*, March 14, 1969.

144 **"far-out science fiction":** *Time*, June 22, 1962.

145 **renewed consideration:** Seamans, *Aiming at Targets*, 98. This discussion of LOR owes much to the definitive work on the subject, James Hansen's *Enchanted Rendezvous*, and also to Baker, *History of Manned Space Flight*, 144–56; Brooks et al., *Chariots for Apollo*, 61–86; *Life*, March 14, 1969; and von Braun, "Concluding Remarks by Dr. Wernher von Braun."

145 **"eyeballing that thing":** Brooks et al., *Chariots for Apollo*, 76.

146 **announcing his support for LOR:** Seamans, *Aiming at Targets*, 43.

148 **between Kennedy and Thomas:** James E. Webb Oral History Interview 1, April 29, 1969, by T. H. Baker, LBJ Library.

149 **needed at the MSC:** Many of the details included in this account of the MSC deal are from Eric Berger's excellently researched "A Worthy Endeavor: How Albert Thomas Won Houston NASA's Flagship Center," *Houston Chronicle*, September 14, 2013.

149 **"someplace out west":** Dutch von Ehrenfried interview, March 19, 2009, JSC Oral History Project.

150 **souped-up Mercury craft:** Robert Gilruth Papers, box 4, folder 19, "Oral History Transcript"; Neufeld, *Von Braun*, 371.

150 **"what instruments went where":** Quoted in Agle, "Flying the Gusmobile."

152 **everyone in the room:** Cortright, *Apollo Expeditions to the Moon*, 50.

152 **"within reason":** Stafford and Cassutt, *We Have Capture*, 40–41.

152 **applied to them too:** Scott et al., *Two Sides of the Moon,* 134.
157 **"it got to be awesome again":** Jack Garman interview, March 27, 2001, JSC Oral History Project; Garman interview, April 14, 2014, Honeysuckle Creek, www.honeysucklecreek.net/interviews/jack_garman.html.
158 **"right through me":** Cernan and Davis, *The Last Man on the Moon,* 58.
159 **"He was never there":** Worden and French, *Falling to Earth,* 70.
159 **but Shepard did:** Thompson, *Light This Candle,* 306.
159 **"You're just an astronaut trainee":** Grissom and Still, *Starfall,* 123.
159 **more complex operations now involved:** Guy Thibodaux interview, October 20, 1999, JSC Oral History Project.
160 **years of development had gone into it:** Launius and Jenkins, *Coming Home,* 112.
160 **"If the rocket got out of sight":** Henry Pohl interview, February 9, 1999, JSC Oral History Project.
161 **"bearing down on you":** Agle, "Riding the Titan II."
161 **jets used ejection seats:** Guy Thibodaux interview, October 20, 1999, JSC Oral History Project.
161 **"My Gemini spacecraft":** Schirra and Billings, *Schirra's Space,* 157–58.
162 **that would include lunar landings:** Launius and McCurdy, *Spaceflight,* 208.
163 **"This nation has tossed its cap":** Quoted in Thompson, *Light This Candle,* 295.
163 **"This effort is expensive":** Bilstein, *Orders of Magnitude,* 71; Ward, *Dr. Space,* 132.
164 **ever saw him do so:** Neufeld, *Von Braun,* 390.
167 **forced to abandon him:** Portree and Treviño, *Walking to Olympus,* 2.
167 **"There was only a sense":** Quoted in Shepard and Slayton, *Moon Shot,* 172.
167 **"the gap is not closing":** Ibid., 174.
167 **an EVA in space:** Oberg, "Russia Meant to Win the Moon Race."

EIGHT: THE WALK, AND A SKY GONE BERSERK

168 **"you have concerns about":** Quoted in Leopold, *Calculated Risk,* 200.

170 **"We're not playing Mickey Mouse":** Quoted in Evans, *Escaping the Bonds of Earth*, 256.

170 **WE ARE 301 MAN-ORBITS:** United Press International, *Gemini: America's Historic Walk in Space*, chapter 2.

174 **military space stations:** Baker, *History of Manned Space Flight*, 215; Slayton and Cassutt, *Deke!*, 216.

174 **"All astronauts are created equal":** Slayton and Cassutt, *Deke!,* 136.

175 **a rookie astronaut:** For valuable insights into Slayton's crew selection methods, see Swanson, *"Before This Decade Is Out,"* 337–38.

176 **who had lived in it:** Kenneth A. Young interview, June 6, 2001, JSC Oral History Project.

177 **became even thicker:** Schefter, *The Race,* 219.

178 **"a capable pilot":** Stafford and Cassutt, *We Have Capture*, 78.

178 **their Gemini spacecraft:** Ibid., 79–82; McMichael, "Losing the Moon."

179 **consequences for the program:** Borman and Serling, *Countdown*, 152.

179 **shaken up by See's death:** Slayton and Cassutt, *Deke!,* 168.

180 **if it ever became operational:** See Hansen, *First Man*, chapters 15 and 16, for more on Armstrong's X-15 experiences.

181 **"consumed by learning":** Quoted in ibid., 31.

182 **"bridge breaking":** Armstrong et al., *First on the Moon*, 144.

183 **"He was a very steadfast person":** Hansen, *First Man*, 127.

183 **Many of his flights were dangerous:** Ibid., 134–36.

184 **"at the margins of knowledge":** *Life*, September 27, 1963.

184 **the selection panel's first meeting:** Hansen, *First Man*, 195.

185 **They turned up the cockpit lights:** *Roundup*, April 1, 1966.

186 **would almost certainly follow:** *New York Times*, March 18, 1966.

187 **two full revolutions per second:** Barbree, *"Live from Cape Canaveral,"* 115. Lovell and Kluger, in *Lost Moon*, page 10, assert

that it was five hundred rpm. In a March 21, 1966, story five days after the flight, the *New York Times* reported: "In the delayed film frames, the rotation appears even faster than the once a second the space agency said had occurred." In the May 2, 1966, issue of *Missiles and Rockets*, page 28, in a story on medical information shared between the U.S. and Soviet space programs, reference was made to "the high roll rate of GT-8, which reached one revolution per 0.8 second." Neil Armstrong, not one to exaggerate, said in a 2008 Discovery Channel documentary entitled *When We Left Earth:* "When the roll rate increased to more than four hundred degrees per second, our vision was beginning to degrade."

187 **"Physiological limits":** "Gemini VIII Technical Debriefing," 60.

187 **gone to worms:** This phrase is used in a press release authored by the two astronauts shortly after the mission and released by *World Book Encyclopedia*, which shared in the rights to the NASA astronauts' stories.

187 **"I gotta cage my eyeballs":** Chaikin, *A Man on the Moon*, 168.

189 **found the culprit:** Due to a problem in the gauge, the crew believed that the propellant for the reentry control system was down to approximately a third when it was actually closer to 50 percent. See *Gemini Program Mission Report: Gemini VIII*, 376.

190 **The crew was healthy:** The following sources were consulted for this account of Gemini 8: Hacker and Grimwood, *On the Shoulders of Titans*, 308–21; Hansen, *First Man*, chapter 19; Scott et al., *Two Sides of the Moon*, chapter 6; Kraft, *Flight*, 253–55; "Gemini VIII Technical Debriefing"; "Gemini VIII Composite Air-to-Ground and Onboard Voice Tape Transcription"; and *Gemini Program Mission Report: Gemini VIII.*

191 **"It was a non-trivial situation":** Abbey-Callaghan file, Robert Sherrod Collection, NASA History Office.

191 **"If we had heard":** Hansen, *First Man*, 271.

192 **save its occupants:** *Missiles and Rockets*, April 4, 1966.

192 **Soviet shot at the moon:** *Time*, March 16, 1966; Hix, "Laika and Her Comrades"; Burgess and Dobbs, *Animals in Space*, 218–19.

192 **"Americans no longer":** *Life*, December 2, 1966.

193 **"almost like working":** Quoted in Hall, *Space Pioneers*, 395.
193 **"all the flexibility":** Cernan and Davis, *The Last Man on the Moon*, 134.
193 **might lose consciousness:** Glynn Lunney, interviewed in the 2008 Discovery Channel documentary *When We Left Earth.*
195 **selected for the mission:** Mark Wade, "Gemini LORV," *Encyclopedia Astronautica.*
196 **1,993 hours in space:** Hansen, *First Man*, 296.

NINE: INFERNO

201 **"We just became anesthetized":** *Dallas Morning News*, April 13, 1967.
202 **his Mercury mission:** Carpenter and Stoever, *For Spacious Skies*, 325.
202 **Apollo program in Houston:** Quoted in Bizony, *The Man Who Ran the Moon*, 139.
203 **specifications provided by Faget's office:** Brooks et al., *Chariots for Apollo*, 113; Davis-Floyd et al., *Space Stories.*
203 **"item by item":** Brooks et al., *Chariots for Apollo*, 113.
204 **true operational conditions:** Ibid., 164.
205 **Soviet space program:** Harvey, "The 1963 Soviet Space Platform Project."
205 **with mixed results:** *Quest* (Fall 1993), 37.
205 **on the lunar surface:** Slayton and Cassutt, *Deke!*, 147.
206 **"a more distant objective":** *Missiles and Rockets*, May 2, 1966.
207 **"who is right and who is wrong":** Russian Scientific Research Center for Space Documentation, *Roads to Space*, 88.
207 **They received no answer:** Pesavento, "A Review of Rumoured Launch Failures," 389.
207 **"long and fatal illness":** Quoted in Burrows, *This New Ocean*, 404.
208 ***Go ahead and land on the moon:*** DeGroot, *Dark Side of the Moon*, 201; Young et al., *Journey to Tranquility*, 285.
209 **the job done right:** Cernan and Davis, *The Last Man on the Moon*, 3; Slayton and Cassutt, *Deke!*, 191; Grissom and Still, *Starfall*, 260.

209 **"a reporter's delight":** *Dallas Morning News*, January 28, 1967.

209 **first flight to the moon:** Leopold, *Calculated Risk*, 220; *Time*, June 11, 1965.

209 **an occasional profanity:** *Time*, February 3, 1967.

209 **"a really great boy":** Quoted in Lovell and Kluger, *Lost Moon*, 24.

210 **Grissom had with Gemini:** MacKinnon and Baldanza, *Footprints*, 240.

210 **former Mercury operations director:** Quoted in Leopold, *Calculated Risk*, 236.

210 **for his thoroughness:** Ibid., 252.

211 **"This is the worst spacecraft":** Al Shepard interview, February 20, 1998, JSC Oral History Project.

211 **"He thought they should be working":** Grissom and Still, *Starfall*, 181.

211 **"They'll fire me":** Young and Hansen, *Forever Young*, 116; MacKinnon and Baldanza, *Footprints*, 240.

212 **"You're going to be in there":** Schirra and Billings, *Schirra's Space*, 183.

213 **Grissom's Block I version:** Shayler, *Disasters and Accidents*, 102.

213 **back in Houston:** Grissom and Still, *Starfall*, 181, 183.

214 **something they didn't have:** *Dallas Morning News*, March 10, 1967.

216 **to complete the job:** Much of this account of the fire and events following it are based on "Report of Apollo 204 Review Board"; Leopold, *Calculated Risk*, 235–60; Lovell and Kluger, *Lost Moon*, 14–20; Shayler, *Disasters and Accidents*, 97–115; and Chaikin, "Apollo's Worst Day." Slayton's actions are based on the account in his book *Deke!*

217 **take up to space with him:** *Washington Post*, January 26, 2017.

217 **She didn't either:** Grissom and Still, *Starfall*, 188–89.

217 **talked to Lowell for a few minutes:** Author interview with Chuck Friedlander, May 7, 2017.

218 **further testing for a while:** Worden and French, *Falling to Earth*, 85.

218 **from the contractor:** Lovell and Kluger, *Lost Moon*, 22–25.

218 **they soon found out:** Chuck Deiterich interview, May 16,

2000, JSC Oral History Project; Lunney et al., *From the Trench of Mission Control*, 330–31.

219 **"It's horrible!":** Murray and Cox, *Apollo*, 204.

219 **congregated in a back room:** Houston and Heflin, *Go, Flight!*, 98–99; Kranz, *Failure Is Not an Option*, 191–202.

219 **"Worst I ever had":** Slayton and Cassutt, *Deke!*, 190.

219 **"nothing would be the same again":** Kranz, *Failure Is Not an Option*, 197.

220 **like Faget could understand:** Borman and Serling, *Countdown*, 173; Murray and Cox, *Apollo*, 215–16. In the latter account, several "senior people in the program" were present, and the gathering occurred "less than two weeks after the fire."

220 **thought of that moment:** Author interview with Chuck Friedlander, May 7, 2017.

TEN: RECOVERY

221 **"We were given the gift":** William J. Cromie interview with Neil Armstrong, February 1969, in Armstrong biographical file, Robert Sherrod Apollo Collection, NASA History Office.

222 **and nodded to her:** E-mail from Chuck Friedlander to the author, September 3, 2017.

222 **"empty-slot flyovers":** Hall, *Space Pioneers*, 397.

222 **"a reasonable possibility":** *Dallas Morning News*, January 28, 1967.

223 **admiration it had a decade ago:** Lambright, *Powering Apollo*, 159.

224 **"These words will remind you":** Kranz, *Failure Is Not an Option*, 204.

224 **"If you are really tough":** Lunney et al., *From the Trench of Mission Control*, 167.

226 **Eight hours later:** Slayton and Cassutt, *Deke!*, 76.

226 **electroshock treatment:** Frank Borman interview, April 13, 1999, JSC Oral History Project; Borman and Serling, *Countdown*, 176.

227 **in a mental institution:** John F. Yardley interview, June 29, 1998, JSC Oral History Project; McMichael, "Losing the Moon."

227 **punning duels with Wally Schirra:** Cunningham, *The All-American Boys,* 100.

227 **a few at NASA took up:** Kraft, *Flight,* 274–75. Shea would deny this later, but by any reasonable measure, he suffered a nervous breakdown, and several of his NASA contemporaries have attested to that.

227 **"a poor administrator":** Borman and Serling, *Countdown,* 178; Kraft, *Flight,* 274–75; George Mueller interview, September 27, 1998, JSC Oral History Project.

227 **"austere budget":** *Dallas Morning News,* March 12, 1967.

227 **had been carried out:** Robert Sherrod interview with George M. Low, November 7, 1969, box 12, file 22, Low interview, Christopher C. Kraft Papers, Special Collections, Virginia Tech.

228 **"Tell him":** Lambright, *Powering Apollo,* 156.

228 **what should have been there:** Burgess and Doolan, *Fallen Astronauts,* 210.

230 **approved 1,341 alterations:** Crouch, *Aiming for the Stars,* 212.

230 **"until we make":** Quoted in Lambright, *Powering Apollo,* 183.

231 **"Be flexible":** Stafford and Cassutt, *We Have Capture,* 108–9; Cernan and Davis, *The Last Man on the Moon,* 164–65.

231 **"include the most spectacular":** Oberg, *Red Star in Orbit,* 91.

232 **lost spacefarers:** Quoted in Brooks et al., *Chariots for Apollo,* 227.

233 **circumlunar flight in the near future:** Collins, *Carrying the Fire,* 282.

233 **some of the firsts involved:** Newkirk, *Almanac of Soviet Manned Space Flight,* 46.

233 **A moon rocket, unquestionably:** Day, "The Moon in the Crosshairs," part 2.

234 **a check for the difference:** Aldrin and Warga, *Return to Earth,* 191–92; Slayton and Cassutt, *Deke!,* 205.

234 **regain its confidence:** Brooks et al., *Chariots for Apollo,* 230.

235 **training habits had become lax:** Collins, *Carrying the Fire,* 270, 282; Stafford and Cassutt, *We Have Capture,* 102; Cunningham, *The All-American Boys,* 130–31; Slayton and Cassutt, *Deke!,* 191.

236 **booted them from his crew:** Cunningham, *The All-American Boys,* 77–80; author e-mail correspondence with Cunningham; French, "I Worked with NASA, Not for NASA." In the last in-

terview, Schirra denied the idea that he would take on a job like caretaker commander, although of course, he might not have wanted to admit it. One possible supporting fact is revealed in this statement from Slayton: "I had put another crew in training for the second manned Apollo without making an announcement" (Slayton and Cassutt, *Deke!*, 166).

236 **"We labored day and night":** Schirra and Billings, *Schirra's Space*, 191.

237 **without his approval:** Borman and Serling, *Countdown*, 181–84; Lovell and Kluger, *Lost Moon*, 32–33; Kraft, *Flight*, 283; Cernan and Davis, *The Last Man on the Moon*, 169–70.

237 **had received since flight school:** Borman and Serling, *Countdown*, 186–87.

238 **scenario he could think of:** Guenter Wendt interview, January 16, 1998, JSC Oral History Project.

240 **"It sounded reckless":** Cortright, *Apollo Expeditions to the Moon*.

241 **put on standby status:** Lewis, *Appointment on the Moon*, 408–23.

ELEVEN: PHOENIX AND EARTHRISE

243 **"To see the Earth":** Archibald MacLeish, "A Reflection," *New York Times*, December 25, 1968.

243 **reentry and splashdown:** Flight Dynamics Controllers, *Oral Histories of NASA Flight Dynamics Controllers*, 169–74.

243 **human mistake in wiring:** von Braun, *Space Frontier*, 203.

244 **usefulness to the agency:** French, "I Worked with NASA, Not for NASA."

244 **"a photo finish":** Quoted in Turnill, *The Moonlandings*, 134.

245 **off instrument readings:** Ibid., 132.

246 **"He was never the same":** Kraft, *Flight*, 292; John Logsdon interview with Robert Gilruth, August 26, 1969, box 12, file 11, Gilruth interview, Robert Gilruth Papers, Special Collections, Virginia Tech.

246 **would be missed:** Bizony, *The Man Who Ran the Moon*, 212–14; Kraft, *Flight*, 292.

247 **"I've had it":** Cernan and Davis, *The Last Man on the Moon*, 178; Quoted in Baker, *History of Manned Space Flight*, 310.

248 **Wally Schirra Bitch Circus:** Kraft, *Flight*, 289. He also said, referring to *Apollo* 7, "We had a failure in the pilot" (Chris Kraft interview, May 23, 2008, JSC Oral History Project).

248 **"pull the plug on Wally Schirra":** Liebergot and Harland, *Apollo EECOM*, 128; Gene Kranz interview, January 8, 1999, JSC Oral History Project.

248 **"Somebody down there":** Quoted in French and Burgess, *In the Shadow of the Moon*, 218.

248 **wasn't as uncooperative:** *Dallas Morning News*, October 16, 1968. The paper also reported that on the fourth day of the flight, Cunningham said, "Well, so far I've been able to resist pretty much getting a cold." See also French and Burgess, *In the Shadow of the Moon*, 215.

249 **threatening mutiny:** Cunningham, *The All-American Boys*, 127–32; Lunney et al., *From the Trench of Mission Control*, 143–50.

249 **"I wasn't going to put anybody":** Slayton and Cassutt, *Deke!*, 219; Cunningham, *The All-American Boys*, 157–58; Liebergot and Harland, *Apollo EECOM*, 130; Lunney et al., *From the Trench of Mission Control*, 173.

250 **"101 percent successful":** Quoted in Evans, *Escaping the Bonds of Earth*, 452; Wilford, *We Reach the Moon*, 171.

251 **EVA and docking thrown in:** French and Burgess, *In the Shadow of the Moon*, 328–29, 337.

251 **"You weren't really *doing* anything":** Ibid., 337–38.

251 **"tightly wound little sumbitch":** Cernan and Davis, *The Last Man on the Moon*, 178.

251 **the moon in their sights:** Chris Kraft interview, May 23, 2008, JSC Oral History Project; Borman and Serling, *Countdown*, 189; Seamans, *Project Apollo*, 108; Kraft, *Flight*, 284. In recent years, there has been skepticism that the Soviet threat to orbit the moon first was the main reason behind NASA's decision to send Apollo 8 around the moon—or even that it was a factor at all. It has been claimed that Frank Borman was the only legitimate source for this, since the Russian threat was mentioned by no one else, NASA official or otherwise, and some have concluded that Borman's account was erroneous.

But there are other mentions by NASA's top administrators of the Soviet threat and how it affected the Apollo 8 circumlunar decision. In discussing the idea, which was broached while Seamans and Webb were in Vienna for a conference, Seamans wrote later in *Project Apollo*, "He [Webb] invited me to his room...to tell me of his recent telephone calls from Tom Paine, NASA's Deputy Administrator at that time. Tom proposed a circumlunar flight for the next Apollo mission. He advised Jim that there were indications of an early Soviet manned mission to the Moon." Christopher Kraft, in his book *Flight,* page 284, quotes George Low as saying, after mentioning the idea to Kraft: "It would ace the Russians and take a lot of pressure off Apollo." And in his JSC Oral History Project cited here, Kraft says of his meeting with Robert Gilruth and George Low in which they first discussed the change: "So they knew the situation with the Russians better than I did, and they said, 'It would also give us a leg up on the Russians, because it appears that the Russians may be trying to do the same thing.' Well, that was in *Aviation Week* anyway, so I knew that was public knowledge....So Gilruth said, 'We ought to call Deke in.' We did. We got him on the phone. He came in. We talked about it for another half hour or so....We went out of that meeting with Deke going to see what he could do in terms of crew training." It's hard to believe that Slayton was not given the same information about the Russian threat, and it's perfectly understandable that he told Borman about it. Borman also repeated his conversation with Slayton and Slayton's mention of the Russian threat at other times, most prominently in the TV special *NOVA: To the Moon*, which aired in 1999. In it, Borman is interviewed with Bill Anders and Jim Lovell, his Apollo 8 crewmates, and in answer to a question from Lovell about what Slayton said to Borman, he replies, "No, he said that the Russians—that the CIA had heard that the Russians were going to launch before the end of the year, and Low was coming up with this plan to send Apollo 8 to the moon, and what did I think about it?" The threat of a Soviet lunar mission may not have been the primary reason for the switch, but it's quite believable that it was the first one Slayton provided to Borman. Finally, several astronauts—Anders, Rusty Schweickart, and Buzz Aldrin among them—are also on the record as stating their belief that the Soviet threat to send a manned spaceflight around the moon was the primary reason that a lunar-orbit mission was moved up in the schedule. See French and Burgess, *In the Shadow of the Moon*, 299 and 337; Aldrin and McConnell, *Men from Earth*, 191.

254 **examined by Surveyor:** Lewis, *Appointment on the Moon*, 256–59.

255 **"It would ace the Russians":** Kraft, *Flight*, 284.

255 **"order-of-magnitude difference":** Chris Kraft interview, May 23, 2008, JSC Oral History Project.

256 **his crew would be ready:** Borman and Serling, *Countdown*, 189; Slayton and Cassutt, *Deke!*, 214–15.

256 **"A one-third chance of success":** *Time*, December 24, 2008.

257 **"I hope none of you take cold":** American Presidency Project, "The President's Toast and Responses at a Dinner Honoring Members of the Space Program," December 9, 1968.

257 **on the way out:** Cernan and Davis, *The Last Man on the Moon*, 179–80.

259 **"The Russians suck":** Liebergot and Harland, *Apollo EECOM*, 131.

259 **they shook hands firmly:** Kraft, *Flight*, 299; Borman and Serling, *Countdown*, 208–11.

260 **"We flew all the way to the Moon":** Quoted in Bizony, *The Man Who Ran the Moon*, 216.

261 **felt vindicated:** Thomas O. Paine Oral History Interview 2, by T. H. Baker, April 10, 1969, LBJ Library.

261 **"To the crew of Apollo 8":** Borman and Serling, *Countdown*, 220.

TWELVE: "AMIABLE STRANGERS"

262 **"Both Neil and Buzz had more":** Author interview, Albert Jackson, August 31, 2017.

263 **"We were old friends":** Cernan and Davis, *The Last Man on the Moon*, 198.

263 **"If you had to ask somebody":** Author interview, Alan Bean, February 8, 2015.

264 **"I didn't want to be responsible":** Armstrong et al., *First on the Moon*, 145.

267 **Lovell deserved better:** Hansen, *First Man*, 338–39.

267 **"the riskiest one to date":** Borman and Serling, *Countdown*, 198.

268 **"Like a kid's balloon at a party":** *NOVA: To the Moon,* PBS, July 13, 1999.

269 **he had work to do:** Hansen, *First Man*, 332.

270 **Aldrin later insisted:** Aldrin, *Return to Earth*, 169–71; Slayton and Cassutt, *Deke!*, 175–76; Kraft, *Flight*, 259.

270 **Buzz kept his seat:** Cernan and Davis, *The Last Man on the Moon*, 156–57.

271 **"He planted his goals":** French and Burgess, *In the Shadow of the Moon*, 123; Aldrin described his father as "stern and disciplined" in Aldrin and Warga, *Return to Earth*, 136.

271 **their own toddler:** Author interview, Maddy Aldrin Crowell, January 2, 2015.

271 **alligator named Agamemnon:** Ibid.

272 **his theories were wrong:** Dean F. Grimm interview, August 17, 2000, JSC Oral History Project.

273 **psychological tests:** In the documentary *Mission Control* (2016), Aldrin blamed "overuse of alcohol" in both his father and mother for his post–Apollo 11 problems, and in his books and other interviews, he has openly discussed, most fully in *Return to Earth*, the suicide and depression in his family, primarily on his mother's side.

273 **for five days:** Aldrin and Warga, *Return to Earth*, 187.

273 **upcoming trip to the moon:** In the 2007 documentary *The Wonder of It All*, Buzz said: "My mother committed suicide a little over a year before my flight to the moon, probably because, uh…she did not want to deal with…well…a contributing factor would be that she didn't want to deal with the notoriety…uh…that would come along with, uh…her son having gone to the moon. There were other unhappinesses in the domestic relationship that were involved in that.

But anyway, there was a basic depressive tendency that exists throughout members of my family. I felt my problem was dealing with a mental depression. Well, it turned out that that really wasn't the problem—it was, uh, something that led to, uh, even more of an inherited tendency of both of my parents for the, uh, overuse of alcohol."

273 **"We've already seen":** Grelsamer, *Into the Sky with Diamonds,* 118.

273 **orbital mechanics:** Cernan and Davis, *The Last Man on the Moon*, 78; Alan Bean interview, June 23, 1998, JSC Oral History Project; Collins, *Carrying the Fire*, 323.

273 **on the subject:** Chaikin, *A Man on the Moon*, 143.

273 **"Aldrin," said one friend:** Quoted in Barbour, *Footprints on the Moon,* 183.

273 **the Mechanical Man:** *Spokane Daily Chronicle,* July 14, 1969.

273 **"I sometimes think":** Ted Guillory, quoted in Armstrong et al., *First on the Moon*, 87.

273 **If a computer could talk:** This profile of Buzz Aldrin is based on many sources, chief among them Aldrin's three autobiographical books (*Return to Earth, Men from Earth, Magnificent Desolation*); interviews by the author and others with Aldrin and his relatives, friends, and acquaintances; the many books written by and about, and interviews granted by, astronauts and others involved with the Gemini and Apollo programs; and various histories of those programs.

274 **"please speak up":** Author interview, Jim Lovell, March 12, 2015.

274 **"wasn't an organization man":** Aldrin and McConnell, *Men from Earth*, 136; Wendt and Still, *The Unbroken Chain*, 71.

274 **his keen scientific mind:** Many Apollo astronauts have written autobiographies, and several of them make clear their unfriendly feelings for Aldrin, who has acknowledged that he wasn't the best-liked man in the astronaut corps. See Cernan and Davis, *The Last Man on the Moon*, 157, 231.

276 **"I felt I had a better chance":** Collins, *Carrying the Fire*, 8.

276 **"could tell a good wine":** Quoted in the *Freelance-Star*, July 16, 1969.

278 **several months in a hospital:** Collins, *Carrying the Fire*, 262–75.

278 **extensive planning and training:** Cunningham, *The All-American Boys*, 209; numerous astronaut interviews and autobiographies.

280 **an occasional cigar:** Author interview, Albert Jackson, August 31, 2017.

281 **would stop the mission:** Lunney et al., *From the Trench of Mission Control*, 240–43.

282 **success of Apollo:** Ibid., 189.

283 **"People talking at once":** Ibid., 243–44.

283 **"Maybe I'm an 'Aunt Emma'":** Tindall wrote more than eleven hundred Tindallgram memos between 1964 and 1970 for both the Gemini and Apollo programs. Most of them are available for viewing at the Apollo Lunar Surface Journal, https://www.hq.nasa.gov/alsj/, the best source of Apollo information outside NASA—and perhaps even better than NASA.

284 **"it just came naturally":** Author interview, Steve Bales, May 10, 2015.

285 **"not a terribly socially endearing thing":** Ibid.

286 **They'd invented it:** John R. Garman interview, March 27, 2001, JSC Oral History Project.

THIRTEEN: A PRACTICE RUN AND A DRESS REHEARSAL

288 **"connoisseur's flight":** Wilford, *We Reach the Moon,* 211.

290 **Twelve mascons:** Lewis, *Appointment on the Moon*, 509.

291 **"based on individual desire":** *Albany Knickerbocker News*, February 26, 1969.

291 **"The decision really":** *Pittsburgh Press*, March 7, 1969.

292 **"equivocated a minute":** Aldrin and Warga, *Return to Earth,* 206.

292 **before other crewmen:** Cernan and Davis, *The Last Man on the Moon*, 231; Aldrin and Abraham, *Magnificent Desolation*, 129.

292 **Mike cut him off:** Collins, *Carrying the Fire*, 346–47.

292 **LM pilot exiting first:** Cunningham, *The All-American Boys,* 215.

292 **with no luck:** Ibid.; Hansen, *First Man*, 365.

292 **"plans called for":** *New York Times*, April 15, 1969.

293 **"devastated":** Koppel, *The Astronaut Wives Club,* 230.

293 **first to walk on the moon:** Kraft, *Flight,* 323.

294 **There were too many functions:** Apollo note no. 430, February 4, 1969, Robert Gilruth Papers, Special Collections, Virginia Tech.

296 **expressing themselves freely:** Cernan and Davis, *The Last Man on the Moon,* 227.

298 **"So, Rocco":** Slayton and Cassutt, *Deke!,* 238; Sherrod, "The General Makes a Decision," untitled manuscript on the history of NASA; memo, June 17, 1969, "Minutes of Apollo Program Meeting, 12 June 1969," in box 53, folder 4, George M. Low Collection, Rensselaer Polytechnic Institute.

FOURTEEN: "YOU'RE GO"

301 **"If we get":** John Hodge interview, April 18, 1999, JSC Oral History Project.

301 **After a late dinner:** Collins, *Carrying the Fire,* 346–47.

302 **on their toes:** Author interview with Al Jackson, August 30, 2017.

302 **"You guys":** Collins, *Carrying the Fire,* 346–47.

303 **"I hate geology":** Armstrong et al., *First on the Moon,* 211.

304 **"The branch chiefs":** Kranz, *Failure Is Not an Option,* 258.

307 **on his SSR console:** There are several versions of this simulation and what happened after it, and all differ slightly in the details. Murray and Cox, *Apollo,* 345–46, provide the basis for this one, supplemented by Kraft, *Flight,* 321–22, and several other accounts by the participants: Kranz, *Failure Is Not an Option,* 267–71; Lunney et al., *From the Trench of Mission Control,* 247–48; Jack Garman interview, April 14, 2014, part 2, by Colin Mackellar, at www.honeysucklecreek.net; Eugene F. Kranz interview, January 8, 1999, JSC Oral History Project; John R. Garman interview, March 27, 2001, JSC Oral History Project; Granville Paules interview, November 7, 2006, JSC Oral History Project; Lunney et al., *From the Trench of Mission Control,* 248; Author interview, Steve Bales, May 12, 2015; Author interview, Jack Garman, September 18, 2014. The four principals who discuss the aborted simulation involving the

1201 code disagree on the date; Kranz and Bales maintain that it happened on the last day of sims for Kranz's White team, July 6, 1969, and Garman and Paules state that it happened about a month earlier, in late May or early to mid-June, although Garman gives two different estimates in two different interviews ("Just a few months before Apollo 11—I'm quite sure it was May or June" in his March 27, 2001, interview, and "About a month before the flight...they did one last one—well, I don't know if it was the last one," in his April 14, 2014, interview). Murray and Cox, who interviewed many of the people involved, conclude that it happened during the last day of simulation for Mission Control—the July date—and they note that Kranz reviewed the simulation logs for that period. In a March 15, 2016, CollectSPACE.com post, SimSup Dick Koos is quoted as saying of the simulation: "It took until July to be able to run it. The final sim was always normal—no aborts." Finally, Richard H. Battin, who helped create the Apollo Guidance Computer at MIT, said in his April 4, 2000, interview for the JSC Oral History Project, referring to the 1201/1202: "Fortunately, they had already experienced that problem...in Houston several weeks before the flight." I believe the preponderance of evidence points to the July 6 date.

307 **"You must think"**: William D. Reeves interview, March 9, 2009, JSC Oral History Project.

308 **overruling a mission rule:** Kranz, *Failure Is Not an Option,* 262.

308 **wouldn't put up with it:** Kraft, *Flight*, 314.

309 **"If you want to abort"**: Aldrin and McConnell, *Men from Earth*, 226; Hansen, *First Man*, 388.

309 **two years to choreograph:** Watkins, *Apollo Moon Missions*, 149.

313 **"to perform better"**: Goldstein, *Reaching for the Stars,* 128.

313 **"We literally trained out fear"**: Bledsoe, "Down from Glory."

313 **"You're not born"**: MacKinnon and Baldanza, *Footprints*, 125.

315 **"Fear is not an unknown"**: Wilford, *We Reach the Moon,* 261.

316 **Most of the other astronauts:** Collins, *Carrying the Fire*, 360. Aldrin later claimed that he believed the odds of a successful landing and return to be slightly more favorable, sixty-forty; see French and Burgess, *In the Shadow of the Moon*, 393.

317 **plenty of others at NASA were:** Briefing for Dr. Paine, July 10, 1969, by George Hage, subject files, Apollo 11, Robert Sherrod Apollo Collection, NASA History Office.

317 **LM's descent or ascent:** "How we will handle the effect of mascons on the LM lunar surface gravity alignments," July 14, 1969, Tindallgram, 69-PA-T-109A, www.collectspace.com/resources/tindallgrams/tindallgrams01.pdf.

317 **if anything unexpected occurred:** Goldstein, *Reaching for the Stars,* 142.

320 **"by a crossfire of searchlights":** Collins, *Carrying the Fire,* 358.

321 **"You're go":** Aldrin and McConnell, *Men from Earth,* 225.

FIFTEEN: THE TRANSLUNAR EXPRESS

322 **"I am far from certain":** Collins, *Carrying the Fire,* 360.

324 **"I could see the massiveness":** "Summary of Flight in Their Own Words," Apollo 11 Mission Account, NASA.gov.

325 **Collins pointed it out to Armstrong:** Collins, *Carrying the Fire,* 363–64.

326 **"If we could solve the problems":** DeGroot, *Dark Side of the Moon,* 234.

327 **would do the job:** Watkins, *Apollo Moon Missions,* 77, 84.

327 **better strategic systems:** Siddiqi, *Challenge to Apollo,* 856.

328 **working on a moon landing:** Ibid., 685–86, 856.

329 **the history of rocketry:** Ibid., 681–86.

331 **Lord's Prayer to himself:** Bergaust, *Wernher von Braun,* 420–21; Neufeld, *Von Braun,* 432–33.

331 **their normal weight:** Aldrin and McConnell, *Men from Earth,* 226.

336 **"It's all dead air":** *Time,* July 25, 1969.

337 **on the front lawn:** Koppel, *The Astronaut Wives Club,* 224–25.

338 **a sigh of relief:** Borman and Serling, *Countdown,* 240–41; Gallentine, *Infinity Beckoned,* 53–60.

340 **not considered essential:** *Life,* September 5, 1964; Mindell, *Digital Apollo,* 123, 294. See also Frank O'Brien's superlative *The Apollo Guidance Computer.*

SIXTEEN: DESCENT TO LUNA

345 **"The unknowns were rampant":** Hansen, *First Man,* 529.

346 **"I've never seen":** Quoted in Harland, *The First Men on the Moon,* 214.

350 **lit up another Kent:** Lunney et al., *From the Trench,* 250; Kranz, *Failure Is Not an Option,* 283–84.

353 **LM's beginning orbit altitude:** Cheatham and Bennett, "Apollo Lunar Mobile Landing Strategy," 177.

355 **"like popping a cork":** Eugene F. Kranz interview, January 8, 1999, JSC Oral History Project.

356 **"It's the same one":** Kranz, *Failure Is Not an Option,* 288.

357 **with no problems:** Kraft, "The View from Mission Control," 882.

360 **control the landing:** Eyles, "Tales from the Lunar Module Guidance Computer."

361 **"Okay, the only callouts":** Houston and Heflin, *Go, Flight!,* 164–65.

364 **everyone started to breathe again:** Flight Dynamics Controllers, *Oral Histories of NASA Flight Dynamics Controllers,* 191.

364 ***What a wonderful name:*** *NOVA: To the Moon,* PBS, July 13, 1999.

365 **clapped Bales on the shoulder:** *Mission Control* documentary.

365 **"Thank you, John":** Hansen, *Enchanted Rendezvous; NOVA: To the Moon.* Sources for this account of the lunar landing include Aldrin and Warga, *Return to Earth;* Aldrin and McConnell, *Men from Earth;* Aldrin and Abraham, *Magnificent Desolation;* "Armstrong Recalls Moon Landing Details"; Armstrong et al., *First on the Moon;* Bogo, "Blasting Off the Moon's Surface"; Collins, *Carrying the Fire* and *Liftoff;* French and Burgess, *In the Shadow of the Moon;* Lunney et al., *From the Trench;* Hansen, *First Man;* Harland, *The First Men on the Moon;* Houston and Heflin, *Go, Flight!;* Kraft, *Flight;* Kranz, *Failure Is Not an Option;* Lewis, *Appointment on the Moon;* Mindell, *Digital Apollo;* Stachurski, *Below Tranquility Base;* Vine, "Walking on the Moon"; Wilford, *We Reach the Moon;* JSC Oral History Project interviews with John Aaron, Richard Battin, Bob Carlton, Jack Garman, Frank E. Hughes, Christopher Kraft, Gene Kranz, and others; author

interviews with John Aaron, Buzz Aldrin, Neil Armstrong, Steve Bales, Bob Carlton, Mike Collins, and Jack Garman; Eric Jones's superb Apollo 11 Lunar Surface Journal.

SEVENTEEN: MOONDUST

366 **"We were lucky":** Vine, "Walking on the Moon."

369 **"Good...good...good":** *Houston Chronicle*, July 21, 1969.

371 **had known about beforehand:** Michael Collins interview by Robert Sherrod, Collins autobiographical file, Robert Sherrod Collection, NASA History Office; Hansen, *First Man*, 488.

373 **Armstrong would later express:** Hansen, *First Man*, 494.

373 **in front of a TV:** Editors of *Life*, *Neil Armstrong*, 76.

374 **the Apollo 11 crew:** Scott et al., *Two Sides of the Moon*, 245–47.

374 **"like being on a sandy athletic field":** Kondratyev et al., *Space Research XI*.

375 **urine-collection device:** Aldrin and Warga, *Return to Earth*, 235.

379 **quickly fell asleep:** Lunney et al., *From the Trench*, 254.

381 **would ever know:** Wilson, "Mercury Atlas 10," 59.

382 **BEST WISHES:** Folder 5, box 38, Michael Collins Papers, Virginia Tech.

382 **The alignment looked good:** Collins, *Carrying the Fire*, 416

387 **last of the revelers:** *Time*, August 1, 1969.

389 **and headed home:** Ed Buckbee, interview with the author, February 11, 2015; e-mail correspondence, December 4, 2017.

EPILOGUE

390 **"Max, they're going to go back there":** Oberg, "Max Faget, Master Builder."

392 **"Sometimes it seems that Apollo":** Cernan and Davis, *Last Man on the Moon*, 344.

BIBLIOGRAPHY

BOOKS

Aldrin, Edwin E., Jr., and Wayne Warga. *Return to Earth.* New York: Random House, 1973.

———, and Malcolm McConnell. *Men from Earth.* New York: Bantam Books, 1989.

———, and Ken Abraham. *Magnificent Desolation: The Long Journey Home from the Moon.* New York: Three Rivers, 2009.

———. *No Dream Is High Enough.* Washington, DC: National Geographic, 2016.

Armstrong, Neil, Michael Collins, and Edwin E. Aldrin Jr. *First on the Moon.* Boston: Little, Brown, 1970.

Baker, David. *The History of Manned Space Flight.* New York: Crown, 1981.

Barbour, John, and the writers and editors of the Associated Press. *Footprints on the Moon.* New York: Associated Press, 1969.

Barbree, Jay. *"Live from Cape Canaveral": Covering the Space Race from Sputnik to Today.* New York: Smithsonian Books, 2007.

Bergaust, Erik. *Wernher von Braun.* Washington, DC: National Space Institute, 1976.

Berman, Bob. *Shooting for the Moon: The Strange History of Human Spaceflight.* Guilford, CT: Lyons Press, 2007.

Biddle, Wayne. *Dark Side of the Moon: Wernher von Braun, the Third Reich, and the Space Race.* New York: W. W. Norton, 2005.

Bilstein, Roger. *Orders of Magnitude.* Washington, DC: NASA, 2011.

Bizony, Piers. *The Man Who Ran the Moon: James Webb, JFK, and the Secret History of Apollo.* Cambridge: Icon Books, 2006.

———. *One Giant Leap: Apollo 11 Remembered.* Minneapolis, MN: Zenith Press, 2009.

Boomhower, Ray E. *Gus Grissom: The Lost Astronaut.* Indianapolis: Indiana Historical Society, 2004.

Borman, Frank, and Robert J. Serling. *Countdown: An Autobiography.* New York: William Morrow, 1988.

Brooks, Courtney G., James M. Grimwood, and Loyd S. Swenson Jr. *Chariots for Apollo: The NASA History of Manned Lunar Spacecraft to 1969.* Mineola, NY: Dover, 2009.

Bruns, Laura, and Mike Litchfield, eds. *Johnson Space Center: The First Fifty Years.* Charleston, SC: Arcadia, 2013.

Brzezinski, Matthew. *Red Star Rising.* New York: Henry Holt, 2007.

Burgess, Colin. *Moon Bound: Choosing and Preparing NASA's Lunar Astronauts.* Chichester, UK: Springer-Praxis, 2013.

———. *Liberty Bell 7: The Suborbital Mercury Flight of Virgil I. Grissom.* Chichester, UK: Springer-Praxis, 2014.
———. *Freedom 7: The Historic Flight of Alan B. Shepard, Jr.* Chichester, UK: Springer-Praxis, 2015.
———. *Friendship 7: The Epic Orbital Flight of John H. Glenn, Jr.* Chichester, UK: Springer-Praxis, 2015.
———. *Sigma 7: The Six Mercury Orbits of Walter M. Schirra, Jr.* Chichester, UK: Springer-Praxis, 2016.
———. *Aurora 7: The Mercury Spaceflight of M. Scott Carpenter.* Chichester, UK: Springer-Praxis, 2016.
———. *Faith 7: Gordon Cooper, Jr., and the Final Mercury Mission.* Chichester, UK: Springer-Praxis, 2016.
———, and Chris Dobbs. *Animals in Space: From Research Rockets to the Space Shuttle.* Chichester, UK: Springer-Praxis, 2007.
———, and Kate Doolan. *Fallen Astronauts: Heroes Who Died Reaching for the Moon.* Lincoln: University of Nebraska Press, 2016.
Burrows, William E. *This New Ocean: The Story of the First Space Age.* New York: Random House, 1998.
Cadbury, Deborah. *Space Race.* New York: HarperCollins, 2006.
Caidin, Martin. *The Astronauts.* New York: E. P. Dutton, 1960.
———. *Man into Space.* New York: Pyramid, 1961.
———. *Rendezvous in Space.* New York: E. P. Dutton, 1962.
———. *Buck Rogers.* Lake Geneva, WI: TSR, 1995.
———. *Red Star in Space.* New York: Crowell-Collier Press, 1963.
Carpenter, Scott, and Kris Stoever. *For Spacious Skies: The Uncommon Journey of a Mercury Astronaut.* New York: Harcourt, 2002.
Carpenter, Scott, et al. *We Seven.* New York: Simon and Schuster, 1962.
Carter, Dale. *The Final Frontier: The Rise and Fall of the American Rocket State.* London: Verso, 1988.
Cernan, Eugene, and Don Davis. *The Last Man on the Moon.* New York: St. Martin's, 1999.
Chaikin, Andrew. *A Man on the Moon.* New York: Penguin, 1994.
Chambers, Mary Randall, and Randall Chambers. *Getting Off the Planet: Training Astronauts.* Burlington, Canada: Apogee Books, 2006.
Clark, Phillip. *The Soviet Manned Space Program.* New York: Orion Books, 1988.
Collins, Michael. *Carrying the Fire.* North Salem, NY: Adventure Library, 1998.
———. *Flying to the Moon and Other Strange Places.* New York: Farrar, Straus and Giroux, 1976.
———. *Liftoff: The Story of America's Adventure in Space.* New York: Grove Press, 1988.
Compton, William David. *Where No Man Has Gone Before: A History of NASA's Apollo Lunar Expeditions.* Mineola, NY: Dover, 2010.
Conrad, Nancy, and Howard A. Klausner. *Rocketman: Astronaut Pete Conrad's Incredible Ride to the Moon and Beyond.* New York: NAL, 2005.
Cooper, Gordon, and Bruce Henderson. *Leap of Faith.* New York: HarperCollins, 2000.
Cooper, Henry S. F., Jr. *Apollo on the Moon.* New York: Dial Press, 1969.
———. *Moon Rocks.* New York: Dial Press, 1970.
Cortright, Edgar M., ed. *Apollo Expeditions to the Moon.* Washington, DC: NASA, 1975.
Crouch, Tom D. *Aiming for the Stars.* Washington, DC: Smithsonian Institution Press, 1999.

Cunningham, Walter. *The All-American Boys*. Rev. ed. New York: Ipicturebooks, 2009.
Daniloff, Nicholas. *The Kremlin and the Cosmos*. New York: Knopf, 1972.
Davis-Floyd, Robbie, Kenneth J. Cox, and Frank White. *Space Stories: Oral Histories from the Pioneers of America's Space Program*. Kindle, 2012.
Dawson, Virginia P., and Mark D. Bowles. *Realizing the Dream of Flight*. Washington, DC: NASA, 2005.
DeGroot, Gerard J. *Dark Side of the Moon*. New York: NYU Press, 2006.
de Monchaux, Nicholas. *Spacesuit: Fashioning Apollo*. Cambridge, MA: MIT Press, 2011.
Dethloff, Henry C. *Suddenly, Tomorrow Came: The NASA History of the Johnson Space Center*. Washington, DC: NASA, 1993.
Dick, Steven J. *Remembering the Space Age*. Washington, DC: NASA, 2008.
———, and Roger D. Launius, eds. *Societal Impact of Spaceflight*. Washington, DC: NASA, 2007.
Doran, Jamie, and Piers Bizony. *Starman: The Truth Behind the Legend of Yuri Gagarin*. New York: Walker, 2011.
Dornberger, Walter. *V-2*. New York: Viking, 1954.
Duke, Charlie, and Dotty Duke. *Moonwalker*. Nashville: Oliver Nelson, 1990.
Editors of *Life*. *Neil Armstrong, 1930–2012*. New York: Life Books, 2012.
Editors of Time-Life Books. *Outbound*. Alexandria, VA: Time-Life Books, 1989.
Evans, Ben. *Escaping the Bonds of Earth*. Chichester, UK: Springer-Praxis, 2009.
———. *Foothold in the Heavens*. Chichester, UK: Springer-Praxis, 2010.
Evans, Michelle. *The X-15 Rocket Plane*. Lincoln: University of Nebraska Press, 2013.
Flight Control Division. *Lunar Module Orientation Guide and Compartment Familiarization*. Burlington, Canada: Apogee, 2010. First published 1968 by NASA (Washington, DC).
Flight Dynamics Controllers. *Oral Histories of NASA Flight Dynamics Controllers*. Lexington, KY: Privately published, 2015.
French, Francis, and Colin Burgess. *In the Shadow of the Moon*. Lincoln: University of Nebraska Press, 2007.
———. *Into That Silent Sea*. Lincoln: University of Nebraska Press, 2007.
Fries, Sylvia Doughty. *NASA Engineers and the Age of Apollo*. Washington, DC: NASA, 1992.
Gallentine, Jay. *Infinity Beckoned: Adventuring Through the Inner Solar System, 1969–1989*. Lincoln: University of Nebraska Press, 2016.
Glenn, John. *John Glenn: A Memoir*. New York: Ballantine, 1999.
Glennan, T. Keith. *The Birth of NASA: The Diary of T. Keith Glennan*. Washington, DC: NASA, 1993.
Godwin, Robert, ed. *Apollo 11: The NASA Mission Reports*. Vol. 2. Burlington, Canada: Apogee Books, 1999.
———. *Apollo 11: The NASA Mission Reports*. Vol. 3. Burlington, Canada: Apogee Books, 2002.
Goldstein, Stanley H. *Reaching for the Stars*. New York: Praeger, 1987.
Gorn, Michael. *NASA: The Complete Illustrated History*. London: Merrell, 2005.
Gray, Mike. *Angle of Attack: Harrison Storms and the Race to the Moon*. New York: W. W. Norton, 1992.
Grelsamer, Ronald P. *Into the Sky with Diamonds: The Beatles and the Race to the Moon in the Psychedelic '60s*. Bloomington, IN: AuthorHouse, 2010.
Grimwood, James M. *Project Mercury: A Chronology*. Washington, DC: NASA, 1963.

Grissom, Betty, and Henry Still. *Starfall.* New York: Thomas Y. Crowell, 1974.

Grissom, Virgil. *Gemini!* New York: Macmillan, 1968.

Hacker, Barton C., and James M. Grimwood. *On the Shoulders of Titans: A History of Project Gemini.* Washington, DC: NASA, 1977.

Hall, Loretta. *Space Pioneers in Their Own Words.* Los Ranchos, NM: Rio Grande Books, 2014.

Hansen, James R. *Enchanted Rendezvous: John C. Houbolt and the Genesis of the Lunar-Orbit Rendezvous Concept.* Monographs in Aerospace History, no. 4. Washington, DC: NASA History Office, 1995.

———. *First Man: The Life of Neil A. Armstrong.* New York: Simon and Schuster, 2005.

Hardesty, Von, and Gene Eisman. *Epic Rivalry: The Inside Story of the Soviet and American Space Race.* Washington, DC: National Geographic, 2007.

Harland, David M. *Exploring the Moon: The Apollo Expeditions.* Chichester, UK: Springer-Praxis, 1999.

———. *The First Men on the Moon: The Story of Apollo 11.* Chichester, UK: Springer-Praxis, 2007.

Harvey, Brian. *Russia in Space: The Failed Frontier?* London: Springer, 2001.

Heiken, Grant, and Eric Jones. *On the Moon: The Apollo Journals.* Chichester, UK: Springer-Praxis, 2007.

Henry, James P. *Biomedical Aspects of Space Flight.* New York: Holt, Rinehart and Winston, 1966.

Hersch, Matthew. *Inventing the American Astronaut.* New York: Palgrave Macmillan, 2012.

Houston, Rick, and Milt Heflin. *Go, Flight! The Unsung Heroes of Mission Control, 1965–1992.* Lincoln: University of Nebraska Press, 2015.

Hunt, Linda. *Secret Agenda: The United States Government, Nazi Scientists, and Project Paperclip, 1945 to 1990.* New York: Thomas Dunne Books, 1991.

Hurt, Harry III. *For All Mankind.* New York: Atlantic Monthly Press, 1988.

Huzel, Dieter. *Peenemunde to Canaveral.* Englewood Cliffs, NJ: Prentice-Hall, 1962.

Irwin, James B., and William A. Emerson Jr. *To Rule the Night.* Philadelphia: A. J. Holman, 1973.

Johnson, Michael Peter. *Inventing the Groundwork of Spaceflight.* Gainesville: University Press of Florida, 2015.

Kelly, Thomas J. *Moon Lander: How We Developed the Apollo Lunar Module.* Washington, DC: Smithsonian Institution Press, 2001.

Klerkx, Greg. *Lost in Space: The Fall of NASA and the Dream of a New Space Age.* New York: Pantheon, 2004.

Kluger, Jeffrey. *The Apollo Adventure.* New York: Pocket Books, 1995.

———. *Apollo 8.* New York: Henry Holt, 2017.

Kondratyev, K. Y., Michael J. Rycroft, and Carl Sagan. *Space Research XI.* Berlin, Germany: Akademie-Verlag, 1971.

Koppel, Lily. *The Astronaut Wives Club.* New York: Grand Central, 2013.

Kraft, Christopher C., Jr. *Flight: My Life in Mission Control.* New York: Dutton, 2001.

Kranz, Gene. *Failure Is Not an Option.* New York: Simon and Schuster, 2000.

Lamb, Lawrence E. *Inside the Space Race: A Space Surgeon's Diary.* Austin, TX: Synergy Books, 2006.

Lambright, W. Henry. *Powering Apollo: James E. Webb of NASA.* Baltimore: Johns Hopkins University Press, 1998.

Landwirth, Henri, and J. P. Hendricks. *Gift of Life.* N.p.: Privately published, 1996.

Lattimer, Dick, ed. *"All We Did Was Fly to the Moon."* Gainesville, FL: Whispering Eagle Press, 1988.

Launius, Roger D. *Apollo: A Retrospective.* Monographs in Aerospace History, no. 3. Washington, DC: NASA History Office, 1994.

———. *NASA: A History of the U.S. Civil Space Program.* Malaber, FL: Krieger, 1994.

———, and Dennis R. Jenkins. *Coming Home: Reentry and Recovery from Space.* Washington, DC: NASA, 2012.

Launius, Roger D., and Howard E. McCurdy, eds. *Spaceflight and the Myth of Presidential Leadership.* Urbana: University of Illinois Press, 1997.

Leonard, Jonathan Norton. *Flight into Space.* New York: Random House, 1953.

Leopold, George. *Calculated Risk: The Supersonic Life and Times of Gus Grissom.* West Lafayette, IN: Purdue University Press, 2016.

Levering, Ralph B. *The Cold War, 1945–1972.* Arlington Heights, IL: Harlan Davidson, 1982.

Lewis, Richard S. *Appointment on the Moon.* Rev. ed. New York: Ballantine, 1969.

Ley, Willy. *Rockets, Missiles, and Outer Space.* Rev. ed. New York: Signet, 1969.

Liebergot, Sy, and David M. Harland. *Apollo EECOM: Journey of a Lifetime.* Burlington, Canada: Apogee Books, 2006.

Lindsay, Hamish. *Tracking Apollo to the Moon.* London: Springer-Verlag, 2001.

Link, M. M. *Space Medicine in Project Mercury.* Washington, DC: NASA, 1965.

Llewellyn, John S. *From the Trenches of Korea to the Trench in Mission Control.* Gonzales, TX: Privately published, 2012.

Logsdon, John M. *The Decision to Go to the Moon: Project Apollo and the National Interest.* Cambridge, MA: MIT Press, 1970.

Lord, M. G. *Astro Turf: The Private Life of Rocket Science.* New York: Walker, 2005.

Lovell, Jim, and Jeffrey Kluger. *Lost Moon: The Perilous Journey of Apollo 13.* Boston: Houghton Mifflin, 1994.

Lunney, Glynn. *Highways into Space.* N.p.: Privately published, 2014.

———, et al. *From the Trench of Mission Control to the Craters of the Moon.* Third ed. N.p.: Privately published, 2012.

MacKinnon, Douglas, and Joseph Baldanza. *Footprints.* Washington, DC: Acropolis, 1989.

Mailer, Norman. *Of a Fire on the Moon.* Boston: Little, Brown, 1968.

McDonnell, Virginia B. *Dee O'Hara: Astronauts' Nurse.* Edinburgh, Scotland: Rutledge, 1965.

McDougall, Walter A. *...The Heavens and the Earth: A Political History of the Space Age.* New York: Basic Books, 1985.

Mindell, David A. *Digital Apollo: Human and Machine in Spaceflight.* Cambridge, MA: MIT Press, 2008.

Mitchell, Edgar, and Dwight Williams. *The Way of the Explorer.* New York: Putnam, 1996.

Murray, Charles, and Catherine Bly Cox. *Apollo: The Race to the Moon.* New York: Simon and Schuster, 1989.

NASA Office of Logic Design. *What Made Apollo a Success?* Washington, DC: NASA, 1971.

NASA Public Affairs. *The Kennedy Space Center Story.* Kennedy Space Center, FL: NASA Public Affairs Office, 1991.

Nelson, Craig. *Rocket Men: The Epic Story of the First Men on the Moon.* New York: Viking, 2009.

Neufeld, Michael J. *Von Braun: Dreamer of Space, Engineer of War.* New York: Knopf, 2007.

Newkirk, Dennis. *Almanac of Soviet Manned Space Flight.* Houston: Gulf, 1990.

Nolen, Stephanie. *Promised the Moon.* New York: Thunder's Mouth, 2002.

Oberg, James E. *Red Star in Orbit.* New York: Random House, 1981.

O'Brien, Frank. *The Apollo Guidance Computer: Architecture and Operation.* Chichester, UK: Springer-Praxis, 2010.

O'Leary, Brian. *The Making of an Ex-Astronaut.* Boston: Houghton Mifflin, 1970.

Ordway, Frederick I., III, and Mitchell R. Sharpe. *The Rocket Team.* New York: Thomas Crowell, 1979.

Orloff, Richard M. *Apollo by the Numbers: A Statistical Reference.* Washington, DC: NASA, 2000.

Pellegrino, Charles R., and Joshua Stoff. *Chariots for Apollo: The Untold Story Behind the Race to the Moon.* New York: Atheneum, 1985.

Petrov, G. I., ed. *Conquest of Outer Space in the USSR, 1967–70.* Washington, DC: NASA, 1973.

Piszkiewicz, Dennis. *The Nazi Rocketeers.* Westport, CT: Praeger, 1995.

Portree, David S. F., and Robert C. Treviño. *Walking to Olympus: An EVA Chronology.* Monographs in Aerospace History, no. 7. Washington, DC: NASA, 1997.

Rabinowitch, Eugene, and Richard S. Lewis, eds. *Man on the Moon.* New York: Basic Books, 1969.

Reynolds, David West. *Kennedy Space Center: Gateway to Space.* New York: Firefly Books, 2006.

———. *Apollo: The Epic Journey to the Moon, 1963–1972.* Minneapolis, MN: Zenith Press, 2013.

Riley, Christopher, and Phil Dolling. *Apollo 11: Owners' Workshop Manual.* Sparkford, UK: Haynes Publishing, 2009.

Rosenberger, Jim. *The Brilliant Disaster.* New York: Scribner, 2011.

Santy, Patricia A. *Choosing the Right Stuff.* Westport, CT: Praeger, 1994.

Schefter, James. *The Race: The Uncensored Story of How America Beat Russia to the Moon.* New York: Doubleday, 1999.

Schirra, Walter M., Jr., and Richard N. Billings. *Schirra's Space.* Boston: Quinlan Press, 1988.

Schmitt, Harrison H. *Return to the Moon.* New York: Copernicus Books, 2006.

Scott, David, Alexei Leonov, and Christine Toomey. *Two Sides of the Moon.* New York: Thomas Dunne Books, 2004.

Scott, David Meerman, and Richard Jurek. *Marketing the Moon: The Selling of the Apollo Lunar Program.* Cambridge, MA: MIT Press, 2014.

Seamans, Robert C., Jr. *Aiming at Targets.* Washington, DC: NASA, 1996.

———. *Project Apollo: The Tough Decisions.* Monographs in Aerospace History, no. 37. Washington, DC: NASA, 2005.

Shayler, David J. *Disasters and Accidents in Manned Spaceflight.* Chichester, UK: Springer-Praxis, 2000.

Shepard, Alan, and Deke Slayton. *Moon Shot: The Inside Story of America's Race to the Moon.* Atlanta: Turner Publishing, 1994.

Siddiqi, Asif A. *Challenge to Apollo: The Soviet Union and the Space Race, 1945–1974.* Washington, DC: NASA, 2000.

———. *Deep Space Chronicle: A Chronology of Deep Space and Planetary Probes 1958–2000.* Monographs in Aerospace History, no. 24. Washington, DC: NASA, 2002.

Sidey, Hugh. *John F. Kennedy, President.* New ed. New York: Atheneum, 1964.

Slayton, Donald K., and Michael Cassutt. *Deke!* New York: Forge, 1994.

Smith, Andrew. *Moondust: In Search of the Men Who Fell to Earth.* New York: Harper Perennial, 2006.

Stachurski, Richard. *Below Tranquility Base.* North Charleston, SC: CreateSpace, 2013.

Stafford, Tom, and Michael Cassutt. *We Have Capture: Tom Stafford and the Space Race.* Washington, DC: Smithsonian Institution Press, 2002.

Stroud, Rick. *The Book of the Moon.* London: Doubleday, 2009.

Swanson, Glen E. *"Before This Decade Is Out...": Personal Reflections on the Apollo Program.* Washington, DC: NASA, 1999.

Swenson, Loyd S., Jr., James M. Grimwood, and Charles C. Alexander. *This New Ocean: A History of Project Mercury.* Washington, DC: NASA, 1998.

Thompson, Neal. *Light This Candle: The Life and Times of Alan Shepard.* New York: Crown, 2004.

Treadwell, Terry C. *Stepping Stones to the Stars: The Story of Manned Spaceflight.* Stroud, UK: History Press, 2010.

Turnill, Reginald. *The Moonlandings.* Cambridge: Cambridge University Press, 2003.

United Press International. *Gemini: America's Historic Walk in Space.* Englewood Cliffs, NJ: Prentice-Hall, 1965.

Vladimirov, Leonid. *The Russian Space Bluff.* New York: Dial Press, 1973.

von Braun, Wernher. *Space Frontier.* New York: Holt, Rinehart and Winston, 1971.

———. *The Rocket's Red Glare.* Garden City, NY: Anchor Press, 1976.

von Braun, Wernher, and Frederick I. Ordway III. *History of Rocketry and Space Travel.* New York: Thomas Y. Crowell, 1966.

Von Ehrenfried, Dutch. *The Birth of NASA.* Chichester, UK: Springer-Praxis. 2016.

Wagener, Leon. *One Giant Leap: Neil Armstrong's Stellar American Journey.* New York: Forge, 2004.

Ward, Bob. *Dr. Space: The Life of Wernher von Braun.* Annapolis, MD: Naval Institute Press, 2005.

Ward, Jonathan H. *Countdown to a Moon Launch.* Chichester, UK: Springer-Praxis, 2015.

———. *Rocket Ranch.* Chichester, UK: Springer-Praxis, 2015.

Watkins, Billy. *Apollo Moon Missions: The Unsung Heroes.* Westport, CT: Praeger, 2006.

Wendt, Guenter, and Russell Still. *The Unbroken Chain.* Burlington, Canada: Apogee Books, 2001.

Wilford, John Noble. *We Reach the Moon.* New York: Bantam, 1969.

Wilson, Andrew. *Solar System Log.* London: Jane's Publishing, 1987.

Wolfe, Tom. *The Right Stuff.* New York: Bantam Books, 1980.

Worden, Al, and Francis French. *Falling to Earth.* Washington, DC: Smithsonian Books, 2011.

Young, Hugo, Bryan Silcock, and Peter Dunn. *Journey to Tranquility.* New York: Doubleday, 1970.

Young, John, and James Hansen. *Forever Young: A Life of Adventure in Air and Space.* Gainesville: University Press of Florida, 2012.

Zimmerman, Robert. *Genesis: The Story of Apollo 8.* New York: Dell, 1999.

ARTICLES

Agle, D. C. "Flying the Gusmobile." *Air and Space* (September 1998).
———. "Riding the Titan II." *Air and Space* (September 1998).
———. "We Called It the Bug." *Air and Space* (September 2001).
"Armstrong Recalls Moon Landing Details." *Aviation Week*, October 13, 1969.
Bledsoe, Jerry. "Down from Glory." *Esquire* (January 1973).
Bogo, Jennifer. "Blasting Off the Moon's Surface." *Popular Mechanics* (May 2009).
———, et al. "No Margin for Error: The Untold Story of Apollo 11." *Popular Mechanics* (June 2009).
Chaikin, Andrew. "Apollo's Worst Day." *Air and Space* (November 2016).
———. "Bob Gilruth, the Quiet Force Behind Apollo." *Air and Space* (March 2016).
———. "How the Spaceship Got Its Shape." *Air and Space* (November 2009).
Chow, Denise. "Mystery of Moon's Lumpy Gravity Explained." Space.com, May 30, 2013.
Cooper, Henry, S. F. "Annals of Space: We Don't Have to Prove Ourselves." *New Yorker*, September 2, 1991.
Day, Dwayne. "Chasing Shadows: Apollo 8 and the CIA." *Space Review*, April 11, 2016.
———. "The Moon in the Crosshairs: CIA Monitoring of the Soviet Manned Lunar Program." 4 parts. *Space Review*, December 14, 2015; December 21, 2015; January 4, 2016; January 11, 2016.
———. "Webb's Giant." *Space Review*, July 19, 2004.
Dille, John. "We Who Tried." *Life*, May 10, 1963.
Evans, Ben. "Open-Ended: What 1967 Might Have Been." AmericaSpace.com, February 7, 2015.
French, Francis. "I Worked with NASA, Not for NASA: An Interview with Astronaut Walter 'Wally' Schirra." CollectSPACE.com, February 22, 2002.
Glenn, John. "If You're Shook Up, You Shouldn't Be There." *Life*, March 9, 1962.
Grissom, Virgil. "If It Goes Wrong I'll Be Responsible." *Life*, June 4, 1964.
Harvey, Brian. "The 1963 Soviet Space Platform Project." *Quest* (Fall 1993).
Hedman, Eric. "The Best Reason to Go to Mars." *Space Review*, September 6, 2016.
Hix, Lisa. "Laika and Her Comrades: The Soviet Dogs Who Took Giant Leaps for Mankind." *Collector's Weekly*, January 23, 2015.
Houbolt, John C. "Lunar Rendezvous." *International Science and Technology* 14 (February 1963).
Kurtzman, Cliff. "Standing on the Shoulders of Giants: What I Learned From Max Faget and Joe Allen." AdAstro.com, October 19, 2004.
Lamb, Lawrence E. "Aeronautical Evaluation for Space Pilots." *Lectures in Aerospace Medicine*, February 3–7, 1964.
Lang, Daniel. "A Reporter at Large: A Romantic Urge." *New Yorker*, April 21, 1951.
Launius, Roger D. "Heroes in a Vacuum: The Apollo Astronaut as Cultural Icon." *Florida Historical Quarterly* 87 (Fall 2008).
Lear, John. "The Hidden Perils of a Lunar Landing." *Saturday Review*, June 7, 1969.
Logsdon, John M. "Selecting the Way to the Moon: The Choice of the Lunar Orbital Rendezvous Mode." *Aerospace Historian* 17 (June 1971).
Mallon, Thomas. "Moon Walker." *New Yorker*, October 3, 2005.
McMichael, W. Pate. "Losing the Moon." *St. Louis*, July 28, 2006.

Oberg, James E. "Max Faget, Master Builder." *Omni* (April 1995).

———. "Russia Meant to Win the Moon Race." *Spaceflight* 17 (May 1975).

Pesavento, Peter. "A Review of Rumoured Launch Failures in the Soviet Manned Program, Part 2: The Lunar Project/1968–1969." *Journal of the British Interplanetary Society* 4, no. 9 (September 1990).

Pyle, Rod. "Apollo 11's Scariest Moments: Perils of the 1st Manned Lunar Landing." Space.com, July 21, 2014.

Reichardt, Tony. "The Luna 1 Hoax Hoax." *Air and Space* (January 2013).

Sawyer, Kathy. "Neil Armstrong's Hard Bargain with Fame." *Washington Post Magazine*, July 11, 1999.

Schanche, Don. "The Astronauts Get Their Prodigious Chariot." *Life*, December 14, 1969.

Shepard, Alan, Jr. "The First Step to the Moon." *American Heritage* 45, no. 4 (July/August 1994).

Siddiqi, Asif. "A Secret Uncovered." *Spaceflight* 46 (May 2004).

Smith, Morgan. "Can You Survive in Space Without a Spacesuit?" *Slate*, August 1, 2007.

Teitel, Amy Shira. "Mercury Astronaut Scott Carpenter and the Controversy Surrounding Aurora 7." *Popular Science*, October 13, 2013.

Vick, Charles, and Dwayne Day. "A Taste of Armageddon." Parts 1 and 2. *Space Review*, January 3 and 8, 2017.

Vine, Katy. "Walking on the Moon." *Texas Monthly* (July 2009).

von Braun, Wernher. "Space Travel and Our Technological Revolution." *Missiles and Rockets* (July 1957).

Wasser, Alan. "LBJ's Space Race: What We Didn't Know Then." *Space Review*, June 20, 2005.

Wentworth, J. Jason. "A History of Surveyor." *Quest* (Winter 1993).

White, Ron. "It's Time to Go." *Quest* (Summer/Fall 1994).

———. "The Right Stuff, the Wrong Story." *Quest* (Fall 1993).

Wilson, Keith T. "Mercury Atlas 10: A Mission Not Flown." *Quest* (Winter 1993).

Young, Anthony. "Apollo 10: 'To Sort Out the Unknowns.'" *Space Review*, May 16, 2016.

ONLINE RESOURCES

AirSpaceMag.com

AmericaSpace.com

CollectSPACE.com

Davis-Floyd, Robbie, and Kenneth J. Cox. 2000. *Space Stories: Oral Histories from the Pioneers of America's Space Program*, a set of interviews resulting from an oral history project carried out under the auspices of the American Institute of Aeronautics and Astronautics (AIAA), the NASA Alumni League, and Johnson Space Center. Interviewees include Paul Dembling, author of the 1958 Space Act; Eilene Galloway, senior specialist in space research for the Library of Congress, 1956–1975; Chris Kraft, Apollo engineer and director of Johnson Space Center; Caldwell Johnson, designer and draftsman for the early space program; Guy Thibodaux, rocket scientist; Max Faget, father of spacecraft design; and Paul Purser, former manager, Langley Research Center. The full texts of the edited interviews appear at www.davis-floyd.com.

Encyclopedia Astronautica (Mark Wade)
Eyles, Don. "Tales from the Lunar Module Guidance Computer." NASA Office of Logic Design, 2004.
Jones, Eric M. "The First Lunar Landing." Apollo 11 Lunar Surface Journal, 1995. Last revised July 7, 2016.
thespacereview.com
Thibodaux, Joseph, Jr. "Reflections of Joseph 'Guy' Thibodaux Jr." www.lsu.edu/eng/docs/HOD/joseph_thibodaux.pdf

MAGAZINES

Aerospace Historian
Air and Space
Astronautics and Aeronautics
Aviation Week
Life
Look
Missiles and Rockets
Newsweek
Popular Mechanics
Quest
Space News Roundup
Spaceflight
Space Review
Time

PAPERS, REPORTS, MEMORANDA, ET CETERA

Cheatham, Donald C., and Floyd Bennett. "Apollo Lunar Mobile Landing Strategy," in *Apollo Lunar Landing Symposium, June 25–27, 1966.* NASA Technical Memorandum X-58006. Washington, DC: NASA, 1966.
Congressional Research Service, Library of Congress. "United States and Soviet Progress in Space: Summary Data Through 1979 and a Forward Look." Report Prepared for the Subcommittee on Space Sciences and Applications of the Committee on Science and Technology, U.S. House of Representatives, Ninety-Sixth Congress, Second Session. Washington, DC, April 1980.
Gemini Program Mission Report: Gemini VIII. Washington, DC: NASA, 1966.
Kraft, Christopher C., Jr. *Robert R. Gilruth, 1913–2000.* Washington, DC: National Academics Press, 2003.
Landis, Geoffrey A. "Human Exposure to Vacuum." GeoffreyLandis.com, August 7, 2007.
Mercury Project Summary, Including the Results of the Fourth Manned Orbital Flight, May 15–16, 1963. Washington, DC: NASA, October 1963.
"Report of Apollo 204 Review Board." NASA Historical Reference Collection, NASA History Office, NASA Headquarters, Washington, DC.

Results of the First U.S. Manned Suborbital Space Flight. Washington, DC: NASA, 1961.

Results of the First United States Manned Orbital Space Flight. Washington, DC: NASA, 1962.

Results of the Second U.S. Manned Suborbital Space Flight. Washington, DC: NASA, 1961.

Results of the Second United States Manned Orbital Space Flight. Washington, DC: NASA, 1962.

Results of the Third United States Manned Orbital Space Flight. Washington, DC: NASA, 1962.

Sheldon, Charles S. "The Soviet Challenge in Space." NASA Technical Memorandum, TM-X-53518. Huntsville, AL, September 15, 1966.

Voas, Robert B. "Project Mercury: A Description of the Astronaut's Task in Project Mercury." Presented at the Fourth Annual Meeting of the Human Factors Society, Boston, MA, September 14, 1960.

von Braun, Wernher. "Concluding Remarks by Dr. Wernher von Braun about Mode Selection for the Lunar Landing Program," June 7, 1962, Lunar-Orbit Rendezvous File, NASA History Reference Collection, NASA Headquarters, Washington, DC.

OTHER

Friedlander, Charles D. and Diane M. Friedlander. *Chuck and Diane Friedlander Memoirs.* Privately published manuscript.

Kraft, Christopher C., Jr. "The View from Mission Control." Unpublished manuscript. Christopher C. Kraft Papers, Virginia Tech.

Sherrod, Robert. Untitled manuscript on the history of NASA. Sherrod Archives, NASA History Office, Washington, DC.

Williams, Walter. "Go!" Unpublished manuscript. NASA History Office, Washington, DC.

INTERVIEWS

John Aaron
Buzz Aldrin
Steve Bales
Alan Bean
Ed Buckbee
Bob Carlton
Jerry Carr
Maurice Carson
Michael Collins
Maddie Aldrin Crowell
Walt Cunningham
Jerry Elliot
Chuck Friedlander
Jack Garman
Dick Gordon
Fred Haise
Bill Helms
Al Jackson
Sy Liebergot
Jim Lovell
Ken Mattingly
Edgar Mitchell
Sam Ruiz
Joe Schmitt
Rusty Schweickart
Reuben Taylor
Tom Weichel
Al Worden

INDEX

ABOUT THE AUTHOR

James Donovan is the author of the bestselling books *The Blood of Heroes: The 13-Day Struggle for the Alamo—and the Sacrifice That Forged a Nation* and *A Terrible Glory: Custer and the Little Bighorn—the Last Great Battle of the American West.* He lives in Dallas, Texas.